MW01295453

PRAISE FOR
THE APOCALYPSE OF WISDOM

"Behind the universally known figure of Jean-Luc Marion lies the largely forgotten Louis Bouyer. Lemna wisely does not try to make this giant of French Catholic ecumenism a forerunner of anybody. His relevance to our current debates is unmistakable on every page of this erudite volume. If you were overwhelmed by the sheer size of Bouyer's corpus, you now have a reliable guide. Lemna does for Louis Bouyer what von Balthasar did for Romano Guardini and de Lubac for Teilhard de Chardin. He provides a synthesis of a brilliant thinker as well as a new path forward for Christian theology in the West, the East, and at the edge of contemporary scientific inquiry into the nature and meaning of thc cosmos."

—PETER CASARELLA, University of Notre Dame

"Catholics seeking resources to surmount the long winter of cultural and ecclesiastical crisis are turning afresh to preconciliar Thomism, and to the great Jesuits and Dominicans of the *Ressourcement* school. But until very recently, few paid serious attention to Louis Bouyer or to the vision advanced by his *Cosmos*. But, none too soon, this is now changing, both in Europe and in the USA. Read Keith Lemna's pellucid work of love, and learn how to read the wisdom of Christ in the pages of cosmic history."

—MATTHEW LEVERING, Mundelein Seminary

"It is hard to say whether this book treats of Bouyer or of cosmology, since the subject matter is Bouyer's cosmology. Lemna is as competent a guide as one could hope for in ascending these heights, for he is a master of Bouyer's thought. Why care about this? Because without cosmology the Church is only a regional sect, one display among many others in the museum of human religions. Cosmology asks whether we can make room for God in his own creation, can recognize his invisible hand everywhere. Bouyer's reputation in liturgy is here deservedly augmented by Lemna's thorough and extraordinary study of the world as gift that reflects the triune life of God in loving praise."

—DAVID W. FAGERBERG, University of Notre Dame

"In *The Apocalypse of Wisdom*, Keith Lemna magisterially displays the unity of Louis Bouyer's œuvre from a hitherto understudied angle while also laying out a vast 'theoanthropocosmic vision' that unites speculative power with sensitivity to the present historical moment. The result is essential reading for both Bouyer scholars and contemplative seekers, who will find Lemna (and Lemna's Bouyer) an indispensable guide to a recovery of a 'cosmicity' that does full justice to the exigencies of faith, reason, and the mythopoetic imagination."

—**ADRIAN J. WALKER**, Catholic University of America

"Decades ago Walker Percy cheekily published *Lost in the Cosmos* to diagnose the malaise haunting our late-modern spirit. Since it seems we are still lost, given our political and ecological trajectories, we do well to search for scouts on the way. Louis Bouyer is one such, setting up biblical and speculative signposts. He spent his life envisioning an integrated and healed cosmos to point the way for theological and ecological renewal. But he is largely ignored in Anglo-American theology. In this clear guide to Bouyer's *Cosmos*, Keith Lemna offers a welcome compendium of Bouyer's thought to enliven Anglo-American sophiology. It is at once crisp commentary and instructive intellectual history on developments key to 20th-century Christian theology. Lemna does a great service filling in a lacuna in cultural memory by introducing us to this generous friend and fellow-pilgrim of so many Christian luminaries, from J.R.R. Tolkien to Jean-Luc Marion."

—**CYRUS OLSEN**, University of Scranton

"Was Louis Bouyer a liturgical scholar? A systematic theologian? A historian of spirituality? A Scripture scholar? The answer is yes. Few figures of the 20th-century theological *Ressourcement* were as adept at grasping and communicating the faith as a living and organic Mystery. We can be grateful to Keith Lemna for making available to us Fr Bouyer's comprehensive and coherent theological vision, one which is simultaneously intellectually and spiritually nourishing."

—**MSGR MICHAEL HEINTZ**, Mount St Mary's Seminary

"Keith Lemna's *tour de force* through Fr Louis Bouyer's work takes the form of a commentary on the latter's *Cosmos: The World and the Glory of God*, providing the Anglophone reader with an introduction to the synthetic genius of this relatively unknown French Oratorian. He deftly explicates the salutary content of Bouyer's *magnum opus,* which the Church and society urgently need today. Our disenchanted view of the world and of ourselves has resulted in a postmodern malaise that has left us in search of we know not what. Bouyer's synthesis is an essential element of the solution to our current existential crisis. Lemna's masterful treatment of the breadth and depth of Bouyer's work lays bare the metaphysical foundation of the cosmos as Trinitarian, making it and everything within it inherently relational—the *analogia entis*. Further, he shows that our epistemological access to this integrated cosmovision comes through created wisdom, connatural to our hylomorphic nature as a seamless unity of spirit and matter and to the relationship between reason and faith. *The Apocalypse of Wisdom* is essential reading for every Christian. It is a must-read for theologians, clergy, catechists, apologists, and everyone else whose vocation it is to understand and convey the ineffable drama of creation and redemption."

—DAVID H. DELANEY, Mexican American Catholic College

"Here are the keys to the thought of the French theologian who has reached out farther and deeper than Teilhard de Chardin in the critical field of cosmology, at the crossroads of science, culture, and faith."

—JEAN DUCHESNE, literary executor for Louis Bouyer

The Apocalypse of Wisdom

The Apocalypse *of* Wisdom

LOUIS BOUYER'S THEOLOGICAL RECOVERY OF THE COSMOS

KEITH LEMNA

Angelico Press

First published in the USA
by Angelico Press 2019

For information, address:
Angelico Press, Ltd.
169 Monitor St.
Brooklyn, NY 11222
www.angelicopress.com

978-1-62138-471-7 pb
978-1-62138-472-4 cloth
978-1-62138-473-1 ebook

Book and cover design
by Michael Schrauzer

In Memory of
Stratford Caldecott

TABLE OF CONTENTS

ABBREVIATIONS

FOR FREQUENTLY-CITED WORKS BY LOUIS BOUYER

CG	*The Church of God: Body of Christ and Temple of the Spirit*
Cosmos (Fr.)	*Cosmos. Le Monde et la Gloire de Dieu*
Cosmos	*Cosmos. The World and the Glory of God*
ES	*The Eternal Son. A Theology of the Word of God and Christology*
Gnôsis	*Gnôsis. La Connaisance de Dieu dans l'Écriture*
IF	*The Invisible Father. Approaches to the Mystery of the Divinity*
LC	*Le Consolateur. Esprit Saint et Vie de Grâce*
MSW	*Mary: The Seat of Wisdom. An Essay on the Place of the Virgin Mary in Christian Theology*
MT	*Le Métier de Théologien: Entretiens avec Georges Daix*
Mystery	*The Christian Mystery: From Pagan Myth to Christian Mysticism*
Sophia	*Sophia ou le Monde en Dieu*

OTHER ABBREVIATIONS

PG	*Patrologia Graeca*
PL	*Patrologia Latina*
SC	*Sources Chrétiennes*
CCSG	*Corpus Christianorum Series Graeca*

INTRODUCTION

WHAT IS THE COSMOS, AND WHAT IS THE place of the human being in it? This perennial, bipartite question, determinative of culture, has been particularly vexing for humanity in its quest for knowledge and meaning amidst the dramatic struggles of the modern age. It has assumed renewed urgency in contemporary thought as a result of growing awareness of and heightened concern for ecological crisis. Some philosophical cosmologists have, in light of this perceived crisis, set forth on a new path in their discipline, beyond the objectivism and mechanist materialism of the Enlightenment. They have called into question the modern disenchantment of the world, the confining of mind and life within the parameters of the self-enclosed human body and brain. Philosophical projects of greater or lesser credibility and with or without overt religious reference now abound that seek to re-enchant nature and to reintegrate humanity with the cosmos.[1] Was modern disenchantment correct to assume that there cannot be a human wisdom that accords with the reality of the world (understood now not as cosmos but as universe without *qualia*), leaving humans able only to project "values" onto the blank slate of an immense field of naturalized beings that has nothing properly cosmic about it? Recent developments challenge both Enlightenment scientism and modern variants of orthodox Christian theology, which are thought to have worked together in unwitting symbiosis to empty the world of its harmonious mind and vitality, reducing it to the status of a dead mechanism or to a collection of inert objects that lie before us to be manipulated, reshaped, or merely consumed in order to placate our voracious appetites. The mechanical way of seeing the world, radically partitioning subject from object, is increasingly recognized to have led to a social ethos that drove

1 Cf. Jane Bennett, *Vibrant Matter* (Durham, NC: Duke University Press, 2010); Michel Onfray, *Cosmos: Une Ontologie Matérialiste* (Paris: Flammarion, 2015); Slavoj Žižek, *Absolute Recoil: Toward a New Foundation of Dialectical Materialism* (New York: Verso, 2015). Žižek is critical of many of these efforts at re-enchantment and seeks to revive a dialectical materialism that he thinks does not collapse subject into object as these other approaches tend to do.

civilizational initiatives whose outcome was the mass destruction of plant and animal life, the eclipse of beauty in culture, and ultimately imminent apocalyptic reckoning. The beauty and order of being, constitutive attributes of the world as cosmos, were lost from view. Can we ever again catch sight of this beauty, is it real, and, if so, can human wisdom ever be made to harmonize with it?

Decades before this awareness of crisis reached its current fever pitch, the French Catholic Oratorian priest and theologian Louis Bouyer (1913–2004) published a monograph on the cosmological question and its ecological consequences that dealt with the theological roots of the issue. It was as deep a treatment of the subject as any single theological volume in the twentieth century and anticipated many of our contemporary concerns. Drawing on religious, biblical, and philosophical wisdom traditions, this theologian was able, in his book *Cosmos: The World and the Glory of God* (1982), to establish an integrated view of humanity in relationship to the world that has the potential to help renew our sense of wonder at the cosmicity of creation and to help heal our broken relationship with the world.[2] Taking with utmost seriousness biblical apocalyptic literature in the context of contemporary, secular, apocalyptic foreboding, he draws our attention to the importance of divine *apokalypsis* or revelation for recovering a sapiential or sophianic understanding of the meaningfulness of the whole and of the common vocation of humanity.

The French writer Gérard Leclerc has recently pointed to this "major book" (as he calls it) by Bouyer as a particularly comprehensive rejoinder to the common view that Christian thought, from Saint Paul onward, is to blame for the a-cosmism and mechanism that has led to ecological catastrophe:

> We would already see [in *Cosmos*] all the current questions. Better yet, they are elaborated in advance and criticized in depth. Nothing of the present difficulties on the interpretation of the

2 The title of this book is taken from Alexander von Humboldt's massive survey of the cosmos: *Kosmos*, 5 vol. (Berlin: 1845–1862). This survey by Humboldt (1769–1859) in fact initiated the modern study of ecology. See Bouyer's *Cosmos*, xi. Bouyer suggests that his own book is meant to show another side of the story to that which Humboldt provides. Bouyer aims to show that modern scientific cosmology "necessarily assumes something beyond its reach." Bouyer, unlike Humboldt, seeks to put the Creator front and center in his cosmology.

> texts of the Bible is ignored. And everything is rehearsed in a journey through the history of thought, until . . . the rediscovery operated by ecology. The objection . . . to Pauline theology is also identified in advance. Paul was extremely sensitive to "the organic nature of the entire creation and of human life inserted into creation, so that only human life can express the ultimate meaning of the created world in its entirety and its unity."[3]

This biblical, "organic," sapiential understanding of creation, which joins together God, humanity, and creation in a single vision of divine Mystery, is the wellspring for Bouyer's "major book." He marshals together a vast array of theological, philosophical, and scientific thought in a uniquely integrative perspective in order to elucidate this vision and gives us, as a result, at once a recovery of Christian cosmology and of the cosmos precisely as cosmos. He shows that Christian thought, on the basis of its biblical organicism, should not be taken to be intrinsically amenable to the modern objectification of nature.

Yet as singular as this book is, it has not been commented on in any depth in English-language publishing. This is a glaring lacuna, especially in these lands that have been, since the age of the Romantic poets, epicenters not only of mechanistic science but also of ecological reaction against modern patterns of life and thought that militate against the poetic perception of reality that can alone give us the cosmos. Bouyer's *Cosmos* charts a sacramental, theological path in response to this reaction that is a needed alternative to materialist projects of re-enchantment or to theological efforts that are insufficiently informed by biblical sapiential and apocalyptic writings. He rediscovers the cosmos as gift of the Creator and as radiant with the Creator's divine light. The primary purpose of this present book is to redress the aforementioned lacuna by illuminating the twists and turns of the path Bouyer charts in *Cosmos* with our own comprehensive map of it in the form of a thorough commentary on it in light of Bouyer's wider writings.

This eminent theologian, although still a little unknown in our day, was one of the greatest and most influential practitioners of his craft in the twentieth century. He has recently been aptly described as both

3 Gérard Leclerc, "Théologie cosmique," in *France Catholique*, June 23, 2015. Available at: https://www.france-catholique.fr/Theologie-cosmique.html. The quotation is taken from Bouyer, *Cosmos*, 89.

the "most brilliant" and the "most dazzling" French theologian of his time—in an age when French theology was a particularly important global source for Christian understanding of the mystery of faith.[4] He produced a wide-ranging oeuvre that was deeply rooted in Scripture, Tradition, liturgy, and the writings of the Church Fathers. The culmination of this work was a nine-volume treatise on systematic theology that was a comprehensive overview of Christian doctrinal themes. *Cosmos* is a capstone monograph in it, summarizing in important ways the decisive first six volumes. Bouyer desired to inculcate a more vivid sense than has been prevalent in the modern age, even among Christians, of the divine presence in creation or of creation in God. Moved by this desire, he critically assimilated throughout the volumes of his trilogies on dogmatics, in a Western, broadly Augustinian-Thomist context, insights from the modern sophiological movement of thought especially as found in the Russian Orthodox theologians Vladimir Soloviev (1853–1900), Sergei Bulgakov (1871–1944), and Pavel Florensky (1882–1937). Like these Russian thinkers, Bouyer related God to the world through the scriptural figure of *Hokmah-Sophia*. This was long before "sophiology" was expounded in the writings more contemporary to our day of Rowan Williams, John Milbank, Antoine Arjakovsky, William Desmond, Stratford Caldecott, Celia Deane-Drummond, Michael Martin, and others.

Bouyer's *Cosmos* not only addresses a topic of increasing importance in a comprehensive way but goes to the heart of the wider thought of this underestimated theologian. The eminent contemporary French philosopher Jean-Luc Marion (b. 1946), who can be aptly described as a spiritual child of Bouyer, has said that in this book especially, among all his many writings, we see that "the theology of Bouyer has a breadth, a space, and a breath that few others have had since the Patristic Age."[5] As an essential part of our commentary, we hope to

4 See Sr. Marie-David Weill, *L'Humanisme Eschatologique de Louis Bouyer: De Marie, Trône de la Sagesse, à l'Église, Éspouse de l'Agneau* (Paris: Les Éditions du Cerf, 2016), 13, backcover. On page 13, in a preface written by Sister Noël Hausman, S.C.M., we read that Bouyer "was the most brilliant of the French theologians of the twentieth century." On the back cover, we read that Bouyer was "the most dazzling [*fulgurant*] of the French theologians of the twentieth century."

5 Jean-Luc Marion, "Cinq Remarques sur l'Originalité de l'Oeuvre Louis Bouyer," presentation at *Groupe de Recherche sur l'Oeuvre de Louis Bouyer*, November 9, 2016. Available at: https://media.collegedesbernardins.fr/content/pdf/formation/etudier-louis-bouyer/2016-11-09-6-intervention-du-pr-j-l-marion.pdf.

show that this book is especially important for recovering and developing the "theoanthropocosmic synthesis" or "theandric humanism" of those Church Fathers who elaborated the Pauline and wider biblical cosmology in confrontation with Greco-Roman philosophy.[6] The loss of this synthetic, biblical vision of creation is surely the root cause of those ills of modern or postmodern life that ecologism seeks to combat, however effectively or misguidedly, and modern Western Christian theology is indeed complicit in this loss of vision.

The link between the human and cosmic is inexorable, particularly in Christ, the New Adam, but modern philosophers and theologians have not always understood this link. Wisdom relinquished its cosmological bearing and was eclipsed by science. Science, coming into the interpretive clutches of mechanical philosophy, no longer saw a cosmos, that is, a harmonious, interrelated whole. Humanity lost its microcosmic integration with nature. Sometimes under the sanction of theologians themselves, the elements of the world were artificially joined together as aggregates, thought to make up together a vast mechanism lacking intrinsic unity. Human *telos* could no longer be grounded in a transcendent purpose that embraced the whole of creation.

For the Church Fathers, on the other hand, particularly in the East, the link between cosmos as genuine cosmos and *anthropos* in Christ was a theme of major importance. This is part of what it means to speak of a theoanthropocosmic synthesis or theandric humanism in their thought and can be seen especially in their transformation of ancient philosophical and medical microcosmisms. For some of the Church Fathers, it was in fact insufficient to refer to humanity as a microcosm or a universe in miniature, as much of Greco-Roman philosophy and medicine did. Humanity, these theologians thought, should be said to be not so much a microcosm as a macrocosm, that is, a great world (*macrocosmos*) in the small one (*microcosmos*).[7] The Romanian Orthodox theologian Dumitru Staniloae (1903–1993) explains that in this Greek Patristic understanding of creation the world will be actualized

6 Cf. Philip Sherrard, *Human Image: World Image, the Death and Resurrection of Sacred Cosmology* (Ipswich, UK: Golgonooza Press, 1992), 10. Sherrard speaks of the "theoanthropocosmic vision" of the Church Fathers. He describes this as a vision that allows us to perceive both the world and ourselves as sacred realities.

7 Cf. Gregory of Nazianzus, *Oratio* 38, *On Theophany*, ch. 2, *PG* 36:324a.

when it becomes "pan-human" or bears the entire stamp of the human in Christ through the human nature he has assumed.[8] It is in fact destined to become "macro-*anthropos*" in Christ by the full realization of his ecclesial Body.[9] They followed and boldly developed Paul's anthropology according to which "only human life can express the ultimate meaning of the created world in its entirety and its unity." In this understanding, *anthropos* lends to the cosmos its very cosmicity. This is not because it projects order onto the world from the *a priori* powers of its created mind, but because God's Wisdom has ordained *anthropos* to be the bridge that unites and reconciles all things together in the Incarnate Lord of creation. This vision of the world is summed up in a Marian and sophianic light in the Western tradition in Saint Hildegard of Bingen's (1098–1179) intriguing icon of the Cosmarchia or mother of creation, the Beloved of God the Father, who carries in her womb the figure of the "Universal Man" standing at the midpoint of creation. As mother of the whole cosmos, she communicates her maternal energies to him. This image indicates the organic unity of *anthropos* and cosmos not only in Christ but in Wisdom.[10]

Louis Bouyer takes up in *Cosmos* these elaborations of the Pauline view, particularly in the sophiological form it was given much later by the likes of Soloviev and Bulgakov, as an essential dimension of his sapiential and apocalyptic recovery of the cosmos, but he lends unique contours to this tradition. He embraces, as we shall see, a more "positive" approach to theology than many of the Eastern theologians whom he admired, and certainly writes with much greater sympathy for the Western tradition as a whole, particularly in its monastic form. In some ways, his book on which we shall center our attention is classifiable as a theological alternative to the great twentieth-century Western cosmologies developed particularly in the Anglophone world by philosophers Samuel Alexander (1859–1938), Alfred North Whitehead (1861–1947),

8 Dumitru Staniloae, *The Experience of God*, vol. I: *Revelation and Knowledge of the Triune God*, trans. and ed. Ioan Ionita and Robert Barringer (Brookline, MA: Holy Cross Orthodox Press, 1998), 4–5.

9 Ibid.

10 This refers to *The Book of Divine Works*, vision 2. See Thomas Schipflinger's comments in *Sophia: A Holistic Vision of Creation*, trans. James Morgante (York Beach, ME: Samuel Weiser, Inc., 1998), 144. Unfortunately, Bouyer never references this icon, which sums up in iconic form much that his theology of creation entails.

and R.G. Collingwood (1889–1943), each of whom is referenced in *Cosmos*, and each of whom, very much in line with nineteenth-century philosophers of nature, sought to reconcile human history with the total process of nature. Bouyer takes these cosmologists into implicit and sometimes explicit account in the text. They are each rigorously mindful of the historical becoming of the cosmos, orienting the attention of readers to the future, to what has not yet come to be, going so far as to say, in the case of Alexander, that God, the entelechy of creation, has not yet Himself come to be but will do so someday with the advance of history.[11] The Oratorian's own cosmology is future-oriented, apocalyptic, anticipatory, but within the limitless parameters of an overarching analogical construal of the eternal and transcendent triune God's relation to creation. He seeks to direct our gaze from where we are to where God is ultimately leading us in His grand design for each of us and for the whole of creation, which is eternally present to Him in the divine Word as Wisdom. The future completion of creation is not a perfection or coming-into-being of God's transcendent life but of created Wisdom in and through the eschatological Church. Human wisdom harmonizes with the cosmos most fully when it is shaped by and assimilates God's revelatory Wisdom disclosed in and through the Word of God. Bouyer shows that cosmology should be a pursuit of wisdom, but that Wisdom is ultimately a gift revealed or uncovered by means of *apokalypsis* in the re-creative interventions of the divine Word and Holy Spirit in history.[12] The theological cosmologist, if true both to the nature of being and to God's revelation, has the task of orienting his or her reasoning according to modalities of insight congruent with Christian expectation for the absolute end of the whole cosmos. This is the path to recovery of the cosmos that has the power to transform the foreboding and despair that too often pass for apocalyptic consciousness in the contemporary period.

11 *Cosmos*, 5.

12 In this book we shall capitalize "Wisdom" only when we are referring directly to divine and created Wisdom, corresponding in some sense to the essence and energies of God. Also, we shall capitalize the personal pronoun "He" in reference to God only when referring to the One God or to God the Father. In reference to Christ or the Holy Spirit, we shall not capitalize the personal pronoun.

OVERVIEW OF *COSMOS* AND ITS META-ANTHROPOLOGY

In an earlier and now-classic text on ecclesiology, *The Church of God*, Bouyer described at length what he intended eventually to accomplish in his future book on theological cosmology in a way that provides a point of departure for understanding it precisely as a theoanthropocosmic synthesis. This latter book would be, he said, a "supernatural cosmology," following in the line of his earlier "supernatural anthropology," *Mary: Seat of Wisdom*, and his book on ecclesiology, which he described as a "supernatural sociology."[13] Both the supernatural anthropology and the supernatural sociology would be given context in light of this cosmology, placed ultimately in the biblical perspective of glory, for which the world is "the effulgence *ad extra*, just as the Divine Word is its glow *ad intra*."[14] In the introduction to *Cosmos*, he defines the word "glory" in not only divine but cosmic terms: "what the Bible calls the glory of God is but the divine radiance of the cosmos when we contemplate it in its infinite vastness and depth, a radiance emanating from all existence, from the entire cosmic being."[15] The human experience of glory in creation points to the Creator who is its source. This glory "is precisely why the created world can be called a 'cosmos,' meaning order and beauty."[16]

The central purpose of his study of the cosmos, he explained, would be to provide a "Christian interpretation of what is called 'the modern scientific vision of the world.'"[17] Bouyer acknowledged that he would thus join the theological *Problematik* identified by the Jesuit paleontologist Pierre Teilhard de Chardin (1881–1955) as crucial for the future of Christian as of all human thought. However, he intended to establish a much broader interpretive horizon for modern cosmology than the Jesuit paleontologist provided. He aimed "to come to grips with the unfolding of Christian revelation and its interpretation, along with philosophical reflection upon the totality of the experience of man when left to his own devices and his own lights, at least as they appear to be."[18] What Bouyer would try to accomplish in *Cosmos* was both a theological hermeneutic and a phenomenology of the cosmos that draw on the whole of human

13 *CG*, xiii.
14 Ibid.
15 *Cosmos*, xi.
16 Ibid.
17 *CG*, xiii.
18 Ibid.

experience, ultimately illuminated by Christian faith and respectful of those to whom divine revelation has not been directly given. Both ancient and modern sources would be taken into account:

> However provisional such an attempt might be in essence—an attempt at a critical synthesis of the experience and (within that experience) of the total awareness of a mankind receptive to biblical faith—an exploration of Christian tradition seems to us much vaster and more profound than it is in our silly cosmology manuals. At the same time, such an attempt ought to be receptive not only to the heritage of the thinkers of antiquity (without whom modern thought would be incomprehensible) but also to the attainments of the great contemporary cosmologies, such as those of Alexander and Whitehead, as well as to the phenomenology of science or of the whole of human existence, in which science is merely a part or a fragment.[19]

These words accurately envision in advance the accomplished text of *Cosmos*. They signal both its grand ambition and its intentionally provisional character. Moreover, they speak of Bouyer's desire to engage modern philosophical cosmologies that reckon with advances in science in the context of a comprehensive anthropology.

In his book *The Invisible Father*, the Oratorian gives an important clue to how he thinks this engagement should be concretely implemented. Theologians, he says there, can effectively become involved with questions pertaining to modern science on the basis of a "phenomenology of faith" that is ever open to the "true *scientia*" of divine revelation.[20] The adoption of a phenomenological approach would enable theology to "harmonize quite naturally with the unexpected perspectives opened up by in-depth explorations of the epistemological problems which have emerged as an inevitable result of science's quite autonomous development."[21] In the first part of *Cosmos*, Bouyer takes a phenomenological starting point of a genetic or genealogical kind, reminiscent of the later work of Edmund Husserl (1859–1938). The world as studied by Bouyer is more akin to the *Lebenswelt* of Husserl than to that of

19 Ibid., 14.
20 *IF*, 304.
21 Ibid.

a pure object considered irrespectively of the intentional modalities of human consciousness to which it is given. It certainly differs from the wholly noumenal being of the mathematically-describable plenum that Galilean science has sought to symbolize and whose potencies modern technology has aspired to unleash in titanic fashion. At various points in his writings, Bouyer describes his preferred method as "positive theology." He defines this in a general way as an attempt to "explain the emergence of detailed foundations of Christian doctrine by tracing their development in the Church's own awareness of these formulations, manifest in the evangelization of the Church."[22] The positive theology of Bouyer is not a mere history of theology. It is infused with phenomenological and hermeneutical concerns: it does not only catalogue historical positions in theology but uncovers the essence of manifestation of religious realities, both natural and supernatural, and elucidates performative human responses, first of all present in liturgy, to the call of the divine Word.[23]

Despite its "positive" character, the cosmology on offer in *Cosmos* cannot be accurately described as devoid of metaphysics or of a "philosophy of nature." It will be one of our fundamental goals to establish the soundness of this interpretation. The reader of the book will not find there, it is true, a philosophy of causality treated in separation from the phenomenology of religious culture and its transformation by divine revelation. The metaphysical dimension of the book is always connected, on the one side, to Trinitarian theology and, on the other, to the widest considerations of anthropology. Cosmic wisdom is not explicated in abstraction from all that is involved in the attainment of human wisdom in history. This speaks to the theoanthropocosmic synthesis, in an updating of the Church Fathers, achieved in the text. The Bouyerian metaphysics is thus akin to the "meta-anthropology" of Hans Urs von Balthasar (1905–1988). The Swiss theologian, a friend and colleague of Bouyer, employed this expression to explicate the philosophical dimension of his own work:

22 *Cosmos*, 131.

23 Cf. Donald Wallenfang, *The Dialectical Anatomy of the Eucharist: An Étude in Phenomenology* (Eugene, OR: Cascade Books, 2017), 26–48. In these pages, Wallenfang, drawing from the work of David Tracy, distinguishes phenomenological from hermeneutical approaches to religious realities. The former is concerned with manifestation and the latter with proclamation.

> It is here that the substance of my thought inserts itself. Let us say above all that the traditional term "metaphysical" signified the act of transcending physics, which for the Greeks signified the totality of the cosmos, of which man was a part. For us physics is something else: the science of the material world. For us the cosmos perfects itself in man, who at the same time sums up the world and surpasses it. Thus our philosophy will be essentially a meta-anthropology, presupposing not only the cosmological sciences, but also the anthropological sciences, and surpassing them towards the question of the being and essence of man.[24]

In this philosophical approach, cosmology is inextricably joined to anthropology, as humanity is understood to be the being in the world that alone brings the being of beings to light. The French Oratorian was prescient to have targeted the theme of wisdom as brought forward by the aforementioned Russian theologians, who also practiced a sort of "meta-anthropology." Sophiology naturally unites cosmology with anthropology, the latter understood with respect to broader considerations of human culture and the history of human thought, ancient and modern. In meta-anthropological fashion, it centers the question of being on the being of man, in view of his reconciling vocation for the whole of creation. Bearing an intrinsically aesthetic character, in line with the scriptural sapiential literature, sophiology is amenable to phenomenological supplementation. Indeed, drawing on insights from nineteenth-century philosophers of nature, sophiologists already joined metaphysical with hermeneutical/phenomenological concerns.

Bouyer, as we shall see, associates in aesthetic and phenomenological mode the gift of Wisdom with the active reception of it by the unifying depth of human perception that he describes alternately as "mythopoetic thinking" and "imagination." He demonstrates that wisdom as we have known it in Western civilization emerges out of the royal myths of antiquity, which ultimately have a ritual basis. Cosmological wisdom is metaphysically understood only in light of its anthropological emergence and disclosure, clothed in the figure of glory. *Cosmos* details the

24 Hans Urs von Balthasar, "A Résumé of My Thought," trans. Kelly Hamilton, *Ignatius Insight*, March 5, 2005. Available at: http://www.ignatiusinsight.com/features2005/hub_resumethought_mar05.asp.

process of this emergence, which is correctively accomplished by the development of human *logos* and by the revelation of God in the world through the Spirit of the divine *Logos*. This exposition is an essential part of the "philosophy of nature" in the text, which is at the same time a theological "phenomenology of the Spirit." The demonstrations that Bouyer provides in this regard in *Cosmos*, as elsewhere in his writings, constitute one of the unique dimensions of his thought, and Balthasar described these demonstrations as "magisterial."[25] The explication and validation of mythopoetic thinking and imagination enables us to recover the world in a sapiential and sacramental way, inseparably connected to the apocalyptic *telos* of human history. At the end of this work, we shall summarize Bouyer's cosmology in this meta-anthropological light.

OVERVIEW OF THE PRESENT BOOK

The present study, in large part an extensive commentary on *Cosmos*, contains twelve chapters, arranged in two main sections, organized according to the structure of this text on which we shall center our attention. The most significant points of each of the chapters of this major book will be expounded. Although we shall draw much on Bouyer's wider corpus in order to help us understand the book, we shall follow the organizational structure of *Cosmos*, so that our exposition and analysis follows Bouyer's own wise ordering of themes without missing anything essential and without becoming too unwieldy. Although this study is largely a commentary on Bouyer's major book on Christian cosmology, it is also an entrée into his wider writings. *Cosmos* is Bouyer's most personal as well as most speculative work and thus opens up the meaning of his other writings. On the other hand, *Cosmos* is illuminated by recourse to them, because much that Bouyer leaves unsaid in *Cosmos*, or under the surface of the text, is explicitly articulated in these other writings, particularly the other volumes in his dogmatic trilogies.

The current study is also an assessment of Bouyer's potential to contribute to the recovery of the cosmos that has been a wider concern of philosophers, theologians, and scientists in recent years, in reaction against the anti-cosmism of modern disenchantment. As we shall see, cosmology was one of the preeminent concerns of Bouyer's entire

25 *MT*, 17.

theological career, from the time of his youth.[26] We can get at the heart of Bouyer's whole corpus by exploring this one theme and concern and see that Bouyer is a forerunner—albeit unfortunately neglected as a resource—especially of theologians who nowadays seek to develop a sacramental ontology of creation.[27]

In the first part of this study, we shall explore the first seventeen chapters of *Cosmos* and their context in Bouyer's thought as well as in wider modern currents of thought. These chapters together constitute more the genetic-phenomenological portion of the monograph. Bouyer's preference for positive theology of a distinctly phenomenological type comes out in full in this first part of the book, where he shows forth, in a presentation that is at once historical and illuminative of universal structures of disclosure, the movement from pagan ritual and myth to Christian liturgy and the scriptural Word in the ancient world, as well as modern developments in thought and culture that derive from this movement. The first seventeen chapters of *Cosmos* trace in a genetic-phenomenological mode the whole of the history of cosmology in Western civilization, from pre-history all the way to the development of scientific theology and the emergence of modern physical science and technology. Bouyer's employment of this approach exhibits great concern for modern, so-called "historical consciousness" but also for the "creative event" in cosmic history that surpasses known possibilities in the world. He aims to show the radical newness of the Christian way of seeing the cosmos and the transfiguring Wisdom of God that God's revelation discloses and imparts in the folly and power of the Cross.

In the first chapter, we shall explore Bouyer's theological cosmology in light of his biography and personal experience and demonstrate the link between his thought and that of Cardinal John Henry Newman (1801–1890), the Inklings, and Bulgakov, particularly with respect to the connection between wisdom and imagination in the recovery and development of cosmology. All of these thinkers are important in this

26 Ibid., 22.

27 Cf. Hans Boersma, *Nouvelle Théologie and Sacramental Ontology: A Return to Mystery* (Oxford University Press, 2009); *Heavenly Participation: The Weaving of a Sacramental Tapestry* (Grand Rapids, MI: Eerdmans, 2011). Boersma shows the concern of 20th-century French theology to recover a sacramental ontology of creation. But, curiously, he makes little mention of Bouyer, who was arguably the most comprehensive of the French theologians on this score.

regard, because they draw profoundly from the deepest traditions of Christian thought on the doctrine of creation. They share the view that explication of this doctrine requires development of human imagination. Bouyer draws together conceptual motifs from these disparate thinkers in a unique fashion. Explication of this drawing-together sets an intelligible context for what he is doing in *Cosmos*. By putting his thought in the context of the better-known doctrines of these thinkers who deeply influenced him, his own ideas become a little clearer.

In the second chapter, following the themes of the first two chapters of *Cosmos*, we shall explore Bouyer's understanding of mythopoetic thinking or imagination with respect to a wider turn to myth in modern thought and to his development of a sacramental gnoseology of divine glory. In his positive theology of creation, Bouyer explores the transfiguration of *mythos* that occurs first by way of divine revelation in the Old Testament and second by way of philosophy. In chapters three through seven of *Cosmos*, he focuses especially on Old Testament revelation in relation to the ancient royal myths. These chapters taken together are a lengthy and minutely detailed portion of the text. Their combined length indicates the importance he accords to scriptural exegesis.

We shall explore them in our own third chapter, setting the context with a brief summary of his wider theology of history, especially as found in his book on the Holy Spirit, *Le Consolateur*.

In chapter four, we shall follow the eighth and ninth chapters of *Cosmos*, where Bouyer shows the relation of philosophical *logos* to the ancient cosmic myths. We shall end the chapter with an exposition of his view that the Christian monk is the transfiguring heir of the philosophical legacy of antiquity.

In chapter five, we shall explore the Christological dimension of Bouyer's cosmology, which he briefly treats in chapters ten and eleven of *Cosmos*. This requires turning as well to his great text on Christology, *The Eternal Son*, where important themes are explicated that he merely hints at in *Cosmos* without description.

We shall follow Bouyer's hermeneutic and phenomenology of modern science in chapters six and seven of our study. Chapter six will explore the darker dimension of Bouyer's history of modern cosmology, as recounted in chapters twelve, thirteen, and sixteen of *Cosmos*. The seventh chapter will treat the more hopeful signs that our author calls forth out of the depths of modern thought in chapters fourteen, fifteen, and seventeen of his text. We shall suggest in this latter chapter that modern sophiology is indeed crucial to his poetic theological recovery of the cosmos.

At this point in the study, in our second part, we shall turn to the more metaphysical chapters of *Cosmos*. We shall give an overview of Bouyer's sapiential way of treating the question of the relation of nature, spirit, and history that was decisive for nineteenth-century philosophy of nature. This, as we shall see, gets to a question that inspires the ordering of the whole book. It is only in the second, shorter, and more speculative part of the book that this question is treated in full. Bouyer refers to this second part as "Retrospections," and it runs from chapter eighteen to chapter twenty-two.

In chapter eight of our own book, we shall focus on the author's eighteenth chapter, where he expounds succinctly his theology of Trinitarian Wisdom. Sergei Bulgakov once said that the central question of cosmology is: "What is the world in God and what is God in the world?"[28] We give our deepest assessment of Bouyer's answer to this question in this eighth chapter, which sets the stage for the final four chapters of our study. We shall in fact draw in summary fashion on the whole of Bouyer's Trinitarian thought in his wider writings.

In chapter nine, we shall explore the Oratorian's angelology as set forth in chapters nineteen and twenty of *Cosmos*, while we shall explore in chapter ten implications of his angelology for how we should understand the relationship between spirit and matter, analyzing more deeply Bouyer's wider philosophy as well as his philosophy of science. This requires us to explore in depth chapter twenty-one of *Cosmos* as well as Bouyer's connection to some very interesting twentieth-century French philosophers of science.

In chapter eleven, we shall focus on the themes in the twenty-second and final chapter of *Cosmos*, which bear on the eschatological Church, the Bride of the Lamb, and the nuptial ontology of creation revealed in

28 Sergei Bulgakov, *The Bride of the Lamb*, trans. Boris Jakim (Grand Rapids, MI: Eerdmans Publishing, 2002), 33. In *The Invisible Father*, Bouyer makes this question even more precise and sets its historical context: "Yet, undeniably, once the Fathers of the West or East had reached this formulation they could only stammer when it came to defining more precisely how created and divine Wisdom could be one without confusion—how, in other words, the inner life of God, in the Trinity of its persons, includes human and cosmic life in itself from all eternity and will unite itself to them when history reaches its term, yet without absorbing them back into itself" (*IF*, 236).

eschatological light through John's Apocalypse. We shall be required to draw from Bouyer's books on Mary and the Church in order to treat these topics in full.

Finally, in our twelfth and summary chapter, we shall reflect on the singular achievement of the Oratorian in setting forth his theoanthropocosmic vision in *Cosmos*. A final assessment of the importance of his turn to Wisdom and imagination will be provided in its dual metaphysical and phenomenological implications. We shall suggest that the philosophical underpinning of his work is aptly described as a sort of "meta-anthropology." From the vantage point of analogical anticipation, the Oratorian sets our sights on the unity of creation, consummated only at the end of history with the Return of the King, the Son of Man, slain before the foundation of the world as the very principle of all creation.

As we have indicated, Bouyer's standpoint in *Cosmos* as throughout his writings is resolutely apocalyptic. This is so in the sense of *apokalypsis* or revelation and also in privileging the Johannine vision of the absolute end as ordering principle for interpreting the entirety of cosmic history. His cosmology is anticipatory of what has not yet come to pass. This is one reason why imagination and the image play such an important part in his thought. As Austin Ferrar (1904–1968) argued, apocalyptic writing, among all the biblical writings, is especially imaginistic, because it concerns "a realm which has no shape at all but that which the images give it."[29] Images and imagination are freed in Jewish and Christian apocalyptic writing (and thinking) from the restrictions placed upon them by the historical actuality to which the gospels are inevitably wedded, although this does not render their productions for that reason contradictory to the historical actuality that is their basis.

In privileging a monastic standpoint or transgressive eschatological perspective on the world, Bouyer's thought can strike the reader as a rather unworldly or otherworldly way of being in the world and of seeing it. We shall maintain here that he is not, in fact, teaching us without good cause that the eschatological standpoint is a requisite to recognizing the provisional character of our very lives in this age. After all, the work of the Body of Christ is not yet complete. Church and world are still too much under the power or influence of the remnants of Belial's kingdom.

29 Austin Ferrar, *A Rebirth of Images: The Making of Saint John's Apocalypse* (Albany, NY: State University of New York Press, 1986), 17.

Bouyer advocates the need for conversion and ascesis in Christian life. He urges that only by following the Way of the Cross can we espy the glory of God that irradiates the cosmos and share the divine gift of charity freely. In exploring the methodological, phenomenological, and meta-anthropological implications of Bouyer's cosmology, while not shying away from what Balthasar described as Bouyer's "dense formulas," we hope not to lose sight of this monastic, biblical, and liturgical understanding of the need for conversion and discipleship in order to recover the world.[30] The cosmology on offer in *Cosmos* is an antidote to worldliness, all so that we can gain full human possession of the world precisely as gift of the Creator meant to reflect the triune life of God in loving praise as "cosmic liturgy." Bouyer once described Newman's theology as especially relevant for our age, which is mired in what he called "general apostasy" from Christian faith.[31] We might well say the same for Bouyer's *cosmology*: for Bouyer as for Newman, we are brought into a new world in Christ (cf. 2 Cor. 5:17). We are brought into the cosmos re-created in and through the worship of the Church. This is the sacramental world, and it not only prepares us for but anticipates our full citizenship in the heavenly Jerusalem. This is not ultimately an "otherworldly" position. The re-created world of Christian hope is not a different world from the one in which we now live but is, instead, our own world "restored to its primitive transparency... in expectation of its total re-creation at Christ's second coming."[32] To speak of the "re-creation" of the world in Christ is, as we shall see, very different from speaking of salvation in a Gnostic way as "de-creation." In this time of general apostasy, of corruption and filth invading or even emanating from the precincts of the visible Church, Christian sympathy for the world has tended to be accompanied by a lack of critical discernment of spirits. Bouyer's cosmology can help the Church to recalibrate, to move away from excessive worldliness and the following of deceptive spirits, by recovering a truly sacramental and therefore anticipatory vantage point on the meaning and nature of the cosmos in its intrinsic ontological dignity as a creation very much in need of re-creation in the Spirit of Christ.

30 *MT*, 17.

31 Louis Bouyer, *Newman's Vision of Faith: A Theology for Times of General Apostasy* (San Francisco: Ignatius Press, 1986), 9–16.

32 Ibid., 168.

SECONDARY LITERATURE

The ascetical and monastic emphasis in Bouyer's writings, oftentimes polemically defended, was not a terribly helpful feature of it for bolstering his reputation among ecclesial authorities on the local level and of progressivist theologians in the 1970s and 1980s.[33] Although highly regarded on a wide scale as a theologian before the Second Vatican Council, and although deeply esteemed by Pope Paul VI before, during, and after the council, Bouyer's work fell into neglect and obscurity for several decades. Few secondary studies of it could be found. This was still the case when we first encountered Bouyer in reading *Cosmos* in 2004. Because this book was so obviously a singular modern treatment of cosmology from a Christian theological perspective, we marveled that it was not part of larger theological discussion on this increasingly important topic. We hoped eventually to produce a study based on the book that shows the significance of cosmology in the wider whole of Bouyer's thought, as well as the uniqueness of this cosmology in the wider whole of historical cosmology generally.

In the meantime, genuinely great studies of the Oratorian's thought began to appear in Europe, beginning with Davide Zordan's *Connaissance et Mystère* in 2008.[34] More recently, Guillaume Bruté de Rémur's comprehensive book on Bouyer's Trinitarian theology, Marie-David Weill's monumental study of his "eschatological humanism," and Dom Bertrand Lesoing's provocative treatise on his ecumenical theology have made significant contributions.[35] Two recent studies are particularly worthy of note, given our own chosen topic. The first is a study by Carolina Blázquez Casado of the cosmovision shared between the Oratorian and the French Eastern Orthodox theologian Dom Olivier Clément

33 Cf. *Le Décomposition du Catholicism* (Paris: Aubier-Montaigne, 1968); *Religieux et Clercs contre Dieu* (Paris: Aubier-Montaigne, 1975) [hereafter *RC*]. With these two books, pugnaciously critical of the postconciliar collapse of French Catholicism, Bouyer sealed his fate as a bit of a pariah among French clerics and the episcopacy, at least until his pupil and spiritual child Cardinal Jean-Marie Lustiger (1926–2007) was appointed Archbishop (and later Cardinal) of Paris in 1981 by Pope John Paul II.

34 Davide Zordan, *Connaissance et Mystère* (Paris: Les Éditions du Cerf, 2008).

35 Guillaume Bruté de Rémur, *La Théologie Trinitaire de Louis Bouyer* (Rome: Editrice Pontificia Università Gregoriana, 2010); Weill, *L'Humanisme Eschatologique de Louis Bouyer*; Bertrand Lesoing, *Vers la Plénitude du Christ: Louis Bouyer et l'Oecuménisme* (Paris: Les Éditions du Cerf, 2017).

(1921–2009), and the second is Marie-Hélène Grintchenko's *Cosmos, Une Approche Théologique du Monde:* Cosmos *du Père Louis Bouyer*.[36] Although these books on Bouyer's *Cosmos* have appeared, we think that its content and place in Bouyer's thought warrant further studies dedicated to it.[37] In our own, we shall emphasize the apocalyptic-imaginative and sophianic heart of the book, which includes an emphasis on the eschatological, theandric humanism of the text and its development of the integrative and organic Pauline view of the whole of creation in the wake of three millennia of Western cosmology. We will show, more than these earlier studies have done, the unique place of Bouyer's cosmology in the context of modern thought taken as a whole. It is important, in our view, not to disconnect his thought from the questions of nineteenth-century nature philosophy that he himself expounds in *Cosmos* and which still have great relevance. The thinkers who raised these questions, whatever their monumental flaws, were visionary in their efforts to try to re-unite God, cosmos, and *anthropos* in the horizon of speculative thought. We hope as well to give a better indication than has been given so far of the largeness of the Oratorian's own speculative mind, which is always tethered to the concrete images of biblical revelation as well as sacred liturgy and, in consequence, can be readily underestimated, as it has been in fact. If Bouyer did not attain real achievement on the speculative level in his work, then it would be of little lasting importance. It would be a futile exercise to engage in extended commentary on it. With the just-mentioned goals in mind, we will explicate more fully than these

36 Carolina Blázquez Casado, *La Gloria de Dios en la Entraña del Mundo: Olivier Clément y Louis Bouyer. Un Estudio en Perspectiva Ecuménica de Dos Cosmovisiones Cristianas. Disertación para la Obtención del Doctorado en Teología Dogmática* (Salamanca, 2014); Marie-Hélène Grintchenko, *Cosmos, Une Approche Théologique du Monde:* Cosmos *du Père Louis Bouyer* (Paris: Parole et Silence, 2015).

37 In an interview with Ignatius Press published on November 12, 2010, we noted that we take Bouyer's *Cosmos* to be an especially important book in his oeuvre. Available at: http://www.ignatiusinsight.com/features2010/klemna_louis-bouyer_nov2010.asp. Unfortunately, we did not yet have access to his *Memoirs* at the time of this interview, and the interview reflects this. See also Keith Lemna, "Trinitarian Panentheism: A Study of the God-world Relation in the Theology of Louis Bouyer," Ph.D. dissertation, The Catholic University of America, 2007. Although our doctoral dissertation focused on cosmological themes in Bouyer, the current study is not a revision of this earlier work. See Bouyer's *Memoirs*, trans. John Pepino (Kettering, OH: Angelico Press, 2015).

earlier studies have done some of the sophiological writings with which Bouyer's theology is connected, particularly those of Bulgakov.

This whole group of European studies sets a solid groundwork for research into Bouyer's thought. We have benefited from reading these volumes and have used them in planning as well as composing the current book. Other books of note on Bouyer from which we have drawn are Jean Duchesne's lovely volume *Louis Bouyer* and the publication of the proceedings of a conference held on Bouyer's thought in Paris in 2014, on the tenth anniversary of his death.[38] *Gregorianum*, the Roman Jesuit theological journal, published an issue dedicated to Bouyer's theology, under the direction of Fr. Jacques Servais, SJ, in that same year.[39] Currently, there is a group of scholars operating under the appellation "Le Groupe Louis Bouyer," who meet tri-annually at the Collège des Bernardins in Paris to discuss the Oratorian's work and who have helpfully contributed to our understanding of Bouyer in their continuing research into his thought.[40]

Bouyer taught in the United States and Great Britain. The recent publication by Angelico Press of his *Memoirs* in English translation has spurred greater interest in his theology in the English-speaking world. He is known to have been a very influential liturgist and writer on spirituality, but the depth and breadth of his theological work has barely begun to be appreciated in these lands. We hope that our own present study will help to foster this appreciation.

38 Bertrand Lesoing, Marie-Hélène Grintchenko, Patrick Prétot, eds., *La Théologie de Louis Bouyer: Du Mystère à la Sagesse* (Paris: Parole et Silence, 2016).

39 See *Gregorianum* 95.4 (2014): 675–799.

40 See *Groupe Louis Bouyer*, https://www.collegedesbernardins.fr/formation/groupe-louis-bouyer. This is the main page for the website for the group. Summaries of the interventions at their meetings are included on this site.

PART I

COSMOLOGY IN HISTORY

CHAPTER 1

Imagination and Wisdom

MARTIN HEIDEGGER ONCE OPENED A LECture on Aristotle by saying that the "personality of the philosopher is of interest only to this extent: he was born at such and such a time, he worked, and died."[1] Louis Bouyer, by contrast, operating from the fundamental conviction that truth and life are inseparable, thought that one's ideas and personality are so inextricably connected that biography is a crucial factor in interpretation.[2] This is reflected in many of his writings, such as his classic spiritual biography of Cardinal Newman, where he placed Newman's ideas in the context of his spiritual journey.[3] Newman himself encouraged the view that biography, where available, is important to theological or philosophical exposition. As he said in *An Essay in Aid of a Grammar of Assent*:

> [I]n these provinces of inquiry [religion and metaphysics] egotism is true modesty. In religious inquiry each of us can speak only for himself, and for himself he has a right to speak. His own experiences are enough for himself, but he cannot speak for others: he cannot lay down the law; he can only bring his own experiences to the common stock of psychological fact.[4]

1 Martin Heidegger, *Grundbegriffe der Aristotelischen Philosophie* [Basic concepts of Aristotelian philosophy] (Marcuse Archive: University of Frankfurt, summer semester 1924), 1. Parts of the present chapter were originally published in Keith Lemna, "Louis Bouyer's Development of Cardinal Newman's Sacramental Principle," *The Journal for Newman Studies* 13.1 (Spring 2016): 22–42. Reprinted here with permission of the publisher.

2 *MT*, 16.

3 We refer to *Newman: His Life and Spirituality*, trans. J. Lewis May (San Francisco: Ignatius Press, 2011). See also "The Permanent Relevance of Newman," in *Newman Today. Proceedings of the [1988] Conference on John Henry Newman*, Wethersfield Institute, ed. Stanley Jaki (San Francisco: Ignatius Press, 1989), 165–74.

4 John Henry Newman, *An Essay in Aid of a Grammar of Assent* (London: Longmans, Green, and Co., 1903), 384.

Without being historicist, these words promote the exploration of psychology and experience in the formation of ideas. They do not entail that ideas in religious matters have no universal validity; rather, certainty in these domains is attained only through unique, unavoidable, personal effort and accomplishment. Contra modern rationalism, Newman held that there is no foolproof, abstract procedure or scientific method for attaining certitude in religion or metaphysics. Theological or philosophical writings can convey universally valid demonstrations, but in matters of inference to truth in these areas of study no theologian or philosopher gives assent on the basis of a single line of purely objective demonstration or communicates an indubitable method for the attainment of truth. Many tacit biographical factors are at play, and the elucidation of these can help one to understand better a thinker's ideas or demonstrations.

In the vein of Newman, who was his most decisive intellectual influence, Bouyer admits that his monograph on cosmology is intensely personal.[5] He confesses it to be, in fact, rooted in his own experiences of the cosmos. In the *Prélude* of the book, he relates experiences that he understood to be foundational for his cosmology and asserts that visions of this sort, which are available to all, are crucial in order to grasp the meaning of his text:

> It seems to me that one can know or recognize the world only if one has had some visions of this kind, as little as they may resemble mine. They are in the reach of everyone, and everyone, without doubt, if he wants, can call them forth from the high funds of his memory. But if someone refuses to do so or to believe it possible, it is better that he close this book immediately: the following will have no meaning for him.[6]

There is a tension in Bouyer's theology in this regard, and it extends beyond his cosmology. On the one hand, he thinks that a theological work exists not to render account of the thought, let alone the "visions,"

5 See Zordan, *Connaissance et Mystère*, 471–78. Zordan recounts at length Newman's personal and intellectual influence on Bouyer.

6 *Cosmos* (Fr.), 12. Unfortunately, neither the *Prélude* nor the *Postlude* were included in the English translation of the book.

proper to its author, but to expound the truth received and proclaimed in and by the Church in the course of the centuries. On the other hand, he recognized, with Newman, the singular character of each theologian's reception of the tradition, which involves uniquely personal, even creative, reception. It is perhaps with this tension in his work in mind that his former pupil, Cardinal Jean-Marie Lustiger, described Bouyer as "the least conformist of theologians and among the most traditional."[7] We can certainly say that this statement is pertinent to his cosmology.

The present work is not a biography of Bouyer. However, out of respect for his convictions, and in the spirit of the foregoing, it is necessary to recount basic biographical points, especially those pertaining to his unique experience, recovery, and development of Christian cosmology. The very idea of biography is conceptually important to Bouyer, inasmuch as imagination and mythopoetic thinking are philosophical underpinnings of this recovery and development. Imagination and story make possible the "real assent" of the intellect, tying our concepts of nature to personal experience and union with the world. Even cosmology requires attention to a mode of thought akin to the biographical, wherein concrete history and experience are addressed.

In this chapter, we shall explore the importance for Bouyer, in his personal and intellectual formation, of the theme of imagination as a means of access to the divine presence in creation or of creation in God. We shall try to develop a broad sense of what he means by imagination in his developed work by exploring his sources. This gets at a topic of great underlying significance in *Cosmos*. We shall see here that this non-conformist thinker is in fact a singular representative of important strands of Christian tradition. He was, as Joseph Ratzinger once said, "a mind of a very special character," but he did not stand alone in clarifying the human discovery of God through imagination.[8] It will be appropriate to begin, in a first section, with an overview of the singular breadth of his theological work and the central significance that he accords to the cosmological theme, particularly in his trilogies dedicated to systematic or dogmatic theology. This first step is fitting given that the full scope of

7 *MT*, 11.

8 Joseph Ratzinger, *Milestones: Memoirs 1927–1977*, trans. Erasmo Leiva-Merikakis (San Francisco: Ignatius Press, 1998), 143.

his work and accomplishments is little known in the English-speaking world. In a second section, we shall talk specifically about his aforementioned religious experiences of the cosmos, and, in a third section, explore the liturgical framework of his cosmological vision. In a fourth section, we shall connect his thoughts on imagination and cosmology to the Inklings and to the modern tradition of English Christian Platonism and conclude, in a fifth section, by introducing his sophiology in connection with that of Bulgakov.

THE AUTHOR'S LIFE AND WORK

Born in 1913, on the cusp of the First World War, Louis Bouyer was eventually to become one of the most important Catholic theologians in Europe in the twentieth century. The author of over fifty books, some posthumously published, and many articles, he was one of the four great post-war theologians in France.[9] First a Lutheran pastor, ordained in 1936, then a Catholic priest and Oratorian, ordained in 1944, the vocation of theologian was central to his ministry, and the sheer volume of his literary output is estimable, especially given the duties and tasks to which he was frequently called. In the midst of a long life of ministry, teaching, and overall ecclesial busywork, Bouyer maintained an exceptionally consistent productivity in writing, which began in the 1930s and ended in the middle of the 1990s, when Alzheimer's disease began to get the better of him. His oeuvre covered many domains: homiletics or catechetics, biblical and Patristic studies, liturgical studies, the history of spirituality, polemical controversy, fundamental as well as dogmatic theology, and even novels.

His professional scholarly career began in earnest as a Lutheran ministry student first with the Protestant faculty in Paris then in Strasbourg, where he completed a licentiate thesis in theology on Saint Athanasius's ecclesiology, which was later published, unchanged, by a Catholic publishing house.[10] After his conversion, he completed a doctoral dissertation at the Institut Catholique in Paris (1945), continuing along this

9 Jean-Robert Armogathe, "Obituary for Louis Bouyer," trans. Adrian Walker, *Communio* 31 (2004): 688–89. The other three "great theologians" were Yves Congar (1904–1995), Jean Daniélou (1905–1974), and Henri de Lubac (1896–1991).

10 *L'Incarnation et L'Église: Corps du Christ dans la Théologie de Saint Athanase* (coll. Unam Sanctam, no. 11, Paris: Éditions du Cerf, 1943) [hereafter *IÉ*].

Athanasian line. It dealt with the early monastic spirituality found in Saint Athanasius's *Life of Saint Antony* and was also published, in 1950.[11] The Athanasian foundation of Bouyer's theology is one among many indicators of his indebtedness to Newman, who held the Alexandrian monk, theologian, and defender of Nicaea as one of his greatest heroes among all the Church Fathers. Moreover, it indicates an attraction to monastic life and thought. Bouyer was converted through the instrumentality of the Benedictine monks at Saint Wandrille Abbey in Normandy, and he lived with them near the end of his life, before the final illness that necessitated his removal to the care of the Little Sisters of the Poor in Paris.

In 1945, Bouyer published a mystagogical account of the liturgy of the Easter Triduum, *Le Mystère Pascal*, a study that proved to be his most influential writing during his lifetime and established him in his early thirties as among the most widely recognized theologians in the Church.[12] The book has been described as "epochal" in its importance and effect, for it was instrumental in helping the modern Church to discover the unity of the Paschal Mystery in a new light as a locus for theological and spiritual contemplation.[13] Bouyer showed that the Cross, Resurrection, and Ascension of Christ constitute together an intrinsic integration of divine action in the world and are, taken together, an essential starting point for Christological and Trinitarian reflection. This emphasis had the effect of highlighting the divine *agape* in a way that many readers of the day found enlightening. The Oratorian did not limit the attention of prayer and theology to the Cross, Resurrection, or Incarnation taken in abstraction from one another. The unity of the divine Mystery, summed up in the expression "Paschal Mystery," communicated through the liturgy of the

11 *La Vie de Saint Antoine. Essai sur la Spiritualité du Monachisme Primitif* (Saint Wandrille: Éditions Fontenelle, coll. "Figures Monastique," 1950). Bouyer's work is notably "monastic" in character, and he lived with the Benedictines at Saint Wandrille near the end of his life.

12 *Le Mystère Pascal (Paschale Sacramentum): Mèditation sur la Liturgie des Trois Derniers Jours de la Semaine Sainte* (coll. Lex orandi, no. 6, Paris: Les Éditions du Cerf, 1945). Published in English as *The Paschal Mystery: Meditations on the Last Three Days of Holy Week*, trans. Sister Mary Benoit (Chicago: Regnery, 1950) [hereafter *The Paschal Mystery*].

13 See R. Cabié, "Quand on Commençait à Parler du Mystère Pascal," *La Maison Dieu* 240 (2004): 7–19. Though Patristic in spirit, the term "Paschal Mystery" is not in fact a Patristic expression.

Church and drawing us into the divine life, became an essential key to the considerations of the Second Vatican Council, as is seen especially in the text of its first major constitution, *Sacrosanctum Concilium*.[14]

A comprehensive theologian, Bouyer's oeuvre has been submitted in recent years to significant exegetical analysis in France, Italy, Belgium, and Spain, some of the productions of which we highlighted in our introduction. The full scope and impact of his writing has just begun to emerge. Jean-Luc Marion claims that he has, since his death in 2004, "compelled recognition more and more as one of the greatest theologians of his century, and not only in France."[15] He was a writer of high literary sophistication, good-natured wit, and, at times, stinging polemical sarcasm.[16] The literary dimension of his work needs to be emphasized to get a full picture of the man. He was well-read in classical and modern drama, literature, and poetry, having obtained a *Licence de Lettres* (Bachelors of Literature) from the Sorbonne. He taught humanities to high school level students for thirteen years at the famed Oratorian school in Juilly, established in 1638, in Seine-et-Marne in France, an assignment which he much enjoyed and gladly carried out. He wrote four novels pseudonymously, one of which, *Prélude à l'Apocalypse*, published under the appellation "Louis Lambert" in the same year as *Cosmos*, has been the subject of important secondary literature in France.[17] Posthumously published and still unpublished manuscripts show the sophistication of his reading of the Western literary canon.[18] The human wisdom that he gained from his lifelong immersion in literary classics is deeply embedded in his properly theological writings and reflected in his elegant style

14 For a discussion on this point, see Dom Hugh Gilbert, *Unfolding the Mystery: Monastic Conferences on the Liturgical Year* (Herefordshire: Gracewing, 2007), 87. It is in fact Gilbert who describes this book as "epochal." It is ironic that this theme was so important for *Sacrosanctum Concilium*, given Bouyer's low esteem for the main contributor to the text of this constitution, Annibale Bugnini (1912–1982), the "architect of the liturgical change in the Roman rite since 1948 and principal author of the Vatican II document on the liturgy, *Sacrosanctum Concilium*" (quoted from *Memoirs*, 219, n. 77).

15 Marion, "Introduction," *La Théologie de Louis Bouyer*, 10.

16 See especially *Memoirs*, 207–30.

17 See Weill, *L'Humanisme Eschatologique de Louis Bouyer*, 276, n. 1.

18 See especially *Lectures et Voyages: Compléments aux Mémoirs* (Paris: Ad Solem, 2016), 19–66 [hereafter *LV*].

of composition, even though his developed theology often implicitly suggests rather than directly references the individual works that constitute this literary wellspring. His very cosmology reflects his literary preoccupations, as we shall see in the following sections of this chapter.

Bouyer's life and work were international in scope. He taught at Strasbourg and at the Institut Catholique of Paris and was an invited lecturer at several European and American universities. He served as an emissary to different continents after Vatican II for the Pontifical Secretariat for Christian Unity. Travels to the Christian East during this time were especially significant to him, and Eastern Christian thought was always at the heart of his theological engagement.[19] He had a comfortable intellectual home at Oxford University, where he could indulge a lifelong fondness for certain strands of Anglophone thought and culture. He was associated with influential and significant scholarly reviews, such as *Dieu Vivant* and *Communio*, for which he made many contributions, and was a mentor of the now-eminent scholars of global reputation who founded the latter journal in its French-language edition, Jean-Luc Marion, Jean Duchesne (b. 1944), Jean-Robert Armogathe (b. 1947), and Rémi Brague (b. 1947).[20] He was a *peritus* in the period during and surrounding Vatican II, and Pope Paul VI, who admired his work and continued to follow Bouyer's new writings even as pope, named him a member of the first International Theological Commission (ITC) in 1969.[21]

19 See *En Quête de la Sagesse: du Parthénon à l'Apocalypse en Passan par la Nouvelle et la Troisième Rome* (Jouques: Éditions du Cloître, 1980). In this volume, Bouyer recounts journeys he took in the 1970s in the Christian East, in the footsteps of Saint Paul and Saint John.

20 Cf. *Memoirs*, 246–47, n. 16.

21 *MT*, 18–19. The first ITC was populated by theologians the scope of whose work, creativity, and influence has no parallel in our contemporary period. The names include Karl Rahner, Hans Urs von Balthasar, Yves Congar, Henri de Lubac, and Joseph Ratzinger. See Bertrand Lesoing, *Vers la Plénitude du Christ*, 245, n. 4. Fr. Lesoing shows that during the fifteen years of his pontificate Paul VI referenced seven volumes in Bouyer's work. He points the reader to the index of the different volumes of *Insegnamenti di Paolo VI* (Rome: Libreria Editrice Vaticana). Four of the works cited were published after Paul VI's election to the papacy in 1963. In 1976, the pope, in a message to the Roman clergy, recommended Bouyer, along with Henri de Lubac and Cardinal Journet (1891–1975), as authors they should read in order to enlarge their doctrinal horizon and nourish their contemplation and preaching.

The crowning achievement of his life's work was the aforementioned nine-volume synthesis of the whole of Catholic doctrine, broken up into three trilogies, of which *Cosmos* is a part. It is necessary to summarize the basic structure and purpose of these volumes in order to get a proper grasp of the place of theological cosmology in his overall writings. The trilogies as a whole distinguish, in the manner of the Church Fathers, "economy" (*οἰκονομία*) from "theology" (*θεολογία*), the former having to do with the history of salvation, or consideration of Christ with his face turned toward creation, and the latter having to do with the eternal Mystery of God, or consideration of Christ with his face turned toward the Father. Bouyer's first trilogy was on economy and the second on theology, although, by the inner logic of the work, taken as a totality, economy and theology are distinguishable therein only while remaining inseparable. Bouyer gives in these volumes an exposition of the whole Christian Mystery on the basis of different but complementary points of view. The six volumes of the first two trilogies were published between 1957, when his classic book on Marian theology, *Le Trône de la Sagesse*, first appeared, and 1982, when *Cosmos* was at last published. He foresaw the ordering of these volumes and described it in his 1957 book on Mary:

> This book is intended as the first of three in which we hope to develop a theology of creation and of the whole economy of grace as designed by God for its benefit. It forms a kind of sketch of a supernatural anthropology, that is, the theology of man and of his destiny in the eyes of God. It is to be followed by a work on the Church and the people of God, in other words, an essay on supernatural sociology. A third essay, devoted to the world, both material and spiritual, from the physical universe to the world of angels, will set forth, as it were, a supernatural cosmology, the theology of the whole creation, considered as a single whole.... If God permits, this trilogy of an "economic" theology... a theology of the plan of God in His works, will itself form an introduction to another trilogy, one that is "theological" in the strict sense. We hope, after tracing the outlines of a "pneumatology," a theology of the Holy Ghost, and of a "Christology," a theology of Christ the Word made flesh, to show, finally, how

> the whole of theology issues in the knowledge and adoration of the Father, "in spirit and in truth."[22]

The actual ordering of the publications varied somewhat from this early, prospective overview. The volumes on theology "in the strict sense" were written and published before the final volume on the economy, *Cosmos*. The volumes on the economy thus enveloped or encompassed the volumes on theology. This should not be taken to suggest that Bouyer understood consideration of the economy to be more important than theology, but that there is a double primacy of the two trilogies with their complementary approaches. In good sophiological manner, "economy" and "theology" in fact intertwine within each particular volume of Bouyer's trilogies. The Church's dogmatic, Trinitarian understanding is inseparable, in each volume, from consideration of the revelation through which we are elevated to know and love God in Jesus Christ by the gift of the Holy Spirit.

Later in life, Bouyer added a third trilogy, which Marion characterizes as a "Christian epistemology": *Mysterion* (1986); *Gnôsis* (1988); *Sophia* (1994). This epistemology is necessary, Marion says, to receive in full the first six volumes on the economy and theology.[23] It answers to our need today for a "gnosis of faith," partaking of the mystery of charity brought to completion in us in communion with divine Wisdom in and through the Church. Marion notes that this third trilogy returns to the point of departure of the first book, to the "supernatural anthropology," where Bouyer describes the Virgin Mary as the created, personal means through which "man becomes the very place of Wisdom."[24] *Sophia*, Bouyer's final and ultimate work, gives the key to the other books. A primary purpose of the whole is to bridge theology and economy and overcome the split between theology and life that is one of the tragedies of Christian existence in the modern age. *Sophia* especially illuminates *Cosmos*, but *Cosmos* is itself a summative volume, as it recapitulates the first two trilogies and prefigures in full the final trilogy. It is, naturally, the central text in his recovery of Christian cosmology and the cosmos.

22 *MSW*, viii-ix.

23 Jean-Luc Marion, "Introduction," *La Théologie de Louis Bouyer*, 13.

24 Ibid. Marion says: "l'homme devient le lieu même de la Sagesse."

Cosmos is, as we have seen, a "supernatural cosmology" and an essay on the "Christian interpretation of what is called 'the modern scientific vision of the world'" that takes into account modern cosmologies and the "phenomenology of science."[25] Interestingly, the publication of the book in 1982 was intentionally delayed, as its redaction was greatly advanced already in the early 1970s. Marie-David Weill suggests that there were two motivations for this delay. The first was that *Cosmos* is precisely the most "intensely personal" of Bouyer's writings. It presents what Bouyer himself called a "synthesis" and so was better left to the end of his first two trilogies. Second, although we cannot give a logical primacy to either economy or theology in Bouyer's theology, economy comes first and last in this respect: humankind can know itself and the cosmos in full only in the light of the revelation of God in Christ.[26] *Cosmos*, the last volume on the economy, thus was published after the trilogy on theology was complete. Bouyer recognized that the cosmological theme is uniquely comprehensive, and that it is reasonable to summarize the whole of the dispensation for our salvation in the light of this widest horizon.

EXPERIENCE OF THE DIVINE PRESENCE

As much as any theologian in the twentieth century, Bouyer gave cosmology a central place in his work. He had a lifelong interest in cosmology and the philosophy of nature, which stretched back to pre-adolescence. In his youth, he wanted eventually to consecrate his life to scientific and philosophical investigation in order to articulate, as he once said, "for the world of today the permanent value of a Christian vision of the universe."[27] He thought that he might someday be a philosophical or scientific cosmologist, but God had another, more important, vocation in mind for him, first to Protestant then to Catholic ministry. Still, a fascination for cosmology never left him, and *Cosmos* is its ultimate fruit.

25 *CG*, xii, xiv. See our introduction, nn. 16 and 18.

26 Weill, *L'Humanisme Eschatologique de Louis Bouyer*, 44–45. See Louis Bouyer, *Rite and Man: Natural Sacredness and Christian Liturgy*, trans. M. Joseph Costelloe (South Bend, IN: University of Notre Dame Press, 1963), 3: "It is only in the mystery of God that light is thrown upon the mystery of man." This passage anticipates, almost word for word, *Gaudium et Spes* 22, the text which John Paul II took to be the centerpiece of Vatican II.

27 *MT*, 22.

It is startling to read in his *Memoirs* how early in his life he took to cosmological issues in a philosophically serious manner. He had a passion for science, and, when he was "barely twelve years old," developed a natural philosophy that was an attempt to explicate a universal understanding of reality based on atomic theory as it was understood at that time.[28] Looking back decades later on this system, he realized that his insights were akin to those of the Greek Pythagorean tradition:

> I was unwittingly following a sort of Pythagorean ideal. Far from implying any materialism, it brought nature itself into the mind by a kind of rational monism in which numbers and figures (seemingly circumscribed to it) made up all of reality in a coherent vision that the absolute Mind gives to Itself in our own minds, which are in every way dependent upon it.[29]

Needless to say, this is a remarkable philosophical perspective for someone on the cusp of his teenage years. This philosophy of nature was, in Bouyer's own estimation, perfectly congruent with the Evangelical Protestant catechism in which he was being instructed in the faith at that time.[30] Like Newman, Bouyer's instinct for faith led him in his youth to see the world in the light of a kind of *sui generis* evangelical Christian Platonism, and he wanted to articulate this way of seeing in a philosophical system.

One of the most formative events in Bouyer's life was the unexpected death of his mother at this young age. Before the time of her death, he lived an idyllic youth with her. She was both his primary affective companion and his first intellectual guide. Understandably, her death tore him apart. He turned inward in a near-pathological manner, and his burgeoning Christian Platonism was, as a result, transformed into an unremitting solipsism. He came to doubt the existence of consciousnesses external to his own and to think that the existence of God and other beings might be nothing more than projections of his own mind. He was advised by a psychologist, the brother of a famous chemist and hero of his, Georges Claude (1870–1960), to interrupt his cosmological studies

28 *Memoirs*, 44. Bouyer describes this as his "personal *Eurêka*."

29 Ibid.

30 Ibid.

and was sent by his father on Claude's advice to a retreat with some old family friends in the little town of Sancerre in the Loire Valley in the French countryside. Bouyer's experience there transformed his life. His solipsism was overcome, and his instinctive grasp of the real existence of an enchanted world was clarified, giving his life's work a lasting impetus.[31]

The Romantic sensibility that pervades his entire monograph on cosmology emerges in some remarkable pages of his *Memoirs* dealing with his youthful experience of the countryside. These highlight his consistent demonstration throughout the *Memoirs* of a poetic or natural philosophical gift for visual perception and memory of physical creation as it is manifested in concrete places. He rhapsodizes in these pages on the encounter that overwhelmed him on his first evening in Sancerre after stepping from the train. "I had been wrested," he reports, "from my reading by the golden softness of that landscape, in whose harmonious lines waterways and woods surrounded peaceful and simple houses casually grouped along riverbeds and tree branches." He took a frugal meal in the middle of the woods by a beach where, on the opposite bank, "the declining sun had not yet let the shade completely absorb the three hills where Sancerre, the first town on the north side, completely dominated the landscape." The river at his feet "slid its emerald mirror between its sandbanks along leafy islands whose reflection shimmered on its water's surface." "All of this," Bouyer continues, "was so serene and sure in its beauty that I felt a sudden peace and was won over forever."[32]

The effect on him of this and other visits to Sancerre was comparable to that of the English Lake District on nineteenth-century English Romantic poets, such as William Wordsworth (1770–1850), for whose work Bouyer felt a great fondness. It was precisely solitary walks in this countryside that enabled him to overcome his solipsism, because he sensed that the world encountered there "called out to human presences and immersed them."[33] These walks in the woods reawakened and clarified his earlier philosophy of nature. They gave it a basis in experience, in Newman's real apprehension or real assent, in the whole of his being, body, intellect, will, and affect. He came to understand that the world is personal in its

31 See especially *Memoirs*, 43–56.

32 Ibid., 43–44.

33 Ibid., 49.

essence, a sign or symbol of a limitless source, the perfect Person. It is, as Bouyer develops in *Cosmos*, the fundamental language in and through which the Creator communicates Himself to us. The Sancerre landscape evoked for Bouyer "supernatural presences behind or within nature itself."[34] These experiences, together with influential readings that formed his interpretive or perceptual lens, most importantly Newman's *Parochial and Plain Sermons* and *Apologia*, instilled in him an understanding that the divine presence is given to us only in and through sensible images. This helps to explain his attraction to Saint Thomas Aquinas's philosophy, even before he first heard it explained by Étienne Gilson (1884–1978) in lectures that he surreptitiously attended in Paris while still a Protestant ministry student.[35] Bouyer recognized his experiences to be a "renewed discovery . . . of the most precisely Evangelical kind of Christianity."[36] This "renewed discovery" was, to his mind, far from any sort of pantheism or superstition but was attuned to the Gospel preached by Christ himself in word and deed.

There is another dimension to his theocosmic experiences, which were not limited to the ones just described. Perhaps influenced by his encounters with and readings of Bulgakov and Newman, Bouyer, later in life, interpreted some of his youthful social engagements with nature in a distinctively sophiological manner. The Russian sophiologists are noted for their profound and richly-described experiences of the divine presence in creation.[37] These experiences brought them to conversion to the Christian faith and were often accompanied by a mysterious feminine presence. In the *Prélude* to *Cosmos*, in a manner that evokes the sophiologists, Bouyer interprets his experiences at Sancerre in the light of his friendship developed there with the young Elizabeth, a cousin of his friends.[38] As a young

34 Ibid.

35 Cf. *MT*, 27; *Memoirs*, 65. It was August Lecerf (1872–1943) who first introduced Bouyer to Aquinas's thought in ecclesiastical Latin classes. Lecerf disabused him of any residual Scotism that he may have harbored from reading the Swiss philosopher Charles Secrétan (1815–1895), whose work was otherwise very influential for Bouyer, particularly with respect to its personalism and sense of the link between history and nature. See *Memoirs*, 63, n. 26.

36 *Memoirs*, 49.

37 See Michael Martin, *The Submerged Reality: Sophiology and the Turn to Poetic Metaphysics* (Kettering, OH: Angelico Press, 2015), 143.

38 *Cosmos* (Fr.), 11–19.

man and Lutheran ministry student, Bouyer wanted to marry Elizabeth, but her mother rejected him, and the marriage was not to be.[39] When they were adolescents, they translated together into Latin lines from Newman on the world of spirits taken from his aforementioned *Parochial and Plain Sermons*. Elizabeth was for Bouyer the very face of the Loire Valley, as, perhaps not incidentally, Cardinal Newman's sister Mary Sophia was, after her death, so often the face of the world for him.[40]

In the *Prélude*, Bouyer speaks of another geographical location, albeit of a very different character, that stirred his capacity for cosmic mysticism. In visits to Contentin, on the northwest coast of Normandy, he perceived the harsh coastland, arid and beaten by waves, as an expression of both the finitude of the world and of the promise for its eventual transfiguration. A crucifix erected there seemed to him to stand as a pledge of cosmic resurrection.[41] On one visit to Contentin, he read with two girls of a family for whom he seems to have been a spiritual father Hans Christian Andersen's *The Little Mermaid* (1837). One of the girls, he reports, perceived the sad reality and the other the solid hope articulated in this classic modern tale. The two girls were, for Bouyer, like a double face of the world present in this coastland region, one tormented and the other serene. They reflected both the luminous and the somber atmosphere of the cosmos as it shows itself in this particular place.[42] In explanation of these experiences in the Loire Valley and at Contentin, Bouyer begins the *Prélude* by saying: "To speak of the world impersonally is impossible."[43] The communal personalities who constitute this world are, he continues, united in groups that "assume or tend to assume a personal face, but which will only be in the end one or another feminine face."[44] For Bouyer, the natural cosmos and natural societies, when personalized in feminine presences, anticipate the bridal character of the eschatological, cosmic Church, brought to completion through the redeeming work of divine grace.

39 Cf. *Memoirs*, 56, 86.

40 See Jan Walgrave, *Newman the Theologian*, trans. A.V. Littledale (New York: Sheed and Ward, 1960), 314. See also Bouyer, "Le Souvenir de Mary Newman," in *Études* 246 (1945): 145–59.

41 *Cosmos* (Fr.), 15–16.

42 Ibid., 18–19.

43 Ibid., 11.

44 Ibid., 12.

Bouyer did not have explicit Marian visions or find himself guided by some mysterious supernatural sophianic figure in the manner of Soloviev. Yet he saw the world as a kind of feminine visage, a manifestation of the "Eternal Feminine," participated in by his female friends, of various ages, whose presence in his life surely gave him great consolation for the life-long spiritual wound that he suffered from his mother's tragic early death.

COSMIC LITURGY

There are two other personal experiences of the world related in *Cosmos* that we must briefly describe. The French-language edition of *Cosmos* contains not only a *Prélude* but a *Postlude*, and it is in this that we are told of these other seminal encounters. These occurred in Athens and Constantinople, and had to do not with the landscape in relation to feminine personages of Bouyer's acquaintance but with the greatest pagan and Christian vestiges of religious cult.

The first experience was elicited on a beautiful September morning in Athens in the 1970s, on a hill that Bouyer playfully suggests may have been the Hill of Muses. In view of the Acropolis and its temple, the Parthenon, Bouyer was struck by a sense of the cosmic vocation of humanity. The human being, he knew from his reading of the Church Fathers, was given the task to complete the Divine Artist's work of creation, and the meaning of human vocation was palpable to him on this morning, in this land still marked by its ancient art and culture. Despite the ambiguities of the Athenian gods and their temples, as well as the failure of the ancient cities to achieve the cosmic perfection that humanity has always sought, Bouyer wondered if the Pallas Athena, the goddess of wisdom present there, does not "evoke a first glimpse of divine Wisdom?"[45] If so, he suggests, the golden numbers of the bundle of columns of the Erechtheion, the temple in the Parthenon dedicated to her and to Poseidon, "indicate it even more secretly."[46]

Our author's experience in Athens pertained to the cosmic significance of what he would call the "pagan sacred."[47] This was prolonged and elevated in his second, later experience, on a visit to the Hagia Sophia

45 Ibid., 378.
46 Ibid.
47 *MT*, 103.

in Constantinople, where a supremely telling vestige of the supernatural sacred remains. Crossing the Sea of Marmara on a misty evening, the Oratorian elevated his eyes from Greek temples onto the arches and abysses of this holy cathedral, dedicated to God's Wisdom by Emperor Justinian in the sixth century. Bouyer perceived the splendor of the divine Wisdom in the evocation of the liturgy as it must have been celebrated there, uniting heaven with earth by the invocation of the Holy Spirit in the *epiclesis* (*ἐπίκλησις*). The divine Wisdom was made manifest in plenitude in this unique liturgical space, but this did not lessen the mystery that it contained. The liturgical cosmos of this supernaturally sacred site maintained the incompleteness of creation in the plenitude of the supernatural revelation that it bodied forth. The Christian cult practiced there was interrupted by historical events. It became a mosque after the sack of Constantinople in 1453. It was not the full achievement on earth of the eternal Eucharist that can come only through the completion of the eschatological Church. Reflection on its continued architectural presence in the midst of the vicissitudes of history heightened Bouyer's expectation for the eschatological fulfillment of the world.[48]

These two experiences of the divine Wisdom in the cultic re-creation of human endeavor, whether pagan or Christian, are crucial for understanding Bouyer's overall cosmovision and recovery of the cosmos. His appreciation for the cosmic dimension of faith and the quest for wisdom were intertwined with his love for and work to promulgate liturgical piety. In the English-speaking world, Bouyer remains known to this day largely as a specialist in liturgy and spirituality. Yet he was attracted to liturgical studies in no small part because he recognized that Christian existence is inexorably cosmic. Human being is being in the world, and humanity has a vocation to shepherd the whole of creation to perfection. For Bouyer, this vocation is accomplished first and foremost through right worship or glorification of God. This is, after all, what it means, in the first instance, to be "orthodox," and this, he realized, cannot be wholly irrespective of the ceremonial particularities of communal faith, even if his liturgical studies were not first and foremost concerned with questions of rubrics. The Oratorian's cosmology is ultimately a theology of cosmic liturgy that

48 *Cosmos* (Fr.), 378–79.

embraces the concreteness of the Church's synaxis in its universal reach.

From the 1940s until Vatican II, Bouyer was a seminal figure in the liturgical movement. He is known for becoming sharply critical of the liberalizing approach to liturgy taken by the influential Centre de Pastorale Liturgique in Paris, whose original mission he helped to inspire, and of liturgical reform in general after Vatican II.[49] His stinging passages in the *Memoirs* that relate his general distaste for the postconciliar liturgical establishment and his special antipathy for the bureaucrat Annibale Bugnini (1913–1982), the chief architect of postconciliar liturgical reform, seems to be the aspect of the book that has garnered the most attention in the English-speaking world.[50]

Bouyer's own realization of the importance of liturgy was the fruit of maturation in faith and growth in theological as well as historical understanding. He discovered over time the cosmic breadth of Christian experience. During years of his visits to Sancerre, in spite of his cosmic mysticism, Bouyer was initially drawn to a type of quietist spirituality or Quakerism which refused the idea that God joins Himself to us through created intermediaries and stressed instead direct, personal contact of God with the individual soul. The cosmic dimension of his discovery of God led him out of this temptation, enabling him to postulate what he calls "a social and incarnate worship" as essential to faith.[51] He came to see that what was good in the individualist spiritual tradition of his Protestant youth could be brought to fulfillment only by being taken into and elevated by the Catholic tradition of mystical, liturgical experience and theology.[52]

In many of his writings, our author attempts to maintain a harmonious balance between individual, social, and cosmic dimensions of experience. The link between these is synthesized and demonstrated in a conjoined theological and philosophical modality in *Cosmos*. With Wordsworth, he understands that the world is so much with us that we cannot be dissociated from it:

49 Cf. *MT*, 63–95. See Matthew Levering, *An Introduction to Vatican II as an Ongoing Theological Event* (Washington, DC: The Catholic University of America Press, 2017), 1–19. These pages contain the best overview in English of Bouyer's work as a liturgist in relation to Vatican II and *Sacrosanctum Concilium*.

50 *Memoirs*, 224–25.

51 Ibid., 51.

52 Ibid., 51–52.

> In spite of what we may think or imagine, we remain inseparable from the world, to which we belong through everything we are, even in our most sublime mystical ecstasies and our loftiest metaphysical speculations. This being, who may be capable of reaching the most perfect vision of the Being of beings, must nevertheless remain in and of the world.[53]

Our very embodiment, individual and social, is essential to our identity. It is, at the same time, intrinsically meant to be actualized in praise. We are created for worship in the composite unity of body and spiritual soul ordered to relation. This lesson is ubiquitous in Bouyer's writings.

Around the time of the 1940s and 1950s, when Bouyer came into his own as a scholar, modern *Religionswissenschaft*, or religious science, took a turn in a quasi-Catholic direction in some quarters, supporting the view that man is *homo religiosus* before being *homo faber*, *homo economicus*, or *homo politicus*. Ritual and myth anthropologists emphasized more and more that ritual worship is the source activity for human sociality, the basis of all human culture, and that it cannot be dismissed simply as a relic of a long-outgrown, infantile stage of human consciousness. Bouyer realized that this turn was propitious for theology, and he incorporated what he took to be the best insights resulting from it as a fundament for his hermeneutic of cosmological science.[54] He took the work of the scholars of religion as decisively important, and his Christian interpretation of the modern, scientific worldview is ultimately a liturgical interpretation aided by ritual anthropology. He used this work to show that it is only through the sacred that the world is made known to us as cosmos. Bouyer made use of insights from religious science in the service of understanding the meaning of modern science as such and as a whole. He understood that scholarly explication of the meaning of ritual, myth, and liturgy can be a key resource for developing anew a

53 *Cosmos*, 12.

54 See *LV*, 12. Bouyer says: "My perception of the capital importance of understanding liturgy in order to enter straightaway into all the meaning and axial views of the dogmatic tradition necessitates bringing me soon to add to these studies [on Scripture and Tradition] that of the modern history of comparative religion. . . ." His focus is especially phenomenological.

coherent Christian cosmology or for plumbing the meaning of science as a human and therefore cosmic achievement. Scientific cosmology, on this view, is a human endeavor inscribed in the wider cultural need to give praise to the Creator and to draw all of creation into the mystery of the Eucharist.

Bouyer tried to synthesize and utilize to this end aspects of the work of several scientific scholars of religion: Rudolf Otto (1869–1937), Joachim Wach (1898–1955), Gerard van der Leeuw (1890–1950), Mircea Eliade (1907–1986), Georges Dumézil (1898–1986), A. M. Hocart (1893–1939), and E. O. James (1888–1972). There was, however, a particular alternative school or line of contemporary religious thought whose influence had greater impact on the development of his hermeneutic of science and the cosmos than any or all of these specialists in religious studies. This alternative school did not operate within the parameters of the *au courant* specialists, although its proponents understood in a profound way the meaning of the Christian Mystery in the light of world religions and sometimes referenced the specialists. Still, it was not, at least in the twentieth century, taken into the heart of mainstream religious studies or theological science. In Bouyer's work, the specialists are taken seriously, but their ideas regarding religion are contextualized in the wider train of thought of this other, in some ways profounder, movement, whose Christian commitment is sometimes explicit, and for which cosmic experience and cosmic liturgy find anthropological reception and expression in the rational power of "creative imagination."

CREATIVE IMAGINATION

This other direction of thought is that of the highly-influential Anglo-Catholic literary circle known as "the Inklings," most famously and importantly represented by J. R. R. Tolkien (1892–1973), C. S. Lewis (1898–1963), Charles Williams (1886–1945), and Owen Barfield (1898–1997). Insights shared by these literary figures, who were also variously philologists, scholars of religion, historians, theologians, Christian apologists, and philosophers, are clearly discernible within the text of *Cosmos*, even though Bouyer does not reference them in this book.

Tolkien's influence was especially prominent in Bouyer's life. The French Oratorian established "writer's friendships" with several notable twentieth-century novelists including Tolkien, whose works he read assiduously.[55] He was first introduced to the *Lord of the Rings* trilogy in 1955, through the mediation of Fr. Patrick Bushell (b. 1912), an Oratorian from London, and he was so impressed that he wrote the first French-language review of it.[56] He exchanged correspondences with Tolkien in the 1960s. In his *Memoirs*, Bouyer recounts, in typical Bouyerian fashion, his "best conversation" with Tolkien, in the latter's "Headington study, on the heights overlooking Oxford." The conversation grew from their shared amusement at some overly-serious and naïve (mostly) American scholars, contributors to a published volume of essays on Tolkien, who took seriously and therefore misunderstood farcical answers that the English mythmaker provided to a questionnaire they had given to him.[57]

In the course of their correspondences and conversations, Tolkien helped to deepen and clarify Bouyer's understanding of the relationship of the Christian Mystery to pre-Christian myths. From childhood, Bouyer was fascinated by Arthurian legends and the *matière de Bretagne* (the "Matter of Britain").[58] He read virtually the whole of the vast literature that concerns the legend of the Holy Grail. His encounter with Tolkien and the Inklings helped him to grow in understanding and capacity to articulate the deep religious meaning of the Arthurian Legend. It illuminated thereby his understanding of the place of the Christian Mystery in the vast scope of world history.[59]

55 *Memoirs*, 177–81. See also Michaël Devaux, "'Le Probleme de l'Imagination et de la Foi,' chez Louis Bouyer: en lisant, en écrivant avec les Inklings," in *La Théologie de Louis Bouyer*, 141–62. This is a very important study by Devaux, based on archival research at Saint Wandrille Benedictine Monastery, where unpublished notebooks and manuscripts by Bouyer are held. In this article, Devaux is able to trace with precision the authors and books that Bouyer read with respect to the Inklings as well as Coleridge. Much of what we say in this section is indebted to Devaux's research. Other writers that Bouyer befriended include T.S. Eliot (1888–1965), Julien Green (1900–1998), and Elizabeth Goudge (1900–1984).

56 Bouyer, "Le Seigneur des Anneaux, une nouvelle epopee?" *La Tour Saint-Jacques* 13.4 (1958): 124–30. Later in life, he returned explicitly to this theme again in *Les Lieux Magiques de la Légende du Graal: de Brocéliande en Avalon* (Paris: OEIL, 1986) [hereafter *LLM*].

57 *Memoirs*, 180.

58 See especially *MT*, 255–63.

59 Ibid., 114.

As with the Inklings, so with Bouyer, symbolism is of the utmost importance for metaphysics and epistemology.[60] Like them, he frequently draws our attention back to the symbol, and the Grail is, for him, the symbol *par excellence*. It is centered on the perennial royal myth, and he interprets the Arthurian cycle in light of the discovery by James Frazer (1854–1941) in the nineteenth century of the fundamental importance of sacral kingship to humanity's entire religious heritage. The ancient king, according to this understanding, was held in honor as a charismatic person uniquely in touch with supracosmic reality who inherited the vocation of an even more ancient figure, the shaman. The trifold power of the king as head of the city-state, as priest, and as prophet was seen as a gift received from the gods.[61]

The king could lead the city, because he was, like the shaman, a wise seer, the breadth of whose vision of the world surpassed that of the common run of humanity. He was in touch with the divine and therefore singularly qualified to organize life in this world. Yet the shaman was able to attain to the status of seer only by succumbing to a "creative malady" or mysterious passage that led to the death of his old, natural life.[62] Frazer understood that the king could, in like manner, save his people only by dying and suffering in their place. The king replaced the shaman, but eventually the shamanic figure emerged anew when the responsibility to govern the city became such an immense burden that it required a splitting of tasks. The shamanic seer now served as a functionary of the king. It is in this context that Bouyer interprets the figures of Arthur and Merlin: the shaman stood by the king as a reminder of the sacral rooting of kingship, just as the wizard Merlin stood by King Arthur as his advisor.[63]

Hugo Rahner showed in a classic study on theology and myth how the Church Fathers interpreted the Greek myths in a Christian light.[64] Bouyer carries out a similar task with respect to the Arthurian cycle in relation to ancient royal myths. The wounded king who saves his people and

60 Ibid., 109–12.

61 *LLM*, 29–32.

62 See *Memoirs*, 47. Here Bouyer speaks of his upheaval after his mother's death as a "creative malady."

63 Ibid.

64 Hugo Rahner, *Greek Myths and Christian Mystery*, trans. B. Battershaw (New York: Harper and Row, 1963).

all of the earth is, for him, a figure for Christ. Arthur is another David, a premonitory sign of the messianic king and of his victory over the malign powers. He is betrayed by Guinevere, Lancelot, and his traitorous nephew, Mordred. Arthur is both the thaumaturgic king and expiatory victim. The Fisher King who emerges in a later phase of the cycle is a figure of the Suffering Servant in Isaiah and evokes the *Ichthus*, the fish that was a symbol for Christ, the Son of God and Savior of the World. The royal talisman of the Grail and the mysterious procession witnessed by Percival in the House of the King are ciphers for Christian liturgy and the Eucharistic cup. The Arthurian cycle exemplarily demonstrates the Christian transfiguration of the pagan sacred:

> So the royal talismans are going to transmute into the Christian sacraments, tending themselves to the vision, to the eschatological realization of their "*res tantum*," of their ultimate reality. This metamorphosis will lead to that of the great symbols of the primitive forest, of the lake, which concentrates on the figure of Merlin. They will come to signify no longer the perpetual rebirths of a life which renews itself only in death in order to tend there anew, but the hope of a life fallen from its origin to attain finally, through this death, lost immortality. Because divine love, Christic, immolated, alone stronger than death, then seized this other love which, by covetousness, tends only to death, in order to rectify it, to recast it in death itself.[65]

Of undoubtedly pre-Christian origin, the Grail signifies "the communication to men by the gods of celestial nourishment that permits them to accede to a superior life."[66] Legends based on this theme lend themselves to Eucharistic interpretation. The Eucharist is the food of life by which humanity is elevated to a properly supernatural communion with the triune God revealed in Christ. The Eucharist heals, perfects, and elevates human nature. It communicates the fruits of our redemption, won through the ordeal of the suffering and death of the Savior on Calvary. Its dramatic-narrative context, the Gospel itself, points to the ultimate

65 *LLM*, 40.
66 Ibid., 115.

"Euchatastrophe," the surpassingly happy ending of the cosmos in the Parousia by the return of the wounded and glorified King of Creation.[67]

The Arthurian myths, on this interpretation, reproduce the original process of development by which the divine Word "shattered then recomposed the myths, with their symbols, indissolubly connected to the ontogenesis as to the phylogenesis of our psyche," ultimately around the figure of Christ.[68] In *Cosmos*, Bouyer aims to spell out with precision the particularities of the process of shattering and recomposing of ancient mythic cosmogonies accomplished by the divine Word through revelation. Turning to his book on King Arthur and the Grail, we get a deeper sense of his understanding of the inexorable importance and significance of religious symbolism in general and of the particular pathway through which the Christic recomposition of ancient religious symbolism was achieved. This is crucial in order to understand his cosmology, because he so strongly holds the view that theological cosmology benefits from explicit recognition that human intellectual endeavor is always actualized by way of the symbol. The universal is made known in and through the concrete. The invisible world is present in real, concrete places in the visible world, and the symbol unites these two domains. God makes Himself known to us as divine Word in and through the words of creation and the religious word, including mythic and ritual symbols, in order to be heard by us. Our intellect requires attunement to the symbol in order to receive divine revelation, and the phalanx of religious symbols present in royal myths provided the matrix utilized by the divine Word, in transfiguring manner, to make himself known to us.

This analysis of royal myths and Arthurian Myth is coherent with the view held by some of the Inklings, from whom Bouyer learned much on this and other matters. It is known that Bouyer read Tolkien's epic trilogy at least five times, *The Hobbit* at least three times, and *Tree and Leaf*. This was not enough to slake his thirst for what the Inklings had to offer.[69] He read secondary literature on Tolkien and works by C.S.

67 J.R.R. Tolkien, "On Fairy Stories," *Tree and Leaf* (London: Harper Collins, 2001), 68–69; *MT*, 115.

68 *MT*, 115.

69 Devaux, "'Le Probleme de l'Imagination et de la Foi,' chez Louis Bouyer," 144–47.

Lewis and Charles Williams as well, including their seminal book on Arthurian Legend.[70] *Cosmos* is not without parallels to Owen Barfield's *Saving the Appearances*, even though Bouyer found this particular book to be rather disappointing.[71] The Inklings helped Bouyer to delve into the very anthropological and gnoseological foundation of myth, including royal myth, and he shares with them a larger metaphysics of the creative imagination that makes mythogenesis possible.

The Western tradition contains a long heritage of reflection on this integrative power of consciousness. Bouyer and the Inklings gladly inherited this tradition, while making their own decisive contributions to it. The Inklings did so both as individuals, each with his own particular insights, and as a loose family of thought—that was not without perduring squabbles! They had immediate predecessors in the nineteenth century in Newman, Wordsworth, Samuel Taylor Coleridge (1772–1834), and Friedrich Wilhelm Schelling (1775–1854). The ancient tradition of Christian Platonism, coming through the Cambridge Platonists of the seventeenth century, influenced these later writers in decisive ways. It is important to recognize at the outset that the appeal that the Inklings and Bouyer make to imagination, as with their nineteenth-century forebears, is not at the expense of reason. Bouyer explains:

> Far from being some sentimental musing, poetic intuition and imagination (as Schelling and Coleridge understood it) should be considered not as irrational but as a higher form of reason. It is an intuitive grasp of the deeper meaning of things and of existence. Opening the way to a truly human existence, it must maintain that focus failing which our life no longer reaches its full potential.[72]

70 Charles Williams and C.S. Lewis, *Taliessen through Logres: The Region of the Summer Stars and Arthurian Torso* (Grand Rapids, MI: William B. Eerdmans, 1974).

71 *Cosmos*, 258, n. 17. Bouyer says of this book: "Perceptive comments, strangely intermingled with insubstantial reveries, may be found in an initially fascinating but ultimately disappointing essay by Owen Barfield, in *Saving the Appearances*." See Barfield, *Saving the Appearances* (Middletown, CT: Wesleyan University Press, 1988).

72 *Cosmos*, 161. See Cyrus P. Olsen III, "Myth and Culture in Louis Bouyer:

Etymologically, the word imagination evokes both *imago*, meaning image or representation, and *imitor*, meaning to imitate. The word *imago* translates the Greek *eikon* (εἰκών), and the English words "image" and "icon" can both translate it. The word "imagination" may be directly derived from the Greek *eikasia* (εἰκασία), which is used in the Platonic dialogues. With respect to its dual evocation, pointing both to *imago* and *imitor*, Mircea Eliade said that for once etymology reflects reality: "The imagination *imitates* the exemplary realities—the Images—reproduces, reactualizes and repeats them without end."[73] For Eliade, imagination is a faculty of *homo religiosus* striving to pattern his life after the exemplary activity of the gods or the divine by mimesis and reproduction, the latter a creative or "re-creative" accomplishment.

Coleridge understood creative imagination or "primary imagination" to be the symbol-making faculty.[74] He defined it as a repetition in the human mind of the Creator's very act of creation: "The primary IMAGINATION I hold to be the living Power and prime Agent of all human Perception, and as a repetition in the finite mind of the eternal act of creation in the infinite I AM."[75] It is through this primary form of imagination, a creative form of mimesis of the archetypal realities ubiquitous in the sensory world, that the human person most fully images God. Coleridge's poetic creations were at the service of imagination and the reawakening of human consciousness to the divine presence in creation. Wordsworth, Coleridge's friend, was in fact the greater poet, and he also defended imagination as a higher form of reason. In his *Prelude*, which the *Prélude* of Bouyer's *Cosmos* invokes, Wordsworth said: "Imagination is reason in its most exalted mood." Coleridge, though, was more fully a theorist of imagination. He developed an epistemology and ontology

On Louis Bouyer's Theology of Participation," in *Gregorianum* 95.4 (2014): 775–98. Olsen captures well the importance of the theme of imagination in Bouyer's thought.

73 Mircea Eliade, *Images and Symbols: Studies in Religious Symbolism*, trans. Philip Mairet (Princeton, NJ: Princeton University Press, 1991), 20.

74 Samuel Taylor Coleridge, *The Statesman's Manual*, in *Lay Sermons*, ed. R. J. White, vol. 6 of *The Collected Works of Samuel Taylor Coleridge*, ed. Kathleen Coburn, Bollingen Series LXXV (Princeton, NJ: Princeton University Press, 1972), 30.

75 Samuel Taylor Coleridge, *Biographia Literaria*, ed. James Engell and W. Jackson Bate (Princeton, NJ: Princeton University Press, 1983), 1:304.

of imagination with the more precise goal of overcoming the disjunction between literalist and Romantic understandings of symbolism that mar human perception of the world in the modern age. These distortions, Coleridge insisted, eviscerate our capacity to think realistically of metaphor and mystery. Literalism, scientific or religious, flatly eschews metaphor and mystery, while certain versions of Romanticism embrace them as all-important but turn them into mere projections of human feeling or sentiment. Neither is able to attain a unified sense of the depth of meaning inherent to any true symbol or to grasp adequately that human intellection is ultimately and irreducibly symbolic.

Some have argued that Coleridge is himself Romantic in the sense just described, a pure subjectivist, but he has many defenders who see him as an advocate of a realist sacramental ontology of creation, with a differentiated, analogical understanding of metaphor and mystery. According to this reading, Coleridge understood the whole of creation to be a symbol of God's being and activity, and he did not think of imagination—which both interprets and creates symbols—in a purely Kantian way, that is, as a constituting power of the transcendental ego. The French philosopher Gabriel Marcel (1889–1973) insisted that imagination is for Coleridge a realistic power of symbolism that allows the universal species to show through the individual, and that he understood the symbols it creates always to participate in the realities symbolized.[76] A symbol is for Coleridge an idea bodied forth through the power of imagination in images. Coleridge is, on this reading, a defender of both the objective reality of ideas and of the *analogia entis*. He speaks at times of the "consubstantiality" of being, but this should be understood to correspond to the traditional doctrine of analogy, the basis of the Catholic theological resolution of the question of the One and the Many, for which all finite beings participate in the unity of *esse creatum* and ultimately in the transcendent *esse divinum* in differentiated ways.[77] However, there was one sense in which he was indeed aligned with wider

76 Gabriel Marcel, *Coleridge et Schelling* (Paris: Aubier-Montagne, 1971), 187–89. Devaux remarks that when it comes to imagination, Coleridge, and Schelling, "Bouyer always cites" this book by Marcel, including in unpublished manuscripts. See Devaux, 153, n. 1; *Cosmos*, 261, n. 1.

77 Cf. Coleridge, *Statesman's Manual*, 29.

Romantic currents of thought and with Schelling: he held that there is an archetypal domain within the subconscious or unconscious human self.[78] Interior archetypes, too, are indispensable to the imaginative production of symbols.

The Inklings turned to imagination in their own way to address the modern problem of disenchantment and anti-cosmism. Their work can at times evoke wider twentieth-century criticisms of modern abuses of Galilean physics.[79] If the Inklings, even at their most rigorous, were not exactly themselves rigorous philosophers, their work nevertheless connected thematically with some decisive twentieth-century philosophical insights, however incidentally so. Philosophers as diverse as Husserl, Alfred North Whitehead, and Michel Henry (1922–2002) targeted in common with the Inklings modernity's annihilating bifurcation between ideal entities chased by physics and the immediate, concrete life-world of human consciousness.

These philosophers each saw the Galilean method of physical analysis as the source of a persistent tendency to perpetrate what Whitehead called "the fallacy of misplaced concreteness." This fallacy confuses abstract for concrete realities. It splits reality into two domains, first, the realm of qualities in the life-world, and, second, the realm of quantities dissected by modern mathematical physics with its geometrical idealism. The former has to do with basic human experiences, such as the scent of a rose or the color of the sky, which are thought by the heirs of Galileo to be acosmic or mere constructions of the human mind. The latter domain of purely quantitative extension and geometrical ideality is thought to be the noumenal underpinning of the now-vanished cosmos, which has become a "meaning-shorn universe."[80] This fallacious bifurcation of reality is the essence of disenchantment.

In a magisterial study of C. S. Lewis, the French scholar Irène Fernandez has expounded at length Lewis's turn to imagination as a remedy

78 Cf. J. Robert Barth, *The Symbolic Imagination: Coleridge and the Romantic Tradition* (Princeton, NJ: Princeton University Press, 1977), 105–27. We follow Barth in much of this paragraph while disagreeing with his description of the *analogia entis* and so alter the presentation on that front.

79 See especially Barfield's *Saving the Appearances*, 92–95.

80 Charles Taylor, *A Secular Age* (Cambridge, MA: Harvard University Press, 2007), 774.

for Galilean bifurcation.[81] Lewis held that this bifurcation exacerbates a weakness that is congenital to human reason. In Lewis's own words: "Human intellect is incurably abstract. Pure mathematics is the type of successful thought."[82] Galilean physics, he thought, simply exploits *in extremis* the inherent woundedness of finite reason and language, for these cannot, in and of themselves, by virtue of their mutual finitude, reconcile subject with object. They cannot fully unite experiential concreteness with universal concepts. If we allow ourselves to be falsely anaesthetized to this woundedness, we are prone to conjecture that concrete experience can be built up entirely on the ground of realities susceptible to full explication by objects that are, in Jean-Luc Marion's words, "poor in intuition," that is to say, reducible *a priori* to constitution by the human mind.

Lewis held that the abstraction of reason renders impossible the total engagement of a thinking being in what it thinks: "The more lucidly we think, the more we are cut off: the more deeply we enter into reality, the less we can think."[83] Experience needs reason and concepts in order to become articulate, but finite language and reason cannot capture the fullness of concrete realities in the net of their abstractions. A higher, more integrative form of reason is needed, less bogged down by the wounds of finitude. This form of reason is the *nous* or *intellectus* of the Church Fathers and great Scholastic theologians, which Coleridge, for his part, associated with imagination, as Bouyer would do much later.[84] This gives context to Lewis's own understanding. Rationalism of the sort that Lewis embraced in his scholarly youth allows no place for intellectual approaches to reality and truth that go beyond the mathematical ideal. It

81 Irène Fernandez, *Mythe, Raison Ardente: Imagination et Réalité Selon C. S. Lewis* (Geneva: Ad Solem, 2005), 127–73. It is not her language. But she does speak, with the phenomenologists, of "Galilean physics."

82 C.S. Lewis, "Myth Became Fact," in *God in the Dock: Essays on Theology and Ethics* (Grand Rapids, MI: Eerdmans Publishers, 1970), 65.

83 Lewis, "Myth Became Fact," 65.

84 Cf. Samuel Taylor Coleridge, *Shorter Works and Fragments*, eds. H.J. Jackson and J.R. de J. Jackson (Princeton, NJ: Princeton University Press, 1995), 2:1268–69. See *LC*, 372. Bouyer says: "What we call the poetic sense is in us like the vestige of the contemplative intellect, and therefore as a remnant inspiration which draws us not toward simple utility but toward the beauty of the world." Further down in the passage he refers to this poetic reason as imagination. Before the passage, he describes *nous* and *intellectus* as intuitive reason.

promotes a detached form of reason and relegates poetry and myth, the products of what Lewis later considered (like Wordsworth) to be a higher form of reason, to a stage of consciousness that has been superseded. This, for Lewis, leaves subject and object in a hopeless condition. They cannot, in principle, attain reconciling communion. In the midst of this epistemological crisis that hobbles modern civilization, Lewis exalted imagination, "born," as Fernandez says, "at the crossroads of flesh and spirit."[85] Imagination, "especially in its mythopoetic function,"[86] is called forth to "realize this prodigy"[87] of healing the rift between the universal concept and concrete experience.

The thought of the Inklings corresponds on this matter, as we have suggested, to a larger tradition of Anglophone Christian Platonism, especially literary, and it is worth noticing how early in his career Bouyer turned his attention to this tradition. In his first published study of Newman, in 1936, he wrote:

> Is it not precisely this quality in English literature which impresses us most? On the one hand: a marvellous realism, a feeling for the concrete, for the seen and heard and touched that can hardly be found anywhere else; and yet, at the center of this very real world, something entirely different from what we are accustomed to see there. On the other hand: a vision of the ideal world which seems to be attained almost effortlessly; and yet the objects in this world, usually represented only by abstractions, are as rich and living and definite—more so in fact—than those things which we see with our eyes, and hear with our ears, and hold with our hands.[88]

This literary tradition overcomes the typically modern form of the fallacy of misplaced concreteness. It "effortlessly" beholds within the immediacy

85 Fernandez, *Mythe, Raison Ardente*, 157.

86 Ibid.

87 Devaux, "'Le Probleme de l'Imagination et de la Foi,' chez Louis Bouyer," 154.

88 Bouyer, "Newman and English Platonism," *Monastic Studies* 1 (Pentecost 1963): 111–31, at 115–16.

of sight and touch to which it is uniquely attentive the "ideal world" that it realizes is even more real and concrete than the merely sensory domain. The invisible, ideal world is "rich and living and definite." It is the realm of "faerie" that Tolkien accessed in his "sub-creations." One thinks of William Blake's (1757–1827) warning with regard to thin philosophical concepts of spirit that "a Spirit and a Vision are not, as the modern philosophy supposes, a cloudy vapour or a nothing; they are organized and minutely articulated beyond all that the mortal and perishing nature can produce."[89] Imagination, in this tradition, joins the abstract with the concrete in a living apprehension, revealing the richly personal character of the ideal world. Mind and its properties are not confined to the walls of the human body, and it does not suffer a thinning out into air by being recognized as spread throughout the length of the cosmos. Mind is conceived as richly intricate in its simplicity. Qualitative existence can be accepted as truly cosmic, as both noumenal and phenomenal at once.

It was Newman, as we have already said, who first introduced Bouyer to the importance of imagination, which the founder of the Birmingham Oratory later called real apprehension and real assent, distinguished from merely "notional" varieties of the same. In *Newman's Vision of Faith*, one of his final studies, Bouyer connects Newman with Coleridge in this regard and explains again that imagination is no mere "fantasy." It is "really creative."[90] The poet, *poietes*, "makes something." The poetic imagination awakens in us a "presentiment," what Saint Paul called "the expectation of all creation" (Romans 8:19). It awakens us to the numinous. Newman realized, Bouyer says, that "the poetic imagination is led by a mysterious instinct best expressed by the Greek poet quoted by Saint Paul ('we are come from the race of gods') or by Wordsworth, Newman's contemporary ('trailing clouds of glory do we come from God')."[91]

89 William Blake, *Descriptive Catalogue*, IV. Quoted by C.S. Lewis in *The Discarded Image: An Introduction to Medieval and Renaissance Literature* (Cambridge University Press, 1964), 133.

90 Bouyer, *Newman's Vision of Faith*, 37.

91 Ibid. See William Desmond, *Is There a Sabbath for Thought? Between Religion and Philosophy* (New York: Fordham University Press, 2002), 132–66. Congruent with Newman and Bouyer, Desmond explains that imagination mediates body and mind. It concretizes ideality. It ranges over creation and is the source of our self-transcending: through it, we discover the other within

SOPHIOLOGY

The Inklings, with their ruminations on imagination, have an important place in Bouyer's thought, even if they are only part of a much larger tradition of intellectual parentage. There is a perennial — not to say static and absolutely uniform — theological trajectory in the history of Christian thought with which Bouyer wanted to associate his theology of the cosmos that gives much further depth to his thinking on imagination. This tradition includes the great classical Christian theologians of the Patristic Age, particularly the Alexandrian Fathers and Cappadocians, and includes Aquinas most of all among the High Scholastics, particularly his *De Veritate*. Like Edith Stein (1891–1942), of whom he became in life "more and more . . . in many regards a disciple,"[92] Bouyer was drawn to the explicit and particularly rich Trinitarian cosmology and participationist ontology of this text, which he saw as a key to interpret the whole of the Angelic Doctor's corpus. Edith Stein, for her part, represented for Bouyer a culminating figure in the Church's mystical tradition, exemplified by Medieval Rhineland mystics and flowing through the Carmelite spiritual masters into the twentieth century.[93] All of these thinkers helped to solidify and deepen his vision of creation, which he describes in summary fashion in the *Memoirs*:

> [T]he material, physical world cannot be separated from an invisible, essentially "intelligible," spiritual world. The material world is, so to speak, the common irradiation of this spiritual world as far as concerns the first-born spirits, the angels; from it the human mind emerges and finds within it not only its medium of communication, but also the awakening of its consciousness. This world, wherein the intelligible and the sensible form a single tapestry, is but a single thought of God. It is eternally present in

ourselves. He describes it as "metaxological," a mediating power in several senses. It is the power at the birth of religion that enables the discovery of God.

92 *LV*, 14.

93 See the whole of Bouyer, *Women Mystics*, trans. Anne E. Nash (San Francisco: Ignatius Press, 1993).

> Him and projected in time and simultaneously in the distinct existence of other consciousnesses.[94]

This vision of creation was to Bouyer's thinking the common patrimony of the deepest liturgical, theological, and spiritual testimonies of the Church, however much different nuances may show forth in its expression through the ages. Bouyer, with a profound grasp of the historical dynamism of human thought, was certainly aware of these nuanced differentiations within a larger spiritual unity.

Our author was introduced to another tradition that gave him new insights into this sacramental worldview and its biblical roots. In his time as a Protestant ministry student in Paris, he encountered eminent representatives of Eastern Orthodoxy, several of whom deeply impacted his theological and spiritual formation.[95] The Russian Bulgakov, himself a theologian of the Holy Grail, especially elicited his admiration.[96] This brilliant Russian theologian would shape the very direction of his life's research. Bouyer marveled at the former Marxian economist from the first time that he heard him preach, in English, at an ecumenical worship service in Paris.[97] Although he became a friend of Bulgakov's arch-rival Vladimir Lossky (1903–1958), whose criticisms of Bulgakov he took seriously, it was on Bouyer's advice that the publishing house Aubier began to

94 *Memoirs*, 55.

95 Ibid., 66–67, 69–70, 72–75, 243. These pages recount Bouyer's encounters with Lev Gillet (1893–1980), a former Benedictine monk who became a priest and "a monk of the Eastern Church." He was received into the Russian Orthodox Church in 1928. Bouyer says of Gillet (on p. 66) that "no one . . . has had a deeper influence on me." "The amount," Bouyer continues, "of theological, or more generally ecclesiastical information I owe him is hard to imagine, not to mention the disparate knowledge in nearly every field."

96 See Bouyer, "La personnalité et l'œuvre de Serge Boulgakoff," in *Nova et Vetera* 53.2 (1978): 135–44. See also "An Introduction to the Theme of Wisdom and Creation in the Tradition," *Le Messager Orthodoxe* 98 (1985): 149–61. See Sergei Bulgakov, *The Holy Grail and the Eucharist*, trans. and ed. Boris Jakim (Herndon, VA: Lindisfarne Books, 1997).

97 *Memoirs*, 58. This was at the Anglican Holy Trinity Pro-Cathedral. Bouyer comments: "This was the occasion for me to discover both the unexpected splendor of Orthodox liturgy and the brilliant personality of Father Sergei Bulgakov, who preached in that correct yet rough-hewn English that was always to be the medium of his contacts with the West."

publish translations of Bulgakov's work into French.[98] He had an admirable, if not flawless, grasp of the main direction of Bulgakov's immense theological and philosophical output. He was inspired by him to pursue an explicitly sophiological or sapiential direction of thought.[99] Bulgakov was the preeminent Russian sophiologist or theologian of Wisdom in the twentieth century, following the path blazed by Soloviev. Although not a "disciple" of Bulgakov, Bouyer drew freely, though critically, from the thought of the Russian master. He ultimately linked his own developed cosmological vision with sophiological dogmatics:

> Later on I would come to recognize it [the world] as the projection of a Wisdom of creation outside of the eternal Word, animated by the divine Spirit who, at the same time, urges it to return to this filial Word to espouse it and come back up with it, in the same Spirit, to the Father, as if in an eternal Eucharist.[100]

Bulgakov said that sophiology "represents a theological or, if you prefer, a dogmatic interpretation of the world . . . within Christianity."[101] We would be justified in marveling with Bouyer at Bulgakov's theoanthropocosmic synthesis. The Russian, Bouyer says, always kept his gaze transfixed on the Paschal Mystery: "All of his theology . . . is only a cosmology and anthropology of the Transfiguration, but showing through from the Cross."[102] Bulgakov explains that "a special characteristic of sophiology" is that it considers "anthropology in its connection with cosmology."[103] Bouyer himself, in sophiological fashion, urges at the very beginning of the first chapter of *Cosmos* that biblical revelation itself should move us to think of the world as inclusive of both cosmos and

98 See Lesoing, *Vers la Plénitude du Christ*, 80, n. 3. Lesoing notes that the first volume published was *L'Agneau de Dieu* [The Lamb of God], in 1944. See *Memoirs*, 76–77. These pages briefly discuss Bouyer's relationship with Lossky, the famous "Neopatristic" theologian.

99 See especially, *MT*, 210–11.

100 *Memoirs*, 56.

101 Sergei Bulgakov, *Sophia: The Wisdom of God*, trans. Rev. Patrick Thompson, Rev. O. Fielding Clark, and Miss Xenia Braikevitc (Hudson, NY: Lindisfarne Press, 1993), 13.

102 Bouyer, "Personnalité et l'œuvre," 136.

103 Bulgakov, *Sophia*, 5.

anthropos. Not only Paul's writings but John's are determinative on the point. "For God," Bouyer quotes John's Gospel, "so loved the world [*τὸν κόσμον*] that he gave His only Son, that whoever believes in him should not perish but have eternal life."[104] "The world," he says, interpreting this text, "appears so closely linked to us so as to be practically synonymous with mankind."[105]

One can glean from reading a little between the lines of *Cosmos* that as far as Bouyer is concerned German Idealists, Romantic nature philosophers, Newman, and the sophiologists were all correct on some level to see the world as inextricably connected with human history. It is, after all, he maintains throughout the text, biblical to see it thus. If we are beings in the world, Bouyer teaches, which divine revelation as much as sound philosophy emphasizes, then the world is not simply an external container that can be readily dispensed with, and its very purpose is intrinsically linked to our destiny in God. A single entelechy inspires the world and human history. This is a properly apocalyptic understanding of the world. In the first chapter of *Cosmos*, Bouyer wonders if "history has a field extending from man to the entire universe."[106] He answers this question throughout the text in an affirmative but layered manner with particular concern to show how divine revelation leads us to answer this question in a new, transcendent light.

Not to be outdone by Bulgakov, Bouyer recognized everything that is at stake in speaking of the world from the perspective of Christian, "dogmatic" interpretation. Theological cosmology, on the shared view of these two sophiologists, should not be mute on questions pertaining to the meaning of human activity, of creativity, of history, and of culture. Bulgakov once explicated, in a manner congruent with Bouyer's later, developed thought, two seemingly contrary but dialectically related standpoints on interpretation of the world that have been prevalent in the modern age.[107] For the first, Bulgakov explained, the world has been seen precisely as a container to be dispensed with, or, even more so, as essentially in-formed by evil. There are biblical passages that can be taken

104 See *Cosmos*, 3; John 3:16.
105 *Cosmos*, 3.
106 Ibid.
107 Bulgakov, *Sophia*, 14–15.

to support this standpoint, such as from John's First Epistle: "Do not love the world or the things in the world, for the whole world is under the power of the Evil One."[108] Christian life, in this first pattern of thought, is seen as a search for escape from the world. Cosmos does not include *anthropos*, and *anthropos*, in the image of God, is fundamentally acosmic, having no more concern for the cosmos than does the misanthropic deity itself. This anti-cosmism or Manichaean option within Christianity is promoted by what Bulgakov calls "the 'pseudo-monastic' outlook."[109]

For the second, the Church and the world are conceived as "two incongruous bodies that cannot be united" except by way of either subjecting the world to the massive power of the Church or by Christian compromise with the world.[110] In the modernist or liberal version of this second, worldly form of Christianity, Christians are beckoned to change their way of thought and life to fit with the demands of the "world"—understood as modern society relating to the cosmos without reference to God. Christian faith and theology have been for so long detached from society as sources of cultural vitality that many Christians themselves have come to hold that human flourishing is the prerogative of secular culture and not the Church. The Church succumbs to the totalitarian logic of secular society. Bulgakov saw sophiology as providing a third way between failed Christian strategies of escape and accommodation, and Bouyer, if not a disciple of Bulgakov, is nevertheless an associate of his in promoting this third way, which they jointly understand to be authentically monastic, apocalyptic without being Gnostic, cosmic without being pantheist, and creative without being historicist.[111]

An important dimension of the sophiological third way of re-uniting Church and world is that it too promotes a turn to imagination.[112] It

108 1 Jn. 2:15 and 5:19. Bouyer often points to the interpretive difficulty of these passages.

109 Bulgakov, *Sophia*, 15.

110 Ibid.

111 Cf. Bulgakov, *Sophia*, 14–21; Bouyer, *MT*, 175–76. Bouyer explores a similar dialectic in the case of Karl Barth (1886–1968) and his followers, especially Bultmann. He says that Barth's acosmic Christianity (which can be associated with the first party in Bulgakov's dialectic) led to de-Christianized Christianity among his followers.

112 See Antoine Faivre, "Theosophy," in *The Encyclopedia of Christian Theology*, ed. Jean-Yves Lacoste (New York: Routledge, 2005), 3:1566–69. Faivre

does so quite explicitly in order to inspire a renewed sense in the broader culture of the divine presence that suffuses all things. We conclude this chapter with a brief listing of three specific ways in which sophiology evinces this turn to imagination, before, in the next chapter, casting a closer eye on Bouyer's own developed phenomenology of imagination or mythopoetic thinking in *Cosmos*.

First, sophiology encourages theological poetry. Can a theologian, qua theologian, when exercising her scientific faculty, also be a poet? Must the two endeavors, theological inquiry and poetic expression, be separated from one another? To the Slavic theological mind, it seems, it is not necessary to keep these two endeavors apart, as if locked in separate cages of human aspiration and vocation. The theology of the Russian sophiologists is often elegant and poetic, even filled with moments of poetic effusion. Bouyer himself said of Pavel Florensky, in describing precisely his foremost speculative writing, *The Pillar and Ground of Truth* (1914), that he was "a poet of undeniably visionary power."[113] Moreover, sophiology, with the unifying reach that is unique to poetic intuition, provides a holistic perspective on creation and on God's relation to the cosmos. The work of the greatest sophiologists is a sign of integrated, poetic personality. They do not separate their speculative work from spirituality, aesthetics, and ethics. They do not forsake the expressive image and the creative power of the symbol-making capacity of the mind. This is a way of theology that encourages an approach to the intellectual life as a total ordering of the soul in harmony with the real as shown forth in the totality of the sensory plenum.

Second, sophiology is self-consciously attentive to the theology of the symbol and the image in iconography. For the Russians, the theology of Wisdom has roots not only in experiences given to them in creation by the mysterious feminine emissary of God who led them forth but in what they took to be explicitly sophiological icons in the Russian

---

shows how important imagination and myth have always been to the Behmenite theosophical tradition that is one of the sources of the thought of Soloviev and Bulgakov. See especially *MT*, 187–208.

113 Bouyer, "La Colonne et le Fondement de la Vérité de P. Florensky," *Communio* [French] 1.2 (1975): 95. The French translation was published that year by L'Age de Homme, Lausanne. Bouyer reviews the translation made by Constantin Andronikof.

tradition. The icons of Sophia in Novgorod, specifically, are, for them, focal hieroglyphs of the presence of the personal Wisdom that suffuses creation and an indispensable locus for theology.[114] Bouyer was aware of this tradition of iconography and gave his own interpretation of it. He held that the icons of Novgorod beloved by the sophiologists refashioned pre-Christian Earth Mother cults. They took the nature imagery of these cults into their symbolism in an elevating way, with reference to Mary, the Church, and Christ.[115] He avers that the interpretation of these images has been ambiguous at times and has subdued a full sense of the Christian transformation of pagan cultus. He detects similar ambiguities in the sophiological theology taken as a whole. In his own "quest for wisdom by the imagination," as Michaël Devaux describes it, he seeks to sift sophiology of any propensity to collapse Christian Wisdom into an insufficiently refined pagan wisdom.[116] Still, to his mind, the speculative work of the Russian sophiologists is creatively yet symbiotically joined to this tradition of iconography in an imaginative unity that can serve as a methodological paradigm for other Christian theologians, East and West.

Third, the very metaphysics of the sophiologists supports the flourishing of imagination. It does so in part by encouraging a new thought of the divine essence in terms of the Father's self-revelation in the dyad of the Son and the Holy Spirit through Wisdom and glory. We might, the sophiologists teach us, think divine being anew in the light of imagination by thematizing divine phenomenality in the *perichoresis* of divine love. This aesthetic rendering of divine life is what Stratford Caldecott (1953–2014) had in mind when he said that Sophia just is "the divine Imagination, hypostatized in relation to each of the three Persons in turn."[117] Wisdom bears an iconic status in God as the "reflection of eternal light, a spotless mirror" (Ws. 7:26). Divine Wisdom, for the tradition of thought that Caldecott expounds, which includes the work of Bouyer, is eternally the showing forth of the unity of divine being in the relations of the Trinitarian persons that will be imaged in fullness on the created

114 Bulgakov, *Sophia*, 4.

115 Bouyer, *Vérité des Icônes* (Paris: Éditions Criterion, 1990), 94.

116 Devaux, "'Le Probleme de l'Imagination et de la Foi,' chez Louis Bouyer," 159.

117 Stratford Caldecott, *The Radiance of Being: Dimensions of Cosmic Christianity* (Tacoma, WA: Angelico Press, 2013), 233.

plane by the perfection of the cosmos in the eschatological Church through the corporate divinization of redeemed and saved humanity. Even more, imagination can be understood to be an effective power in God as related to creation. God brings into being and perfects *esse creatum* by "the eternal act of creation in the infinite I AM." We think that Bouyer understands Wisdom in this way, and our eighth and twelfth chapters will expound this interpretation. In the latter, final chapter, we shall see that with respect to created Wisdom sophiology transforms the theme of imagination as it was utilized in nineteenth-century philosophy of nature. Overall, sophiology inspires us to think the God-world relationship by means of attentiveness to the unified *Gestalt* of cosmos and history in and through the biblical figures of Wisdom and glory. It draws deeply not only from the Wisdom books but from Saint John's apocalyptic images, especially that of the Bride of the Lamb. God's Imagination and our own, made in His image, become a metaphysical consideration in a unique way in the sophiological trajectory of thought, challenging us to seek real apprehension of the divine unity, omnipresence, and all that is entailed in our hope for the eschatological perfection of creation. Bouyer's cosmology is rooted in both apocalyptic imagination and the theology of Wisdom. These are, if distinguishable, inseparable in the end. The Oratorian's joint, albeit diffusive, deployment of these themes discloses their essential kinship.

CHAPTER 2

Myth and Knowledge of the World

THIS PERSONAL BACKGROUND WE HAVE JUST provided can help us to understand better a central argument in *Cosmos* which Bouyer begins to develop already in the first two chapters: that myth and the mythopoetic function of creative imagination are essential to the progress of both rational thinking and divine revelation. Myth, on this view, is not a mere preamble to later developments of human thought and culture that can be tossed aside, once humanity is sufficiently progressed, as a relic of an outgrown stage of human consciousness. Our overall understanding of history requires recognition of continuity as well as discontinuity between the mythic mode of human consciousness and movements of reform that tend to accuse the mythopoetic mind of being the fabricator of idols. Even in ages when movements of reform are at their most puritanical pitch, humanity still comes face to face, at special times, in dreams, revelatory experiences of nature, or in poetic inspiration, with the fact of the permeability of the individual mind to vast and mysterious realms of spiritual existences and ultimately to the divine, transcendent source of being. Moreover, modern, scientific directions of thought have not been entirely immune to inducements to re-enchantment. In his appeals to sophiology, to the Inklings, to certain specialists in the study of religion, and to all who are sympathetic to these mystical or sacramental directions of thought, Bouyer finds a community of specifically modern thinkers who can be called upon to help to rediscover the enchantment of the world in a renewed imagination informed by living faith.

In this chapter, we turn to a more direct exegesis of *Cosmos*, focusing on the first two chapters of the book, which will involve further discussion of the manner in which Bouyer weaves together these modern sources into his work in order to recover a biblically apocalyptic way of seeing the world. The introduction and first two chapters of *Cosmos* deal with the issue of gnoseology directly. This is Bouyer's preferred starting point in discussing cosmology. These chapters call our attention to the fundamental place of religion and the sacred in the development of human cosmic awareness. We take three especially significant themes as crucial in order to get to the heart of these chapters: first, the importance of religion, ritual, and myth in shaping knowledge; second, the importance of mythopoetic thinking or creative imagination and the sacred as a constitutive power of human consciousness; third, the religious depths of the mystery of language. We shall explore these three themes successively, in separate sections. The first theme pertains more to chapter one, entitled "The World as Question," and the second theme more to chapter two, entitled "The World as Object." We present the third theme as a way of setting up much that is to come in subsequent chapters of this study.

MYTH, RITE, COSMOLOGY

The reader encounters from the outset in the first main chapter of *Cosmos* a sequenced series of 57 questions regarding the world, which Grintchenko has comprehensively summarized and expounded in her treatise on the book. She notes that Bouyer's starting point evokes Thomas Aquinas's "tour of the horizon" (*tour d'horizon*) of theological objections in the *Summa Theologiae*, and she explains well the significance of the sequencing of the questions. "The explicit or implicit responses that we give to these questions," she says, "and the manner by which we perceive them determines our relation to God and our understanding of the world in its complexity and unity."[1] Bouyer holds that the divine Word modifies our questions, especially because it introduces us to the radically new knowledge that the world is a free creation *ex nihilo*. Recognition of this requires that we take into account "the relationship between the knowledge of the world arising directly from the actual experience of man as he is living in the concrete world, and the revised and completed knowledge given by the Word of God [to Israel and the Church]."[2] The last question raised in the fifth section of the chapter is of the utmost importance in order to grasp how Bouyer relates our knowledge gained through natural experience with that which is given through revelation.

1 Grintchenko, *Cosmos, Une Approche Théologique du Monde*, 41.
2 *Cosmos*, 7.

It follows from a more basic level of reflection on whether there is an antinomy between spirit and matter. He puts one and the same query in two different ways:

> From that level of reflection [on the relation of spirit and matter], we reach (or return to) more complex or deeper questions, although they are but subtler forms of earlier ones: in speaking of the world, do we have in mind an object facing us, so to speak, or a reality which includes us? In other words, can we count on the common presupposition of nineteenth-century German idealist philosophers, that there is a basic opposition between Nature and Spirit (*Geist*), with man seen as the emergence of Spirit (*Geist*) into the world of Nature, while still in unavoidable tension with the world?[3]

Grintchenko rightly suggests that the problematic signaled by this question orients the whole of the book. Bouyer surely knew that it strikes the informed theologian or philosopher as evocative of a certain tradition of Patristic thought that sees humankind as a microcosm and *metaxu* of creation.[4]

In his earlier book, *The Invisible Father*, Bouyer makes clear that he thinks there are certain directions in modern philosophy that the theologian should take into account, whatever the ambiguities of the representative philosophers who have set them. These include especially Hegel's historical philosophy, Husserl's phenomenology, and Wittgenstein's linguistic philosophy.[5] One would not, however, be well advised to consult his writings to find a thorough reading of any particular modern philosopher, even these three whom he marks out as especially important. There is little textual exegesis of their work in his, and he largely rejected

3 Ibid.

4 To our knowledge, Bouyer does not himself use the word "metaxu." However, the word corresponds to his understanding of the vocation of humankind. Christopher Dawson, one of Bouyer's favorite historians, used the expression to describe the Patristic understanding of human nature as bridge between spirit and matter. See Bradley J. Birzer, *Sanctifying the World: The Augustinian Life and Mind of Christopher Dawson* (Front Royal, VA: Christendom Press, 2007), 25–26.

5 *IF*, 302–3. See Zordan, *Connaissance*, 759.

the solutions they offered to the problematic of nature and spirit just now identified. Bouyer does see this problematic as crucial to human thought, and he chooses to address it head-on, but largely not by turning directly to the writings of the philosophers themselves. Initially in *Cosmos*, he turns directly for philosophical support to alternative channels of thought for insight, to thinkers who operated under a scientific guise and who orbited on the peripheries of the German philosophical tradition. The canonical German philosophers themselves seem to have been entirely too speculative in an *a priori* way for his liking, although, as we shall see in later chapters, the sophiological dimension of his work connects him very deeply to their entire line of questioning.

In the first two chapters of *Cosmos*, twentieth-century phenomenology comes to the fore in his response to the problematic, but it is less the strictly philosophical phenomenology of Edmund Husserl and his followers than the "scientific" phenomenology of religion that developed in the line of Otto, van der Leeuw, and Eliade. This strand of religious science is indebted in turn to the philosophy of myth and revelation in the later writings of Schelling and to the work of Friedrich Creuzer (1771–1858) before him, providing another indirect route linking Bouyer's thought to that of the nineteenth-century.[6] Bouyer connects this phenomenological turn with certain insights found in depth psychologists and in French sociologists of religion such as Émile Durkheim (1858–1917) and Lucien Lévy-Bruhl (1857–1939). He uses concepts from the more recognized (at least in his time) specialists in religion, psychology, and sociology to show that every human intellectual endeavor, including cosmology, has its ultimate origin in religion and in humanity's experience of the sacred. This demonstration implies that the *metaxu* of human subjectivity through which spirit and physical nature can be reconciled is embodied in a religious context by an historical accomplishment of human freedom. We know the world or exercise our power of rational consciousness thanks only to the cultural instigations of religious ritual and myth. Human awareness is given through the gift of a mysterious inspiration, and only reflection on the religious media through which this inspiration is transmitted enables us to understand the process of

6 Cf. Julien Ries, *L'«Homo Religiosus» et Son Expérience du Sacré* (Paris: Éditions du Cerf, 2009), 8–9.

our own cultural development as well as the development of the world. Human knowledge, whether natural or supernatural, is essentially religious and traditional. Before getting into the issue of how it is, precisely, that the divine Word modifies our questioning in later chapters of *Cosmos*, Bouyer first clarifies that human knowledge of the world is historically constituted in a religious context.

After raising the aforementioned 57 questions in the first five sections of chapter one, our author quickly moves through a series of points in three final sections on 1) the relationship of rational reflection to revelation, 2) the link between knowledge and tradition, and 3) the link between scientific, mythic, and revealed knowledge. He suggests ultimately that even scientific knowledge of the cosmos was made possible through religious tradition. Much that is implied without being directly stated in this first chapter presumes work that he had already done in earlier books, such as *The Invisible Father*. In order to understand this first chapter in depth without having to list the 57 questions that fill up the first five sections, it is helpful to turn directly to this earlier text, which is a treatise on God centered on the first person of the Trinity. Bouyer insists in this book as in *Cosmos* that we can discover God only in and through our experience of the world, and, inversely, that we can discover the world in its truest meaning only through knowledge of the divine, transcendent source of being. He bases his approach on Newman's distinction between real apprehension/assent and notional apprehension/assent: "Abstract reasons for believing in God have never been the source of any man's faith."[7]

Bouyer shows in this earlier book that if we are to understand the origin of our discovery of God or the cosmos we have to recover the sedimented or "primitive" foundation of human consciousness and culture in religion and experience of the sacred.[8] Use of the word "primitive" rightly raises hackles today, but it does not have to assume the pejorative connotation it did throughout most of the modern age. The recurrent temptation of much modern thought, which postmodernism

7 *IF*, 3.

8 Ibid., 6. Bouyer shows awareness here of Husserl's technical language regarding "sedimented" meanings and uses the Husserlian expression in this context.

has not definitively vanquished, is to equate the primitive with childishness, savagery, or barbarism, and to think that modern civilization has matured beyond it to a properly scientific consciousness. Bouyer, in phenomenological vein, attributes a different status to it. For him, the primitive has to do with the inescapable foundation of consciousness and culture, the irreplaceable substructure of these that supports all later superstructures.[9]

Many nineteenth-century scholars of religion exhibited an especially brutal, condescending attitude toward the primitive. If they could admit that the origin of human society seems to be in religion, they nevertheless attempted to explain this fact on the basis of linguistic or conceptual confusions attendant to humanity's first childish attempts to interpret the causes and meaning of nature. The study of religion in this reductionist trajectory of thought equated the primitive with a fall into religious superstition. Perhaps we could go so far as to admit that the primitive was a necessary stage of consciousness, but, even so, it is obvious that we have long since overcome it through the slow and steady progress of human civilization to its acme in the modern secular state.

Scholars of religion of this reductionist stripe have been motivated to search for the origin of religion in what is not religious.[10] They see religion as a transitory phenomenon, and, if present from the beginning, not for that reason *sui generis*. Religion is understood by these scholars to be abnormal and a-religion to be normal. The theories of Max Müller (1823–1900), one of the founders of the study of the history of religion, are representative in this regard, and Tolkien targeted them for rejoinder in his famous Andrew Lang Lecture.[11] According to Müller, religion originates from confusion caused by the personification of nature at an elementary stage of human linguistic development. The primitive mind, he argued, projects onto the world sentiments regarding its own actions and attributes, thereby ascribing to nature a personal character that it

9 Ibid.

10 *IF*, 3–20; *Rite and Man*, trans. M. Joseph Costelloe (Notre Dame Press, 1963), 14–37; *Sophia*, 9–18. Bouyer interprets the history of *Religionswissenschaft* in line with Mircea Eliade in *The Quest: History and Meaning of Religion* (Chicago: University of Chicago Press, 1969), 1–36. For more recent confirmation, see Julien Ries, *L'Homme et le Sacré* (Paris: Éditions du Cerf, 2009), 206–91.

11 We refer to the already cited *Tree and Leaf.*

lacks in itself. Müller postulated that thought is one thing and language or speech another. The latter gives form to thought through metaphor in myth in order to direct the imagination. The primitive mind takes metaphor literally, by a mistake that is natural to the power of language, from which mythopoetic creation, "a disease of language," issues forth.[12] This equation of myth with disease set off Tolkien.

The general outlook on history prevalent in religious science in the nineteenth century is encapsulated by the evolutionary viewpoint of Auguste Comte (1797–1857), one of the founders of sociology. He distinguished theological, metaphysical, and positive stages of human intelligence and argued that humanity progresses from a first, childish, religious stage of development as a race, to a second, abstract, and metaphysical way of seeing it. In a third and final stage of consciousness, Comte argued, humanity advances beyond the theological and metaphysical epochs and at last progresses to the properly scientific, positive condition of intelligence. At this final stage, humanity is content to observe facts from which it discovers the general laws of nature and controls the cosmos for its material profit.[13]

Like many scholars of religion in his time who embraced a more Romantic current of thought, Bouyer chided this sort of reductionist and evolutionary mindset. What were these nineteenth-century scholars trying to do but to explain away religion rather than to understand it as it really is in its irreducible essence? These theorists accorded religion little more than an irrational, epiphenomenal status. It seemed to Bouyer and those of like mind that with the immense growth in

12 See Ernst Cassirer, *The Philosophy of Symbolic Forms*, vol. 2: *Mythical Thought*, trans. Ralph Manheim (New Haven, CT: Yale University Press, 1955), 21–22.

13 Cf. Christopher Dawson, *Progress and Religion: An Historical Inquiry* (Washington, DC: The Catholic University of America Press, 2001), 23, 24, 156, 171, 184. Dawson is one of Bouyer's sources. See Bouyer, *LV*, 15–16. Bouyer says here that the study of history and culture was always of great importance to him, but other theologians thought that it was a waste of his time. The only exceptions, he claims, were Henri de Lubac and Hans Urs von Balthasar. He calls Dawson "the great English historian of Christian civilization." There are striking methodological parallels between Dawson and Bouyer, particularly in how each man construed the task of natural theology. In this, they were jointly, deeply influenced by Newman.

anthropological data from the nineteenth and twentieth centuries the weight of persistent facts should pressure practitioners of religious science to reject these reductionist assumptions. A shifting assessment of the origin of religion coincided with the first stirrings of postcolonialist thought and postmodernism in the West. If this breakthrough was hardly, from the hindsight of our own day, with the rise of the New Atheism and evolutionary psychology, of permanent, universal effect, practitioners of the science of religion in the middle of the twentieth century more and more recognized the speculative, *a priori* nature of Enlightenment-influenced theorizing.[14] Religion could now be seen in its true form and meaning. The discipline of the phenomenology of religion emerged with some influence.

Those who studied religion as phenomenologists recognized that religion shows itself in its phenomenality or in the essence of its manifestation as irreducible to other factors, and, furthermore, that it not only constitutes the foundational layer of humanity's natural societies and cultures but that it is a permanent reality of human existence. It is not derived from some earlier and more fundamental cultural substratum, such as linguistic confusion, society, or, as Karl Marx (1818–1883) thought, economics. There is no more basic level of human experience that precedes it, from which it arises. It is, in fact, the source of all subsequent culture and consciousness. Religion is that which gives meaning to human existence. As Eliade put it, "the beginnings of culture are rooted in religious experience and belief."[15] Bouyer likewise spoke of religion as both genetically and perennially the "most crucial thing in man's life."[16] Religion is "life lived fully," in attunement with cosmic being:

14 Cf. Douglas Headley, *Living Forms of Imagination* (New York, NY: T&T Clark, 2008), 116–17. Headley explains that there are two major types of theory of religion. One is Romantic and the other rooted in an Enlightenment model. The latter is rationalist and tries to explain religion in the way just described, that is, on the basis of what is not religious or "natural." The former sees religion as *sui generis* and criticizes the Enlightenment model for being incapable of accounting for the meaning of religious symbolism. The latter model dominated in the nineteenth century even in the period of Romanticism.

15 Eliade, *The Quest*, 9.

16 *IF*, 41.

> It [religion] is human life brought consciously into harmony, and harmonizing in consequence the individual and society and the cosmos as a whole. To close one's eyes to this is to miss the primary datum of the religious phenomenon and to condemn oneself to total incomprehension.[17]

These scholars saw religion as the primitive cultural activity that gives humankind the very cosmos in its harmonious unity. It puts humanity in touch with the mysterious, transcendent Other who is the source of beings and being, the groundless ground of unity without which there is no cosmos, both *mysterium tremendum* and *mysterium fascinans*, absolutely transcendent and thereby able to be, as Bouyer says, invoking Saint Augustine, "closer to each and everything, each and every being, than they can possibly be to themselves."[18]

Religion, as many different schools of *Religionswissenschaft* agree, takes concrete form in ritual and myth. The phenomenologists acknowledged these to be the primordial practices or forms of culture from which all subsequent forms of human culture derive. Following in some respects Dom Odo Casel (1886–1948), Bouyer saw the developing ritual anthropology to be of considerable importance, capable of being utilized in the theology of liturgy. He attended in his writings to a dispute that has sometimes arisen over the relation of ritual to myth at the origin of human culture. Nowadays, some argue that these are ultimately conjoined activities, but that ritual does have a priority of sorts.[19] Bouyer, the theological liturgist, argued that myth, when most vital, is wedded to ritual embodiment. He insisted that ritual activity brings cosmic

17 Ibid.

18 Ibid., 5.

19 Cf. Robert Bellah, *Religion in Human Evolution: From the Paleolithic to the Axial Age* (Harvard University Press, 2011), 135–36. Bellah says: "For over a hundred years the argument as to which came first, ritual or myth, went on without resolution. It was one of those arguments that many felt would be best abandoned because irresolvable. If scholars like Donald and Deacon are right, however, the argument is at last over. Ritual clearly precedes myth." This seems to accord with Bouyer's analysis of the dialectic of religion and religious awareness, although the point is not decisive. Both Bellah and Bouyer argue the case that ritual and myth are usually intertwined in human religion and that myth serves a liturgical function.

attunement to the body and is the source of the reflective intelligence that gives us the capacity to attain "the widest and deepest vision of reality."[20] Its function includes that of orienting human intelligence cosmically, for "ritual is not one human activity among others but that in and by which man consciously coincides... with what one can call the universe's lines of force and aligns his own life on the axle of the life of the cosmos."[21]

Myth, in this view, is an activity distinct from ritual but emerges in order to shed light on it. In so doing, it sheds light on the meaning of human being and the cosmos. Some scholars in the nineteenth century held that myth was aetiological. It is, they thought, a primitive, outmoded attempt at causal explanation of the cosmos. Others held that it was not aetiological but liturgical, giving poetic voice to the mystery of human rituals. Bouyer argues, as an anthropological correlate to his great motif of cosmic liturgy, that myth is aetiological or cosmological precisely *because* it is liturgical. There is no fundamental opposition between liturgical interpretation and cosmology. In giving meaning to the rites, myth illuminates the cosmos.[22] It gives poetic form to the inchoate vision of divine, cosmic, and anthropic unity that is at first ritually given and, at the same time, releases ritual from the temptation to lapse into magic. Ritual practice is prone to veer in this direction because humans, experiencing for the first time in ritual activity their freedom of action, are tempted to see themselves as the masters of the rites and able thereby to control the gods. Myth serves as a reminder that the rites are first and foremost actions of God or the gods and not magical, primarily human actions by which man wrests control of divinity. It nevertheless faces its own ambiguities and temptations. It can become idolatrous. In unleashing creative imagination, the myth-makers may project the human image directly back onto its prototype, confusing divinity for the humanity of its representations.[23] Nevertheless, cosmology, the knowledge and study of the cosmos, originates in religion,

20 *IF*, 18.

21 Ibid., 12.

22 Ibid. See also *Sophia*, 10–15. Bouyer follows E.O. James in this interpretation.

23 Cf. *Cosmos*, 19.

in myth and ritual.[24] This is true even of modern scientific cosmology, specifically in relation to the transfigured religion of Christian revelation and the Eucharist.

It is not only phenomenologists of religion who can help theologians embrace religion in its *sui generis* character and grasp the meaning of its ritual and mythic symbolism. Depth psychologists and sociologists may have a thing or two to offer as well. Bouyer invokes the depth psychologists in chapter two of *Cosmos*, but he treats depth psychology and sociology in conjoint manner and in greater detail in chapters two and three of *The Invisible Father*.[25] He especially takes the "discovery" of the unconscious mind by certain thinkers in these disciplines to be a sound and lasting contribution to human knowledge as it provides, in his view, a necessary correction for thin, rationalist, Cartesian accounts of the ontology of the soul.[26] Sigmund Freud (1856–1939) is taken by our author rightly to have identified processes by which the conscious mind surrenders self-dominion to forces within its own hidden life. On the Freudian view, the unconscious mind contains only material given to it by the conscious mind, frightening things that the latter cannot embrace and which end up surreptitiously affecting it. The unconscious mind is the domain of complexes, of repressed memories, which Bouyer vividly characterizes as forces "agglutinating, organizing, and eventually crystallizing themselves in the psychic underground."[27] These subterranean forces of the mind are active, wild, and creative presences imposing themselves on our consciousness over time, robbing us of freedom and autonomy.

24 Ibid., 8–11.

25 *IF*, 37–52. At least in the Catholic theologian Bouyer's case, the sociologists helped to establish on a rational basis what was already known.

26 Ibid., 24. Bouyer was always very interested in depth psychology. He frequently mentions his involvement with the work of the Swiss-Canadian psychiatrist and author Henri Ellenberger (1905–1993). Bouyer says that he was associated with Ellenberger's "great book" *The Discovery of the Unconscious* (New York: Basic Books, 1970), and even suggested its title to the author. See *LV*, 13. See also *Memoirs*, 47, 91. The "discovery of the unconscious" in the modern age originates long before the depth psychologists, going back to Jacob Böhme (1575–1624) himself. It is a central sophiological theme, carried through Schelling in the nineteenth century. See Bouyer's early work, "Développements Récent de la Psychologie en Suisse," *La Vie Intellectuelle* 15.12 (1947): 98–117.

27 *IF*, 24.

Psychoanalysis, as Bouyer explains it, aims to return to the past of the soul, which, if it remains hidden in deceptive disappearance, attenuates the self-mastery of the conscious mind. The practitioners of this discipline recognize the importance of anamnesis or remembrance in overcoming the subterranean control of the complexes over our consciousness in order to bring about its reintegration. This tells us something deeply important about ourselves that modern rationalism had forgotten, namely, that we forget our past only at the cost of a loss of self-integration:

> If there is one fantasy absorbing us moderns, it is that of pure futurity. We would fain believe that the future, an untrammeled, creative future, is everything, and in order to enter upon it we are prepared cheerfully to sacrifice our entire past.... But the first lesson Freud can give us is that such a sacrifice cannot be made.[28]

Freud's discussion in his later works of the Oedipus complex and of the link between *Eros* and *Thanatos* contains important insights that Bouyer thinks can provide impetus, if properly critiqued and readjusted, on the path to a nuptial understanding of the meaning of the cosmos. Freud recognized the mysterious conjunction between suffering, death, and love. He realized, on the one hand, that human beings long for their own fecundation, while, on the other, that the deepest, most perennial forms of religious symbolism show that this longing is tied to a fascination with death. From the unconscious wellsprings of the human psyche, linking us to the depth and breadth of the cosmos, we seem to have a premonition of the need for sacrificial *agape* in order that life may truly flourish.[29] Recognition of all that is implied in the link between love and death is explicatory of the symbolism of ancient myths, surpassed by divine revelation, and enables a deeper anthropological penetration into the meaning of the cosmos. We can see better in reflecting on the cosmic mystery of suffering love how the Christian Mystery surpasses the ancient mysteries by leading us "towards a resurrection of our whole being, body and soul, in a transfiguration of the

28 Ibid., 25.

29 Ibid., 32; *Rite and Man*, 44. This seems to be the point Bouyer wishes to convey when he invokes Freud on the so-called "death instinct."

whole cosmos."[30] Cosmic resurrection is the ultimate fecundation of life, but it comes only through the ultimate sacrifice of God on the Cross, the decisive event of divine love in our midst. This divine intervention, which liberates the world from death precisely in and through death sacrificially accepted, is remotely prefigured in the whole of creation in its cycles of death and new life and in the ancient myths that attune human consciousness to the mystery of these cosmic cycles with which human life is conjoined.[31]

Freud hardly gave us the last word on the unconscious mind. Bouyer thinks that his maverick pupil, Carl Gustav Jung (1875–1961), added an important dimension to our understanding with his demonstrations that the unconscious mind has a supra-individual character.[32] The unconscious, Jung realized, contains far more material than Freud allowed, material that could never have been put there by or come through the individual's conscious mind, content that appears in memory or in dreams but that does not derive from the unique experiences of the individual. This information or content is given in symbols, in archetypes that, by way of structural similarity, link the individual to great religious minds in history. Bouyer thinks that there are important insights here, but that Jung was misguided in his attempts to clarify or develop his discoveries, for he did not distinguish sufficiently the archetypes of the unconscious as psychic data from cosmic signs of divine presence that are given to the human being from outside and as other in hierophanies or manifestations of the sacred. He immanentized myth in a Gnostic way and thus failed to recognize myth's genuine significance:

> Myth's prime achievement is not that of reflecting any unalterable structure of the psyche or, derivatively, of the universe. It is that of proclaiming, incessantly and insistently, from the very moment of its birth, that the human spirit can never be closed

30 *Mystery*, 35.

31 *Rite and Man*, 44, 123–50. The first page referenced here is directly about Freud on *Eros* and *Thanatos*, which he says will form a basis for his later analysis of the paradoxical conjunction of death and life. He explores this theme in relation to pagan mysteries and Christian sacraments in pages 123–50. See also *Cosmos*, 27.

32 *IF*, 33–34.

> upon itself, but only exists and only experiences itself in reference to something beyond itself.[33]

Jung's theory of the collective unconscious calls to mind sociological theories of the social constitution of consciousness, and Bouyer turns to these in chapter three of *The Invisible Father*, particularly to Lucien Lévy-Bruhl's thesis that individual consciousness is not inherent to human nature but developed at a stage subsequent to that of the religious, collective consciousness of the earliest societies. This thesis was meant to account for the totemism that Émile Durkheim thought was the origin of religion. "Totemism" describes a form of sacred consciousness present in certain clans or tribes that identifies the human individual with the whole of society and with the natural world. In totemism, the "collective consciousness" of the clan or tribe is assimilated to an animal symbol thought to represent an ancestor who survives in all the members of the group. These French sociologists suggested that individual consciousness does not yet emerge at the totemic stage of human culture.[34]

Although the theory of the totemic origins of religion and consciousness in its details is rejected nowadays—and was already by the middle of the twentieth century—it hints at the truth of the organic nature of human society. Bouyer explains:

> He [Durkheim] succeeded in showing effectively, on the basis of the totemistic cult, how that collective spirit which every genuinely organic society requires for its birth and survival, supposes that all the members of society refer themselves to an enigmatic being, finding themselves in him even though he be 'other,' as much in relation to all as to each.[35]

The totem itself, as a religious phenomenon, is neither universal nor primordial, but it is one among several religious symbols that carry out reference to an "enigmatic being" characterized by both remoteness and sympathy, dual attributes of *Das Heilige* as described by Otto. This being

33 Ibid., 35.
34 Ibid., 40–41.
35 Ibid., 39–40.

is both "wholly other" and "unbelievably close" to us, and we recognize ourselves as one, as a social whole, only in reference to our relationship to it in and through ritual worship and myth.[36]

There is, however, Bouyer argues, no need to postulate on this basis a collective consciousness that is later shattered into atoms of consciousness, as Lévy-Bruhl did. One cannot add or divide consciousnesses, which are always, Bouyer insists, substantial unities.[37] There is, in some sense, a social dimension to consciousness, and our author takes the work of the sociologists to be congruent with that of Jung in showing its reality. Lévy-Bruhl's expression "collective consciousness" and Jung's "collective unconscious" are taken nevertheless to be unsatisfactory in capturing this unified sociality of mind. In sophiological fashion, Bouyer prefers to speak of the "maternal" character of society, from which individual consciousnesses emerge in the fullness of personal integrity.[38] In this nuptial framework, it is true to say that society and tradition are the source of individual consciousness, but that individual consciousness is not pulled out from the whole in the way that a shapeless blob might be extracted from a gelatinous mass. Rather, the individual is always, intrinsically, informed by a "reciprocity of relations" shared by the total community.[39] This is a perennial ontological reality. "Collective consciousness," if the term has any validity, is the shared consciousness of every mind in intrinsic relationality to other minds.

This analysis is crucial to understand the first chapter of *Cosmos*. A key passage gets to the heart of the argument there:

> At first sight, there is nothing more personal, indeed more individual, than our elementary knowledge of the world and its maturation into a more thoughtful form of knowledge. This happens as a result of the progressive emergence of the multitude of questions we have identified, through which our basic experience of the world challenges our intelligence. Nevertheless, however valid and primary this aspect may be, we

36 Ibid., 40.
37 Ibid., 41.
38 Ibid., 42.
39 *Cosmos*, 9.

> should not overlook its complement, the collective or, more specifically, social aspect. Indeed, for each of us, the discovery of the world . . . is in fact inseparable from the discovery of human language, and so from the intercommunication of minds. This discovery is a necessary concomitant of our awakening to thought and life, so that it proves actually impossible to separate our most personal experience of the world from our experience of personal life.[40]

The Cartesian tradition has been naïve with regard to human sociality, particularly when it comes to understanding the manner in which modern scientific cosmology is achieved. Our consciousness awakens, as properly individual, only in the maternal womb of our sociality and history. This "womb" is a shared communion of consciousnesses constituted by language.[41] Science is itself a social endeavor, although it requires the achievement of individuals capable of sifting what has been given and developing it. The interplay of society and individual is characteristic of any and all traditional and therefore human achievements. Tradition, contra certain inadequate characterizations of it, is an organ of progress, not a shackle condemning us to rigid stasis.[42] It is through Tradition that what is given to us in the social whole is carried forward into new terrain in and through the individuals who receive it and make it their own. Cosmology, as a traditional endeavor of humankind, is a constitutive feature of the unified *telos* of the human race, joined together in a totality that is meant to be a harmonious unity, synchronically as well as diachronically diversified yet integrated. The ritual and mythic past is

40 Ibid.

41 Ibid. Throughout the trilogies, Bouyer frequently invokes the linguistic turn and, in *Cosmos*, references Wittgenstein and the importance of "language games," although he does not seem to have read Wittgenstein very deeply. His understanding of the origin and social function of language seems to be much closer to Tolkien, for whom asking "what is the origin of stories . . . is to ask what is the origin of language and mind." See *Tree and Leaf*, 17. This will be discussed in greater depth in the final section, below. This approach to language is very much akin to that of Owen Barfield.

42 *ES*, 59–63. These pages contain a discussion of oral tradition that show that the tradition of handing on the divine Word carries the fullness of the divine Word from the beginning but that tradition is developmental and progressive.

ever-present as the sedimented origin of the whole of human culture in development and of modern science as part of the whole.[43]

In the final section of chapter one of *Cosmos*, Bouyer explains that mythic knowledge "grows from the deep roots of individual and collective experience of the world or, more accurately, of individual experience within the human community."[44] Both revealed knowledge and philosophical knowledge develop on the basis of mythic knowledge and continue to adhere to it even when they seem to oppose it most fervently:

> They [revelation and philosophy] transform our consciousness of the world, and what might be called our consciousness as beings in the world. Nevertheless, neither can totally disregard mythic consciousness nor, without risking disintegration, break away from a fundamental link to that mode of consciousness which is refined and corrected by science or totally transfigured through revelation.[45]

There is a special relationship that obtains between the consciousness and tradition of specifically modern science, on the one hand, and that of divine revelation given to Israel and the Church on the other. Modern science could not have emerged if divine revelation had not cleared a path for it, and Bouyer's phenomenology and hermeneutic of modern science locates its origin in the widespread development, over centuries, of settled cultural instincts made possible by biblical revelation accepted *en masse*, which itself reconfigured and elevated the great mythic themes. He explores this topic in greater depth in chapter twelve of *Cosmos*, whose essential points we shall highlight in our sixth chapter.

MYTHOPOETIC THINKING AND THE SACRED

The themes of myth, Tradition, philosophy, and revelation are further assessed in chapter two of *Cosmos*, in the context of a discussion centered

43 *Cosmos*, 8–10. This whole paragraph is our paraphrase of the section entitled *Knowledge and Tradition*. In our paraphrase, we draw on other Bouyerian sources.

44 Ibid., 10.

45 Ibid.

on how it is that human beings, as individuals and as societies, come to know the world. The chapter has an epistemological flavor, but it is not epistemological in the Cartesian sense, in that the author does not wonder how it is we escape from the inner cabinet of the mind with its mediatory, internally-generated representations of the world. Bouyer sides with philosophical realists for whom such a picture of the mind is nonsensical, in that it fails in its very premises to recognize the nature of consciousness as inherently intentional, in the sense meant by Husserl, which is to say that it is always ordered to realities that transcend it.[46] Consciousness just is, as the etymology of the word (taken from the Latin *con-scire*) suggests, a "knowing with," and if the unconscious mind has the social character just described, it too cannot be confined within the precincts of interiority of the individual self.

The Oratorian theologian argues in chapter two that our experience has a sensory dimension but is not reducible to some purely sensory domain or set of impressions, because human intellect penetrates even into the sensory realm. This coheres with the gnoseology of the Angelic Doctor, for whom the human substance has one substantial, intellectual form, the spiritual soul, the lower functions of which are suffused with a downward-penetrating, intellectual capacity. Aquinas held that "the soul is in the body as containing it, not as contained by it."[47] The body is, we might say, "metaphysically surrounded" by the spiritual soul, which informs it ubiquitously. It is thus illusory to think of the human senses as if they could be detached fully from the rational mind. From the standpoint of an intellectualist account of human sensory perception akin to that of Aquinas, Bouyer criticizes the Humean tradition for holding that sensory experience gives us pure facts existing in the sense organs before the mind can get any grasp of them. Philosophers who take this point of view do not, he argues, realize that their construal of sensory experience is itself a rational abstraction. He does not use the expression, but Bouyer targets with this critique the fallacy of misplaced concreteness that we discussed in the previous chapter. He argues that

46 See especially Bouyer, *Introduction to the Spiritual Life*, trans. Mary Perkins Ryan and Michael Heintz (Notre Dame: Christian Classics, 2013), 32–40.

47 Aquinas, *ST* I.52.1: "anima enim est in corpore ut continens, et non ut contenta."

philosophical materialists mistakenly take sensory experience to exist prior to or outside of the intellectual mind and therefore the world with which human sensibility accords in its presumably pure state to be constituted solely of mindless matter. Philosophical materialists try to bring the multiplicity of being as given to us in experience into unity by the conceptual instrumentality of matter, but matter is not itself a material image given to us in and through sense experience. It is an abstraction achieved by the rational mind working in unity with sense experience.[48]

Bouyer's anthropology and gnoseology aims to return us "to the nucleus of our total being which the Bible calls 'the heart.'"[49] He argued in a relatively early book on mystagogy that the unity of human intellectual experience in fact resides on the affective and aesthetic plane:

> All our sensations are originally bound up together in the unity of an experience in which the sensory only exists inside the affective.... But the affective itself is unified in it by what we have distinguished as the sense of the true, the beautiful, the good, which again is only one single spontaneous intelligence perceiving itself in and through its common perception of its body and the universe in which it is immersed.[50]

He said this long before he began to make the theme of imagination and mythopoetic thinking more thematic in his later writings, but this passage clearly anticipates this later concern. Bouyer's thinking on this front is reminiscent of those perennialist philosophers and theologians who stress that the act of human perception is not achieved by rational interpretation that relates the interior image in the senses to an exterior object but is, instead, given through the unifying, immediate, and intuitive act of the *intellectus*.[51] However, Bouyer does not separate perception

48 Cf. *Cosmos*, 12–13. See also Guillaume Bruté de Rémur, *La Théologie Trinitaire de Louis Bouyer*, 32–34. Rémur helpfully connects the second chapter of *Cosmos* with earlier writings of Bouyer.

49 *Newman's Vision of Faith*, 18.

50 *Christian Initiation*, trans. J.R. Foster (London: Burns and Oates, 1960), 19.

51 See especially Wolfgang Smith, *The Quantum Enigma: Finding the Hidden Key* (San Rafael, CA: Angelico Press, 2005), 21–22.

as given through the intellective act from human embodiment in time and history. If mind and sense ultimately constitute a unity in intelligent affective experience or in imagination, they do not issue in reflective knowledge of the whole all at once in a static and definitive way. Indeed, because human sense and intellect are so deeply intertwined we cannot expect the latter to operate irrespective of time, and our author's deeply cosmic anthropology links the progress of human knowledge in its unity in imagination (or *intellectus*) to the spatio-temporal constitution of the human body. In chapter two of *Cosmos*, Bouyer describes our individual and social intellectual achievements as being genetically constituted, dependent upon history and time-consciousness. Drawing directly from Alfred North Whitehead, Bouyer argues that human consciousness moves from an initial condition of unified experience characterized as a confused jumble to self-conscious understanding of the unity-in-difference of the world.[52] He aligns himself with Whitehead in arguing that sense and intellect interact through a triadic form of process that culminates in the deployment of symbolic intelligence, which is more or less effective in giving expression to the whole of our experience, depending on the character and achievements of the individual or society through which this process is freely actualized. This triadic movement of consciousness is a universal characteristic of the individual human mind but also characterizes the maternal matrix of societies, cultures, and civilizations. We first possess, Bouyer says, a confused perception of a block of the whole, an "indiscriminate perception of unknown chaos."[53] This is "shattered" in a second stage as we move through time and history. We have successive, punctual moments of integrative vision that memory allows us to compare, freezing them at a "second level of consciousness."[54] A third stage of perception is the ever-provisional endpoint, a conscious apprehension or perception of reality as unity-in-diversity, and it is at this stage that reflective cognition comes into its own.

52 *Cosmos*, 13. Bouyer indicates in this note his indebtedness to Whitehead for this chapter. He refers to Whitehead's *Process and Reality* (New York: Macmillan, 1929), and *Modes of Thought* (New York: Macmillan, 1938). We think that Bouyer shows evidence of having read Whitehead directly and perhaps even a little more deeply than he did other modern philosophers.

53 *Cosmos*, 13.

54 Ibid., 14.

This third stage of consciousness shows itself culturally in the life of ritual worship and mythogenesis.[55] Myth is humanity's original, conscious articulation of the unity-in-difference of reality. It is "a synthetic elaboration of our experience of the world, re-established in unity through the integration of successive and discrete views of reality into one intuitive and all-inclusive vision, so that the world may be formally acknowledged in its primordial unity, rather than being just mysteriously sensed."[56] It is thus that myth gives us our first "explicit consciousness of the world."[57] Myth gives us the world as cosmos, because it gives it to us as a unity reflecting a transcendent source. Myth articulates human intuition of the cosmic irradiation of glory, the light that God "shines upon all things . . . without which we cannot sense, reflect upon, and finally see his own light face to face."[58]

Individual minds are joined in this dialectic through which reflective awareness emerges by a communion of language whose condition of possibility is the "original integration of individual consciousnesses, and of the contents of each individual consciousness," which serves "as a basis for our intuitive perception of multiplicity and unity, or rather of multiplicity as a sign of reality's living unity."[59] This integration is on the model of a child in a mother's womb, wherein the child is always a distinct individual, yet absolutely dependent upon its mother.[60] The world in itself is a language, a congeries of expressive signs and symbols, and archetypal presences abound, both beyond the visible plane of the cosmos and within the deepest interiority of the human self. As a "language shared by the minds immersed in it," the world manifests the transcendent Spirit of God, through whom the divine Word is communicated to us.[61]

55 Ibid.
56 Ibid.
57 Ibid.
58 Ibid., xii.
59 Ibid., 15.
60 *IF*, 42–43. Bouyer insists on the importance of "open societies" for true religion. This is to say true religion would not stifle individual consciousness, freedom, and conscience, absorbing the individual into the womb from which he or she arises.
61 *Cosmos*, 15.

The maturation of the human mind is accomplished through the historical movement from *mythos* to *logos* or from, as Bouyer says, "organically linked knowledge in its primary stage" to "the same knowledge sifted by discursive reasoning."[62] The third level of consciousness just described is first expressed in *mythos* and then in philosophical *logos*. Both the *logos* of philosophy and the *Logos* of divine revelation sift *mythos*. The latter, which met up with the *logos* of philosophy especially in the time of the Church Fathers, eventually branched into modern scientific cosmology. Bouyer holds that *mythos* is not ever itself entirely lacking in *logos* but enshrines its own type of *logos*, a "participatory logic," one of inclusion rather than of analytic separation.[63] If philosophical reason goes too far in rejecting this participatory logic, it undermines its own conditions of possibility for existence. In addition to pointing us to the transcendent source of cosmic being and unity, the *logos* inherent to *mythos* operates from a sense of harmonious relation between human spirit and cosmic reality:

> Mythopoetic thinking approaches cosmic reality first through a sure instinct that there exists a spontaneous harmony between our spirit and that reality, then through the very quality which allows our spirit to grasp reality, not only from a specific and superficial viewpoint, but by means of a deep sympathy with its fundamental evolution.[64]

Mythopoetic thinking and creative imagination are, perhaps paradoxically, at the origin of classical philosophy. The *logos* of the Greek tradition drew its inspiration from mythopoetic wonder at the unity and connaturality of mind and being.[65]

There are some parallels between our author's turn to mythopoetic thinking and better-known strands of Catholic thought that revolted against philosophical rationalism in the first half of the twentieth century. We refer specifically to the philosophy of action of Maurice Blondel

62 Ibid., 12.
63 Ibid., 19.
64 Ibid., 23.
65 Ibid., 20.

(1861–1949) as well as to the process philosophy of Henri Bergson (1859–1941) and those he influenced. Bouyer shares with these thinkers to some extent a propensity to relativize the intellectual concept. Blondel excoriated modern rationalists for failing to root abstract concepts and ideas in the context of the historical dynamism of the concrete, freely-acting subject. Bergson, for his part, stressed that reality is life, *élan vital*, the thrusting force of creative evolution mounting to ever-higher plateaus of created being. He held that if science wants to touch life as it really is it cannot do so by means of abstract concepts elicited in the mind through the work of discursive reason, which chops the living, organic, flowing unity of the real into static pieces or fragments. Yet its practitioners and defenders often confuse isolated, abstract fragments of reality's living whole for the ultimately real. Bergson urged that philosophy needs renewal by turning to reflective intuition, to direct encounter with freedom, spirit, and the endless, flowing process of novelty that characterizes the world in its actual being.[66]

The challenge to rationalism issued by Bergson and Blondel was influential to some Thomists who would themselves wield an immense influence on Catholic thought in the 1950s and 1960s, particularly in the line of Pierre Rousselot (1878–1915), killed in battle at Ésparges, in World War I. Thomists who followed down Rousselot's path turned the thought of Blondel and Bergson in an intellectualist direction. They proposed a reading of Aquinas that stressed *intellectus*, the higher, intuitive, synthetic reason that alone enables the proper functioning of discursive human rationality. Like Blondel, they argued that faith and Christian action are needed in order to reach a synthetic view of the unity of the world. They combined Thomist intellectualism with Blondel's philosophy of action and Bergson's evolutionist intuitionism. Bouyer was hardly in complete disagreement with Thomists in this line of thought, although he exhibited from time to time a distinct distaste for Karl Rahner's version of this tradition, which he thought too greatly historicized human nature and naturalized thereby the supernatural presence of God in creation.[67] His recourse to mythopoetic thinking is not without

66 See Gerald McCool, *From Unity to Pluralism* (New York: Fordham University Press, 1989), 44–46.

67 Cf. *IF*, 78–79.

connection with this recasting of Thomism in reaction against eighteenth- and nineteenth-century rationalism and materialism.[68] Moreover, this tradition of thought was the first to discover the philosophical importance of Newman, as the English Oratorian's "real assent" was thought to accord with Bergson's "intuition."[69]

However, the turn to mythopoetic thinking that Bouyer urges has a more specific and immediate genealogical backdrop. His more direct partners in conversation are nineteenth- and twentieth-century Romantic scholars as well as later scholars of religion, depth psychologists, and literary figures, especially the Inklings, who exhibited interest in the mythopoetic substratum of consciousness and creative imagination. Many of the important twentieth-century scholars of religion whom we briefly mentioned in the first section above attempted to narrow in on what defines mythopoetic thinking as a distinctive form of human thought. Bouyer's definition of mythopoetic thinking is his own, but it synthesizes and theologically "corrects" a larger discussion that includes not only the Inklings but these other scholars of religion. It does not come out of nowhere, and he was not talking to himself alone in turning to myth. His concern in this regard is not simply idiosyncratic, and if it bears obvious connection to larger Catholic interventions in epistemology in the twentieth century, its true intellectual context is this other discussion in religious studies, so important for Catholic theology of liturgy in the twentieth century—in the likes of Casel and Romano Guardini (1885–1968)—itself a legacy of nineteenth-century Romanticism.

Hans and H. A. Frankfort, in their classic study *The Intellectual Adventure of Ancient Man*, provide an eminent example of this wider discussion.

68 Ibid., 74–82. Bouyer gives a generally positive assessment of the Transcendental Thomists in these pages, with the exception of Rahner. See also *Mystery*, 1. Here, Bouyer connects Casel, Bergson, and the Transcendental Thomist Joseph Maréchal (1878–1944). The latter two led Bouyer to the conclusion that there is a "genuinely Christian experience which . . . leads to a personal meeting with God, a union with God in Christ."

69 For a discussion of the history of Newman's reception in France, see Toby Garfitt, "Newman at the Sorbonne, or the Vicissitudes of an Important Philosophical Heritage in Inter-war France," *History of European Ideas* 40.6 (2014): 788–803. See also Bouyer, "Newman's Influence in France," *The Dublin Review* 435 (1945): 182–88. Bouyer argues here that Newman influenced the history-minded Augustinian Thomism that developed in the line of Rousselot.

They tried to describe the thought patterns enshrined in the archaeological fragments left by ancient cultures in terms of mythopoetic thinking. They argued that the fundamental tendency of this form of thinking is to construe the world in terms of "I-Thou" relationality rather than the subject-object duality that characterizes modernity. Ancient thought, they held, was inextricably wrapped up in imagination, and its records are recalcitrant to abstraction from their imaginative forms. The ancients refused in their speculative thought, such as it was, to distinguish the realm of nature from the human realm. Viewing natural realities in personal terms, the ancients did not think of nature as constructed on the basis of universal laws but of unique instants in a web of "dynamic reciprocal relationship."[70] Each presence in nature, each concrete moment, could be experienced as a "Thou," an event motivated by will rather than governed by anterior, universal laws.

The Frankforts argued that "primitive man" immediately experienced the world as animate, as a "Thou." The world was not understood to be an object confronting a subject, as with modern, scientific thought, nor was it perceived as a fundamentally inanimate realm which the primitive human mind peopled with the ghosts of the dead, as E.B. Tylor (1832–1917) proposed with his influential reductionist theory of the origin of religion. Ancient speculative thought did not project in *a posteriori* fashion the characteristics of personality onto the world but immediately perceived the world as personal and revelatory of purpose and will.[71] This view runs contrary to Tylor's animist thesis, and follows Lévy-Bruhl to some extent, although the Frankforts, like Bouyer, reject Lévy-Bruhl's characterization of the ancient mind as "pre-logical." Nevertheless, the Frankforts agree with Lévy-Bruhl that the ancient mind immediately perceived the world in a synthetic way and did not give social importance to discursive reason. Early societies collectively represented the world to themselves in a participatory manner, one that seems absurd to our

70 Henri Frankfort, H.A. Frankfort, John A. Wilson, Thorkild Jacobsen, and William A. Irwin, *The Intellectual Adventure of Ancient Man: An Essay on Speculative Thought in the Ancient Near East* (University of Chicago Press, 1977), 5. Without mentioning this specific book, Bouyer once said that Henri Frankfort is one among several figures in the comparative of history of religion who captured his "critical but passionate attention." See *LV*, 12.

71 Frankfort, *The Intellectual Adventure of Ancient Man*, 5–6.

modern, scientific mentality. Moderns and premoderns, according to this understanding, do not receive the same facts of perception and then reason about them differently. Rather, our very fundamental perceptions contradict one another. Odd to modern eyes, perceptual identifications and differentiations are rife in cultures dominated by what Lévy-Bruhl called *participation mystique*. The mythopoetic mind might, for instance, place the sun and a white cockatoo in the same class of beings while simultaneously distinguishing both of these very sharply from a black cockatoo. It selects out apparently incongruous elements for attention from the totality of what is given in perception.[72]

Mircea Eliade's account of the mythopoetic mind gives us another classic example in the wider discourse to which we refer.[73] The Romanian scholar located the roots of the production of myth in humanity's experience of the sacred, which is, he argued, "an element in the structure of consciousness, not [as Hegel thought] a stage in the history of consciousness."[74] This is to say that the sacred is always with us, even in the secular age. Myth is for Eliade, as for C. S. Lewis, always "true story," and it is to be distinguished from mere mythology or fable. Myth is true because it "narrates a sacred history."[75] Myth-making is a distinctive activity of the human mind, not to be confused with poetry, literary creation, or even epic mythologizing.[76] Eliade thought that myths of creation bear special significance. These relate how beings and being came to be through the deeds of "Supernatural Beings."[77] The act or acts of creation took place in a sacred time that is renewed in religious ritual. In the ritual context of mythic recital, profane time is abolished, and the deeds of Supernatural Beings are reiterated. Deeds done *In illo tempore* are thought to be effectively done again and not merely commemorated. Religious action

72 Cf. Barfield, *Saving the Appearances*, 28–35. Barfield discusses Lévy-Bruhl's thesis in this regard in a balanced way. He draws from Lévy-Bruhl's *How Natives Think*, trans. Lilian Ada Clare (New York: G. Allen & Unwin, 1926).

73 Cf. *MT*, 225–27.

74 Eliade, *The Quest*, i, preface.

75 Mircea Eliade, *Myth and Reality*, trans. William R. Trask (San Francisco: Harper and Row, 1963), 5.

76 The Frankforts also distinguished the properly mythopoetic way of thinking in this manner. It reflects a common phenomenological distinction utilized by these scholars.

77 Eliade, *Myth and Reality*, 5.

is always at the very least proto-sacramental, its symbols thought to be capable of effecting anew and making really present what they signify. Eliade sought to rediscover the perennial truth of myth and with it of symbolic thinking and imagination. Myth is true, he insisted, in that it alone has the symbolic and imaginative power to make us aware of the unity of reality in its greatest depth and breadth. Myths are unique in their ability to relate the fundamental story of being in its direct significance to human life and to open the path of escape from the latter's profane and degraded historicity.[78]

With the Frankforts, Bouyer realized that mythopoetic thinking reaches into the immediacy and exteriority of our sense perceptions, and his phenomenological and Thomist account of the unity of sense and intellect bears on this. He rejected with these thinkers the modern idea of a pure sensation that is supposed to submit to subsequent interpretations that may be either premodern and "pre-logical" or modern and "logical." He held that the dialectic of knowledge, in the threefold structuring of human consciousness described above, is universal, and this is a basis for his dismissal of Lévy-Bruhl's claim that the primitive mind is perceptually "pre-logical."[79] He stressed instead that different cultures and societies direct with different types of focal attention and orientation of importance the integrative powers of the mind and the imagination that we all share in sharing human nature. This decision affects our very mode of perception.[80]

With Eliade, Bouyer recognized the sacred as a perennial "element in consciousness"—if this is meant phenomenologically as consciousness directed intentionally toward what is given in the world as a religious essence. Bouyer expressly said that "the sacred is an element of human language, inherent to and subsisting in human nature."[81] Like Eliade, he spoke of "hierophanies," mysterious acts or manifestations of the Wholly Other, to which human consciousness is constitutively ordered. For Bouyer, the sacred is not opposed to nature. Jean Duchesne well explains Bouyer's view:

78 Ibid., 8–20. Cf. Eliade, *Images and Symbols: Studies in Religious Symbolism* (Princeton University Press, 1991), 9–21.

79 Cf. *IF*, 10–11; *Sophia*, 10–12.

80 *Cosmos*, 6, 24–25.

81 *MT*, 104.

> "To consecrate" or "to sacralize" signifies . . . not to increase on the earth the domain of the divine, but to concentrate it and define it in the dimensions of space, of time and of matter, in order to enter into relation with it by means of a language where the community accomplishes itself by concrete actions and not only words and concepts, in some places and in some moments conformed to the physical as well as mental and spiritual nature of the human being.[82]

The sacred is not an addition to nature, whether cosmic or human, but an indicator of its deepest reality, a demarcated entryway into living relation with reality as it should be, in peaceable communion with the divine.

It is necessary to mention that Bouyer has a theological purpose in view in defending myth and mythopoetic thinking in the way that he does. This purpose, at least as it pertains to *Cosmos*, is to fight against demythologizing currents of thought that evacuate the Christian experience of any specifically cosmological dimension, in the manner of the theology of Rudolf Bultmann (1884–1976). The title of chapter two of *Cosmos* is, as we said, "The World as Object of Human Experience." Bouyer teaches us in this chapter that the world is never in fact given to us as a pristine object. Bultmann thought just the opposite, at least in that he thought modern science has at last unveiled the cosmos precisely in its pure objectivity. In *Le Métier de Théologien*, Bouyer argues that this Bultmannian reading of science is a mythic reading in the pejorative sense of the term: it fails to receive the "true epistemological position which corresponds to the stage of the development of science in our day."[83] Bouyer suggests that this stage was best identified by Whitehead, who recognized that science, like all forms of human intellectual endeavor, is fundamentally a symbolic enterprise. Bultmann's objectifying of the world pits scientific conceptualism against mythic symbolism, while Bouyer argues, in defense of the biblical symbols that are the source of Christian cosmology, that myth and science are not radically opposed. It is not as if one wallows in the murk and mire

82 Duchesne, *Louis Bouyer*, 30. Cf. *Rite and Man*, 78–94; *Cosmos*, 26–27; *IF*, 6–9.

83 *MT*, 113.

of refractory symbols while the other bathes in the crystalline pure waters of objective concepts. Myth and science both trade in symbols and symbolic intelligence. The difference between myth and science is that their symbols are typologically distinct: "in their actual utilization, [they] only attain the truth in an imperfect and stammering fashion, and constantly to be corrected."[84] The mythic image and the scientific concept each possess a symbolic function, and each, it seems, benefits from a mutual operation of corrective interpretation.

MYTH AND LANGUAGE

At points in our discussion in this chapter, we have broached the subject of the linguistic constitution of human consciousness, a prevalent theme in twentieth-century philosophy as well as in Romantic and Traditionalist currents of thought that Bouyer takes seriously, while rejecting their excesses. The Oratorian alters Eliade's statement that the sacred is a permanent element in consciousness by speaking of it as a permanent element of language. He does not readily separate language and thought. He holds that religion awakens thought by awakening language, and he reminds the reader in *Cosmos* that the Greeks describe thought and language in a single word, *logos* (*λόγος*).[85] His reflections on myth and human knowledge of the cosmos are wedded to an assimilation of elements of the linguistic turn in philosophy that he himself associates with Wittgenstein but that goes back to earlier figures. In concluding this chapter, we shall focus briefly on Bouyer's linguistic theology, or theology of language, a subtle, often underlying presence in the pages of his volumes on theology and economy and especially his summative volume on cosmology.[86]

Our author explicitly acknowledges in chapter two of *Cosmos* the epistemological importance of the linguistic turn in unsettling the naïve prejudices that flow from the objectification of the world operated by scientific naturalism.[87] In the full scope of his work, Bouyer was able to contextualize this turn with reference to the entire tradition of Western

84 Ibid.
85 *Cosmos*, 18.
86 We shall return to this topic in chapter 11.
87 *Cosmos*, 16.

thought, particularly the shared theology of the Church Fathers, such as one finds it in Saint Maximus the Confessor (c. 580–662), for whom the order of the cosmos is an embodiment (*ἐνσωμάτωσις*) of the divine Word or *Logos*. "Always and in all," Maximus said, "God's *Logos* and God wills to effect the mystery of His own embodiment."[88] Bouyer follows the Confessor's doctrine of divine exemplarism according to which all things in creation have meaning as individuals and taken together as a whole by participation or sharing in the life of the divine *Logos*, the Creator of all things.[89] All created realities are *logoi* of the *Logos*, words of the divine Word. Bouyer, as we shall see, places them more specifically in the Wisdom of the *Logos*.

Language emerges in biblical revelation as a fundamental and primordial reality, present in the divine life itself. From the beginning of the Book of Genesis to the end of the Book of Revelation, God shows Himself in the economy of creation and salvation as active, creative, and re-creative Word. Given that God presents Himself in biblical revelation as Word, it should hardly be surprising that certain theologians would see the twentieth-century linguistic turn as fortunate. In *Cosmos*, Bouyer does indeed herald Wittgenstein's discovery of the philosophical significance of language, but there was another twentieth-century Jewish philosopher of language, not referenced in *Cosmos*, who grabbed his attention relatively early in his theological career and whose work he expounds in greater depth than Wittgenstein's. We refer to Martin Buber (1878–1965), a philosopher deeply imbued with the insights of Hasidic mysticism, who made language a central theme in his existentialist phenomenology. In *Rite and Man*, Bouyer gave a helpful synopsis of Buber's philosophy of the word, connecting the insights of this Jewish mystic to the anthropology of ritual and myth in a way that is pertinent to our present discussion. He suggested there that Buber's analysis of language in *Ich und Du* (*I and Thou*) helps us to understand language as the "primordial action" of the human being.[90] Buber demonstrates that the "I" enters into the world as an effective agent through the spoken word.[91]

88 *Ambiguum 7*, *PG* 91:1084c–d.
89 See *Sophia*, 123–24.
90 Martin Buber, *Ich und Du* (Leipzig: Im-Insel Verlag, 1923).
91 *Rite and Man*, 54–55.

Speech is not, in Buber's understanding, purely an imposition of the self upon the world, for the exertion of self through language entails the existence of another who hears. The existence and action of the "I" who speaks implies the existence of the receptive "thou." The world is not simply a collection of things or "its" but an effective interaction of persons. Speech, Buber holds, is fundamentally the action of an "I" in search of an ultimate and transcendent "Thou." Bouyer agrees with Buber that speech is a sacred action: it emerges from within the soil of religious ritual and through the myths and prayers that give expression to the rites. The sacred word is, in fact, first of all prayer and is then articulated in myth.[92] Because religious ritual is the fundamental action that unifies uncreated and created being in man, the word that illumines it sheds light on the totality of existence. The origin of the word must be in an ultimate "Thou" who exists as its hidden, transcendent Giver and Receiver. "It seems then," Bouyer summarizes, "that the first words of man are a cry towards God as man's first profound reaction to the hierophanies."[93]

The question of the inner link between language and myth has been explored by several other philosophers of note. It continues to be a topic that elicits great interest among philosophers.[94] The Inklings plumbed the question in both their scholarly works and their literary creations. Tolkien, a great philologist, sought, in mystical vein, to penetrate the inner meaning of language, and he recognized the importance of "faerie stories"—myths—to achieve this end. He spent his life engaged in laborious, practical study of languages, motivated by his love of Norse mythology. The study of language and myth was a kind of religious vocation for him, as Philip and Carole Zaleski explain:

> Behind these practical [philological] studies lay powerful, intertwined, and potentially contradictory beliefs: that language provides a key to the rational, scientific understanding of the world and that language is more than human speech, that it

92 Ibid., 100.

93 Ibid., 56.

94 Most recently, see Charles Taylor, *The Language Animal: The Full Shape of the Human Linguistic Capacity* (Cambridge, MA: Harvard University Press, 2016), 70–79.

> claims a divine origin and is the means by which God created the cosmos and Adam named the beasts.[95]

Bouyer shares many of these beliefs with Tolkien, which are not, we think, self-evidently "potentially contradictory." He demonstrates their biblical warrant and argues that biblical revelation assumes into itself and elevates the mythopoetic intuition that sees the world in terms of the relationality of "I-Thou" rather than "I-It."

In *The Invisible Father*, Bouyer argues further for the religious origin of language and for recognition that language is a gift that presumes the existence, however much in absence, of a Thou who is the ultimate Giver of all gifts. Traditionalists are not without good reasons for stressing that society is founded in religion through the gift of language:

> Every religious tradition represents language as a gift of the gods that makes society possible and continues to hold it together like a thread. Conversely, Genesis sees in the fragmentation of speech into mutually incomprehensible languages a curse from heaven upon a sinful society.[96]

Consciousness, whether individual or social, cannot develop without language—"language, being both the fruit of the efforts of all the individuals of a society and yet, prior to that, the very principle of their formation, is as it were the expression of a society's collective or, rather, maternal soul."[97] If society is held together by language, it is nevertheless true that individual consciousness cannot flourish without the language through which the "I" acts on the world. Disruptions of language deform human consciousness. Language, as traditional, is both a stable structure enabling the cohesion of society and an inexhaustible, unpredictable wellspring for development, insusceptible of being delimited by the societal imposition of absolute, authoritarian control. It has its roots in the subsoil of the unconscious mind, which is, in turn, rooted in the human body. Our

95 Philip and Carole Zaleski, *The Fellowship: The Literary Lives of the Inklings* (New York: Farrar, Strauss and Giroux, 2015), 24.

96 *IF*, 47.

97 Ibid.

very physiology is, Bouyer says, "the epiphany of the innate physical structure of the universe which remains without a voice until it finds it in the common speech of human minds."[98]

This quotation recapitulates themes that we expounded in the first two sections. It brings us back to the motif of humankind as *metaxu*, the reconciler of spirit and nature who naturally joins these together in the composite unity of body and spiritual soul. For Bouyer, human nature is by its very ontology potentially redemptive.[99] He draws this idea from those Church Fathers for whom the First Adam was a potential redeemer of a world that was already thrown into disruption and disharmony by the fall of many angels. We shall explore this topic in depth in chapters nine and ten, but an overview of the theological narrative that Bouyer assumes in his work on cosmology is helpful at this point to orient us as we move forward. His thinking on this is especially aligned with that of Saint Maximus the Confessor, who held that spiritual and material planes of being co-inhere in a sympathetic unity centered on humankind.[100] The Confessor teaches that humanity has been given the task to struggle by ascetical purification and contemplation in order to unite bifurcated existences: 1) the Uncreated and created; 2) the intellectual and the sensible; 3) heaven [sky] and earth; 4) paradise and the inhabited world; 5) male and female.[101] He held that this reconciling work can be accomplished only by Christ the New Adam really assuming our nature.

Bouyer first developed aspects of this Christological, cosmic anthropology in his book *The Meaning of the Monastic Life*. He references there Job 38:7—"When the morning stars sang in chorus and all the sons of God shouted for joy"—to suggest that creation was a gift of the superabundant generosity of the triune God meant to be a chorus of praise directed to the Creator, from whose stream of divine being in agapeic sharing all that is not-God came to be:

98 Ibid., 48.

99 *Cosmos*, 107.

100 See Andrew Louth, "The Cosmic Vision of Saint Maximos the Confessor," *In Whom We Live and Move and Have Our Being: Panentheistic Reflections on God's Presence in the World*, eds. Philip Clayton and Arthur Peackocke (Grand Rapids, MI: Eerdmans Publishing Company, 2004), 184–96, at 187.

101 Ibid., 191–93.

> It was indeed under the image of an immense choir resounding with the divine glory in the unanimity of love orchestrated by the Word, that Christian antiquity represented the primordial world to itself. In this wholly spiritual universe, at the beginning all was song. To the hierarchy of created powers within this unity corresponds the sympathy and symphony of the cosmic liturgy in which the countless myriads in that festal assembly of which the Epistle to the Hebrews speaks, glorified the Creator with a single voice.[102]

The harmony of this liturgy was corrupted. Dissonance set in as a consequence of the prideful rebellion of some of the most exalted ministers of the divine will and leaders of the choir, including the most exalted of all, whose fall was first, Lucifer. But God admits of no regrets in creating. He turns this great treason, this fall, into an opportunity to realize an even greater glory. He gives the gift of His Spirit to matter in a new way, instilling a creative life in it that renders it autonomous from the power of the fallen angels to which it had once been tied. He lifts up, as Bouyer teaches, a new spirit in the world, a spirit clothed in flesh: the First Adam, natively redemptive, whose being is the precondition for the gratuitous, salvific gift to creation of the Second Adam in the fullness of time.[103]

Adam was a "substitute angel" whose task was to restore the harmony of the cosmic temple. He was to guide the celestial hierarchy to a renewed ascension in song of praise to the Creator. The failure of the First Adam in this mission did not stifle the divine Wisdom but presented a new opportunity for divine action in the New Adam. In Christ, the New and Final Adam, "cosmic liturgy is not merely restored but reunited to its divine exemplar."[104] In Christ and through the Spirit creation ascends to a fresh and ultimate plateau, sharing directly in the life of the Father. Bouyer provides an extended quotation from Saint Gregory of Nyssa that summarizes the vision of creation and redemption that he wishes to recover:

102 Bouyer, *The Meaning of the Monastic Life*, trans. Kathleen Pond (New York: P.J. Kennedy and Sons, 1955), 29.

103 Ibid., 30–31.

104 Ibid., 33.

> The whole of creation is but one single temple of the God who created it. But, sin having intervened, as the voice of those overcome by evil became silent, as joy no longer resounded in the heavens and the harmony of those who celebrated this liturgy was destroyed; and as the human creature no longer took part in the sacred festival of hypercosmic nature—the trumpets of the apostles gave forth their sound, like those trumpets of rams' horns of which the Law speaks, for they were preparing for the coming of the true Unicorn, the monogenous Word. In accordance with the power of the Spirit, they have thus provided a sustained echo for this word of truth, so that the ear of those whom sin had hardened might be opened and there might be once more a single festival celebrated in harmony, with tabernacles of the earthly creation placed next to the sublime and supereminent Powers which stand around the heavenly altar.... Then, lift up your hearts, that they may enter the choir of spirits, taking David as master, conductor and leader of the choir, and sing with a single voice with him the sweet phrase which I repeat: blessed is he who cometh in the name of the Lord.[105]

This is a musical story of the cosmos, a narrative of song, Christ-transfigured myth. It glories in the unity of God, cosmos, and *anthropos* in which "the whole creation is but one single Temple of God." It assumes an exalted understanding of language in line with divine revelation itself. Bouyer's development of this story in *Cosmos*, which we shall explore in a progressive manner in this study, enables us to see new connections between the ancients and the moderns, the Church Fathers and the Inklings, the Church's first theologians and the best of the modern mythmakers as well as philosophers of language.

Language, in this story, is the instrument through which the cosmos is given harmonious order. Humanity's use of language has cosmic consequences. Have not the postmodern philosophers themselves had an inkling of this? Man is, as Heidegger said, the "Shepherd of Being," and his vocation is exercised in language, which is, as Heidegger also said,

105 Ibid., 37. Taken from Gregory of Nyssa, *Oratio in diem natalem Christi*, *PG* 46:1128–29.

the "House of Being." One cannot, as Max Müller sought to do, separate thought from language. No more than one can separate except through deceptive abstraction sense from intellect in the human soul can one separate language from things in the entirety of the cosmos. Things are in words. The *logoi* of creation are the all-encompassing meaning of material realities. Biblically, man is given the power to name the beasts. This does not bring them into existence but does give them a new, elevated voice in the cosmic choir of praise. Not just any type of language is repository to this immense power of naming. As Heidegger suggested, power of this sort is exercised only in and through the poetic word:

> This naming does not consist in something already known being supplied with a name; it is rather when the poet speaks the essential word, the existent is by this name nominated as what it is. So it becomes known as existent. Poetry is the establishing of being by means of the word.[106]

Poetic speech is fundamental, primary, and even co-creative. Tolkien seems clear on this point, and Bouyer has great sympathy for the idea. On this understanding, prose is derivative, secondary. However, we might well go so far as to say, if we follow the story of the cosmos as primordial chant, that poetry itself is derivative of music.[107] The song of praise would be, as Gregory of Nyssa implies, the fundamental language of the cosmos. Poetry could well be recognized as a vestige of it. Prose would be a vestige of a vestige. Tolkien articulates this understanding in *The Silmarillion* (1977). Bouyer's explication of cosmic liturgy, already given in his early book on monasticism, before he read Tolkien, confronts us with such a view. The human vocation has to do, then, with getting language right. This means that the cosmic

106 Martin Heidegger, "Hölderlin and the Essence of Poetry," in *Existence and Being* (Chicago: Regnery, 1949), 304. Quoted by Peter Kreeft in *The Philosophy of Tolkien: The Worldview Behind* The Lord of the Rings (San Francisco: Ignatius Press, 2005), 153–62, at 157. We are following here, in this paragraph, the preceding paragraph, and the next, much of Kreeft's profound, although succinct, discussion. Needless to say, he does not mention Bouyer or the figure of the monk.

107 Kreeft, *The Philosophy of Tolkien*, 157.

choir of praise needs to sound out anew in a single, harmonious voice. The human being has been given the vocation, in the New Adam, to direct this choir. This is no easy or automatic task. It requires preparation and training in the grace of God through asceticism and mystical practice. Bouyer first recovered this great metanarrative and humanity's inherently redemptive nature in a book on monasticism, because he thought that it is monks who would lead the way in bringing humanity's common vocation to successful completion. Monasticism just is, for Bouyer, integral humanism.[108] He sees monastic life, which in the early Church carried the torch of martyrdom into the new context of imperial Christianity, as paradigmatic of Christian life. Christian monks are not set apart as if simply their own class of being but are at the head of the renewed human choir.[109] If true to their vocation, they show the path for all Christians. They are the first guardians of the cosmic language recapitulated in Christ, the stewards of liturgy. They are the keepers of the Church's eschatological memory, scholars who plumb the depths of meaning inherent to the myth that became fact in Jesus of Nazareth. Newman distinguished Saint Benedict's poetic way of teaching from the scientific way of teaching for which Saint Dominic is the exemplar and from the practical way of teaching for which Saint Ignatius of Loyola is renowned.[110] The entire weight of Bouyer's work suggests that we have lost the proper balance here and have become hardened to the logic of Saint Benedict's poetic way. The other two ways have become primary and foundational. These other ways and their associated activities are important and monks—as all fully integrated human beings—likewise are called to engage in them. However, the ways of Saint Dominic and

108 Bouyer's thinking in the regard is very Eastern. John Paul II described this view approvingly in his apostolic letter *Orientale Lumen* 9: "Moreover, in the East, monasticism was not seen merely as a separate condition, proper to a precise category of Christians, but rather as a reference point for all the baptized, according to the gifts offered to each by the Lord; it was presented as a symbolic synthesis of Christianity."

109 See Jean-Charles Nault, "La Contribution à une Théologie de la Vie Monastique," in *La Théologie de Louis Bouyer*, 249–62. Nault shows that Bouyer's eschatological humanism is clarifying of the meaning of monastic life.

110 Newman, "The Mission of St. Benedict," from *The Atlantis*, January, 1858. Available at http://www.newmanreader.org/works/historical/volume2/benedictine/mission.html.

Saint Ignatius of Loyola depend upon that of Saint Benedict, not the other way around. Science and praxis are founded in poetic intuition and contemplation. Contemplation is the animating principle of worthy activism. The loss of recognition in the Church of the foundational character of the monastic and poetic path for all forms of religious life has contributed to the modern loss of the cosmos.

CHAPTER 3

From Myth to Wisdom (I): The Old Testament

HUMAN BEINGS ARE, AS ROBERT BELLAH ONCE said, "the product of all previous human culture."[1] Provided that this is not taken to mean that human beings are reducible to their cultural heritage, even if global in scope, one could derive this very maxim from Bouyer's hermeneutic of tradition and of the place of modern science in the vast sweep of human history. Science and the scientist, as human beings, are in no small part the products of the totality of what has come before in history and culture, from the mysterious roots of human language and consciousness in religious pre-history, to the dawning of civilization with the first emergence of the city-states in the ancient Near East, to the special history of Israel, called out by God in Abraham, culminating in the birth of Christ and the formation and expansion of the Church as his Body. Bouyer develops such a hermeneutic, contextualized by a theological understanding that emphasizes the human vocation or task in the cosmos to assume the mantle as leader of a renewed universal choir of praise to the Creator after the fall of some of the most exalted figures in the angelic hierarchy.

The theological understanding or vision operative in the whole of Bouyer's work is deeply Trinitarian, and this informs the theology of creation and history that is implied throughout much of *Cosmos* and finally explicated in the second part of the book. Throughout his writings, Bouyer draws deeply from the immense treasury of Patristic tradition, particularly although not exclusively from the Christian East. His presentations are suffused with a Trinitarian creationism inspired by the Church Fathers and their greatest successors, although this largely remains under the surface in the first, genealogical part of *Cosmos*. Nevertheless, the Oratorian ultimately shows that creation is being led to the

1 Bellah, *Religion in Evolution*, 228.

consummation of its unified purpose as eschatological Bride *in Spiritu, per Filium, ad Patrem*.[2] With the coming of the celestial Bridegroom, foretold in John's Apocalypse (Rev. 21:2), the final revelation of God's glory will show forth from the Spirit of God in the wedding of the creature to the Creator in the heart of the Father's celestial abode. Even now, the Holy Spirit fills the entirety of cosmic history with what Bouyer calls his "long kenosis," just as he brings eternally the divine life to the perfection of being as Gift of Love.[3] His grace alone allows entrance into the great Mystery of triune life. The theological context here is no less Pneumatocentric than Christocentric. Bouyer in fact originated an important line of "Spirit Christology" for which Pneumatocentrism and Christocentrism are inseparable from a vision of the unity of divine being eternally present in the fontal divinity of the Father.[4]

Bouyer unites this theological vision with a sapiential rendering of the wider meaning of religion and the sacred. The idea of a "religionless Christianity" was anathema to him. He articulates his Trinitarian theology of creation and history in union with an account of the transfiguration of the sacred accomplished by the revelation of the divine Word, and he insists that if religion and the sacred constitute the necessary substratum of human culture, Christian revelation, the self-communication of God in His Word to humanity, does not leave the pagan or natural sacred and religion where it finds them:

> Religion is natural to man, but equally naturally his religions tend to idolatry. What Voltaire said is only too true, that, if God made man to his own image, man makes God to his. There is probably no one, however apparently irreligious, who does not adore someone or something. But there are few, even among the most religious, who do not, to some extent, project themselves into their object of worship.[5]

2 Cf. Bouyer, *Meaning of the Monastic Life*, 75–114.

3 *LC*, 367. See also *IF*, 268–70. These latter pages detail the Spirit's eternal fruitfulness in the divine life.

4 See Hans Urs von Balthasar, *Theo-logic III: The Spirit of Truth*, trans. Graham Harrison (San Francisco: Ignatius Press, 2005), 53–55. Balthasar endeavors in this volume to develop this line of thought opened up by Bouyer.

5 *Christian Humanism*, trans. A.V. Littledale (Westminster, MD: The

Neither the divine *Logos* nor the human *logos* are satisfied with the native situation of religion and the sacred after the Fall, as the history of civilizations attests. Both divine revelation and philosophy do in fact "demythologize." As important as myth and mythopoetic thinking are to human culture and the advance of cosmology, they did not give us the perfect truth of divine, cosmic, and anthropic being from the very first moment that (postlapsarian) human consciousness awakened in and through the sacred word. The pagan sacred was transfigured through the advance of wisdom, both divine and human. Bouyer recasts the whole problematic of faith and reason in light of a rich description of this advance. In *Cosmos*, he marks three of its decisive stages: 1) the Old Testament movement from Wisdom to apocalypse, 2) the advent of Greek philosophy, and 3) the union of Wisdom and Word in the Incarnation of Christ.

In this chapter, we shall focus on the first stage of this advance, exploring its nuances in six sections, detailing: 1) the Trinitarian theology of history presupposed in *Cosmos*, 2) the decisive importance of the royal myths and sacrifices, 3) the Old Testament transformation of pagan cosmogonies centered on the royal myths, 4) the development of Wisdom and apocalyptic literature out of the royal myths, 5) the apocalyptic ordering of Wisdom to the unveiling of the celestial city of God as the end-point of history and the cosmos, mediated by the figure of the Celestial Man, and 6) the Jewish mysticism of *Shekinah* and *Merkabah*. We shall cover essential themes from chapters three through seven of *Cosmos*, following the ordering of these chapters while providing a larger context and analysis. In the next two of our own chapters, we shall explore, successively, the second and third stages of the advance that Bouyer describes.

THEOLOGY OF HISTORY

The modern, scientific worldview could not have arisen without preparation through the long history of worldviews and cosmologies that preceded it in the movement of human thought from pre-history, to ancient civilization, to the rise of the Catholic Church and its various modern divisions, to present-day "planetization" or globalization.

Newman Press, 1969), 9.

Modern science is, to turn again to Husserlian concepts, a piece or moment in the unfolding *telos* of the entirety of this history in its unity. We can surely agree with those present-day Catholic thinkers who hold that the theologian should not, in the current context of contemporary pluralism, tepidly shy away from the aspiration to put forth, however cautious she may be to avoid turning history into a deductive science, a "Grand Narrative" that describes fundamental motives or features of this advance. A thorough assimilation of Sacred Scripture, in its liturgical and ecclesial setting, will not at any rate permit such a strategy of evasion. Philosophers and theologians who proclaim the end of plausible master narratives may well be duplicitous or self-deceived, affirming what they deny in the very denial.[6] Bouyer provides such a narrative, although not without circumspection, in that he is always heedful of the surpassing meaning of the Mystery of Christ as the principle of creation and its history. This narrative is richly informed by Trinitarian faith in God's containment of all finite being and beings in development. The unity of the narrative attests to our author's advocacy of the unique function of the mythopoetic mind, which encourages the perception of the world and history in unity. It is a narrative open to reception of the events of divine grace that may disrupt settled patterns of historical existence or slowly unfolding potentialities of natural process. It is narrative suffused with theodrama, heedful of individual human actions and contexts in relation to God.

The history or drama that Bouyer recounts has much to do with the Christian transfiguration of the pagan sacred, and the author begins a long exposition of this, rooted in historical scriptural exegesis, in chapter three of *Cosmos*. This exposition runs, in a first stage, for five dense chapters, whose main themes we are drawing together here. Yet in order to gain a better grasp of Bouyer's presentation in these chapters, it is necessary to give a cursory overview of elements of his Trinitarian theology

6 Cf. Alasdair MacIntyre, *Three Rival Versions of Moral Enquiry: Encyclopedia, Genealogy, and Tradition* (South Bend, IN: University of Notre Dame Press, 1990), 196–215. MacIntyre gets to the root of the problems attendant to rejection of master narratives. Both MacIntyre and Charles Taylor have confronted postmodernism with alternative master narratives that strive to take into better account than historicizing genealogists the continuity of moral agency as a historical reality necessary for all intellectual discourse.

of history, which is only presumed in the first chapters of *Cosmos* but was fully articulated in *Le Consolateur* (1980), the ultimate volume in his trilogy on theology, published two years before *Cosmos* appeared.[7]

Our author holds that the sacred has a history that is universally, intimately connected to the global progress of the history of salvation. In *Le Métier de Théologien*, he explains that although "the sacred" is intertwined with religion it is distinguishable from "the religious."[8] The religious, he insists, refers to the mysterious reality that gives unity to existence and has its source in God, in the transcendent One of pure freedom and generosity.[9] The sacred, on the other hand, is "the remnant sign . . . of the presence and activity of God without which nothing else would exist."[10] The religious cannot be fully contained by the sacred. Nevertheless, to refuse the sacred is to refuse God the possibility to disclose Himself to us in the world He has created. Sacred symbols are the vehicles of God's self-revelation. J. R. R. Tolkien argued that Christianity was a myth but not one created by the imagination of human poets. It contains, he explained, all the characteristics of a myth but was fashioned by God Himself.[11] The divine Imagination, on his view, is at work in Christian revelation in a unique and unprecedented way. Bouyer thinks in similar manner that God is like a divine Poet who says things in the economy of salvation that had never before been said, using sacred words already present in the total history of human religion.[12]

Drawing on the Church Fathers, the French Oratorian sees creation as a unified, developing economy or series of economies of grace, transforming the sacred, though this process is hardly accomplished in smooth, continuous trajectories of upward movement. In *Le Consolateur*, in an exegesis of his favorite Patristic theologians, Bouyer explains that the history of the cosmos is ubiquitously moved, directly or indirectly, by the healing, redeeming impulsion of the *Logos* and Spirit of God toward

7 See *MT*, 221–22. Bouyer himself implicitly suggests the close connection between *LC* and *Cosmos*.

8 *MT*, 102–3.

9 Ibid.

10 Ibid., 103.

11 See Fernandez, *Mythe, Raison Ardente*, 72.

12 *MT*, 104.

the completion of charity in the eschatological Church.[13] God wills to draw humankind, as individuals and all together, into the Eucharist of the Church, conforming the elect to Jesus Christ in the memorial of His death and Resurrection, so that they can give back to the Father through the Son in the Holy Spirit the very love of God with which they have been loved from all eternity.[14] God thus extends to humanity His divine paternity in His only Son, in whose crucified flesh all things are recapitulated. Humanity is washed clean and made able to assume, as Bouyer says, a "virginal maternity," a new receptivity to the divine that alone opens human nature as well as distinct persons in their vast multiplicity to fulfillment.[15] There is a constant historical movement, a dialogue in freedom between God and His creatures, leading through the Christic transmutation of the pagan sacred—its rites, prayers, and myths—to the nuptial communion of the final apocalypse that will at last consummate history and creation.

Throughout the entire expanse of history, Bouyer teaches, God communicates His inspiration to the human race through the Word in the Holy Spirit, first through "mythopoetic inspiration" that touches our *nous* or *intellectus*, then through the "prophetic inspiration" in which there is a unique piercing by the Holy Spirit of the human heart, leading at last through "verbal inspiration" in connection with the Incarnation.[16] Bouyer's theology of inspiration in these pages of *Le Consolateur* evokes the views of Newman, who, at least from the time of his book *Arians of the Fourth Century* (1833), commended the Alexandrian Church Fathers of the third century for their practice of the *arcani disciplina* or "discipline

13 *LC*, 365–79. These are the pages for chapter 18, our focus here.

14 Ibid., 370. See especially his discussion of Saint Maximus the Confessor on this page.

15 Ibid., 366.

16 Ibid., 372–412. These categories are used more loosely than what one finds in Bouyer's *Dictionary of Theology*, trans. Charles Underhill Quinn (New York: Desclée Co., 1965), 240, under "Inspiration." There Bouyer says: "In the most accurate sense of the word, such as it is understood today by theologians and in the documents of the magisterium of the Church, divine inspiration pertains exclusively to the text of Holy Scripture, which means that it must be looked upon as having God himself as its chief author, or, in other words, as being the Word of God not only in its content but even in its expression." It is "verbal inspiration" in the categories used here that is closest to this meaning.

of the secret" and all that it implies about the economy of salvation. In their famous catechetical school, they proportioned the presentation of doctrine to the capacity of catechumens to receive it. Not everything was divulged to everyone all at once. They recognized with Saint Paul that not everyone is ready to receive solid food in matters of doctrine from the very beginning (1 Cor. 3:2). Newman understood that this discipline was a method to translate the content of divine revelation to the stage of advance in the spiritual path of the neophyte. It was not a false gnosis, a power play, a holding back of secrets in order to bolster the authority of bishops, priests, and catechists. The Alexandrians modeled their pedagogical approach on divine revelation itself, which was accomplished by a progressive economy or economies (including what Newman called a "pagan dispensation") and in conjunction with pervasive, if fleeting, communications of inspiration. God did not give the full content of divine and created mystery to His People in one fell swoop. The biblical Word was given in the direct history of salvation by means of successive covenants, a school of prophets, a progressive deepening and sometimes shifting of the message, and types whose meaning was at last unveiled only in the Mystery of Christ, the ultimate antitype.

The economy of grace was not, however, limited to the unique history of the covenantal People of God. It extended to the origin of the human race. Newman took as his own the "sacramental principle" of the Alexandrians, according to which there are "various Economies or Dispensations of the Eternal." These theologians and philosophers held that "Nature was a parable: Scripture was an allegory: pagan literature, philosophy, and mythology, properly understood, were but a preparation for the Gospel." As John Keble (1792–1866), one of Newman's mentors, put it, even the Greek poets and sages were prophets, for "thoughts beyond their thoughts to those high bards were given."[17] Newman held to the "divinity of Traditionary religion," which is to say that all knowledge of religion comes from God through the divine Word.[18] There are, according to

17 John Henry Newman, *Apologia Pro Vita Sua* (New York: Longmans, Green, and Co., 1908), 17–18.

18 Ian Ker, *Newman on Vatican II* (Oxford: Clarendon Press, 2014), 128–31, at 128. The quotation comes from Newman, *Arians of the Fourth Century* (New York: Longmans, Green, and Co., 1908), 1, 3.5.

this view, national traditions of religious truth everywhere on earth, and these originate from paradisaic revelation. "Natural" knowledge of God has a source that goes beyond the immanent moral sense of conscience or the power of reason to trace the divine presence in creation from the phenomena of the visible world. The traditions of religious truths among all humankind stem from specifically historical economies of revelation going back to a "primitive revelation."[19] All religious knowledge, of which there are fragments and glimmers found everywhere in human culture, comes from God in history. This does not mean that the Church is not necessary for salvation. Indeed, the "special" revelation that God has given to the covenantal people He has called out in Abraham and fulfilled in Christ and the Church corrects and surpasses the primitive or general revelation that He gives universally and disparately, even though the latter is historical and not simply cosmic in character.

This Traditionary understanding of religion, natural and supernatural, aligns in some respects with Bouyer's *ressourcement* of the Church Fathers, which drew especially on Saint Irenaeus's Trinitarian rendering of the economy or economies of the world according to which the divine Word and Spirit are everywhere present in history as the Father's "two hands." "Through the Word," Saint Irenaeus said, "all His creatures learn that there is one God, the Father who controls all things and gives existence to all."[20] For Bouyer as for Saint Irenaeus, the Spirit always works inseparably with the Son in this task of educating the human race, ordering all things to achieve communion with the *Incarnate* Word in the fullness of time. A

19 Cf. Gregory Solari, *Le Temps Découvert: Développment et Durée chez Newman et Bergson* (Paris: Les Éditions du Cerf, 2014), 187, n. 9. Solari details the understanding of tradition found in the influential work of Josef Geiselmann, *Die Heilige Schrift und die Tradition* (Fribourg: Herder, 1962). He suggests that this "Traditionary school" expounded by Geiselmann and represented by Newman and the Tübingen theologians of the nineteenth century was consecrated by the Catholic magisterium in the conciliar texts of Vatican II, particularly chapter 2 of the constitution on divine revelation, *Dei Verbum*. The key advance made, according to Solari, was to link Tradition with revelation via the principle of divine economy as understood by Newman. The first three chapters of the text by Geiselmann have been translated into English. See Geiselmann, *The Meaning of Tradition*, trans. W.J. O'Hara (New York: Herder and Herder, 1966), 74–75. In these pages Geiselmann discusses Newman.

20 Saint Irenaeus, *Adversus Haeresies*, 4.20.6–7.

special economy or dispensation of God is given in history, through the patriarchs and prophets of Israel, in order to prepare the way for the coming of the Incarnate Word. History, in this view, is the space-time of salvation, first promised then brought to realization through the progressive work of the Trinitarian missions in the *oikonomia*.[21]

The descent of the Spirit is an incubating force throughout history and "pre-history," overshadowing humanity, awakening the human universe from the torpor and chaos into which it had fallen with the sin of Adam and Eve, through whom (along with the fallen angels) death entered the world (Rom. 5:12). Humanity is progressively healed, restored, and elevated so that it may be presented anew to the Word of God, the Bridegroom, the Celestial Man.[22] God's communication to humanity of prevenient grace in the sending of the Holy Spirit prepared it to receive the new life given ultimately through the transpierced heart of the Savior on the Cross. Bouyer holds that each person reborn in Christ brings Christ to birth in his or her own life by the grace extended to him or her through the maternal mediation of the Catholic Church. We are children of the new Eve, Mary, whose maternity is prolonged in us through baptism.[23] The revelation that is transmitted to us in the Church is a consummating transfiguration of prophetic inspiration, enabling us to be really adopted into Christ, divinized and elevated into a condition of the true gnosis of faith, moved by love. Bouyer suggests that there are prophets outside of God's direct dispensation given to the covenantal people of Abraham, yet they give only ephemeral visions of the transcendent, divine generosity.[24] They possessed only obscure intimations of the *Logos* and the Spirit.[25] They did not constitute continuous schools of prophecy in the way that the biblical prophets did, moved by the singular if implicit idea of preparing the way for God's definitive gift to us of His transcendent life in history.[26]

21 Cf. Zordan, *Connaissance et Mystère*, 60; Grintchenko, *Une Approche Théologique du Monde*, 274–75. These two commentators on Bouyer thus link his thought to Saint Irenaeus.

22 *LC*, 366.

23 Ibid., 367.

24 Ibid., 372–76. See also *IF*, all of part II, 85–142.

25 *LC*, 19–35, 365–80. See Balthasar's discussion, *Theo-logic III*, 53.

26 See Newman *An Essay in Aid of A Grammar of Assent* (London: Longmans, Green, and Co., 1903), 10:6.

This Trinitarian theology of history implies a Trinitarian cosmology. For Bouyer, the God who saves us in the flesh of Christ is the very one who created all things and sealed the first creation with His signature.[27] Creation is given being in and through the divine Word and is formed out of the chaos spoken of at the very beginning of the Book of Genesis (1:1). The existence of the Trinity is wholly independent of creation, but God's triune life is reflected everywhere, to the extent that it can be said that the Trinity prolongs itself in creation.[28] Bouyer is fond of invoking in this connection Meister Eckhart's distinction between the *bullitio* of God and the *ebullitio* of creation.[29] God is eternally "boiling" with life, and this life "spills over" (freely) in creation and salvation history.[30] All of creation is an echo of the divine Word, and it is traversed by a vital flux awakened by the Holy Spirit, the very breath of life in God.[31] The material world is itself, accordingly, only an expression or materialization of the spiritual world, a language, as we have seen, and is thus truly an animated world.

The Father works in creation like a potter. His two hands, the Son and the Spirit, mold the world through the education of human freedom.[32] The Word espouses us as exterior to us, giving us form in His visible Body, the Church, while the Spirit awakens us from within.[33] We are drawn into the Son by the Spirit who is in us. We can freely respond to the solicitations of God that come to us in and through the world, but this response requires the gift of God's own freedom to us in the hypostatic love of the third person of the Holy Trinity. The Word does not cease to speak to us in all things, just as his Spirit does not cease to sustain our will. The language of the Word can seem illegible to us, though, and it is very difficult for us to embrace the presence of the Holy Spirit. History is characterized by progress and regress in this regard,

27 *LC*, 368. Bouyer follows the Eastern Patristic tradition, especially the Cappadocians, on this point.

28 Ibid., 369.

29 *Sophia*, 144; Bouyer, *Women Mystics*, trans. Anne Englund Nash (San Francisco: Ignatius Press, 1993), 65; *IF*, 268–71.

30 We shall explore this in greater depth in chapter 8 of the present study.

31 *LC*, 369.

32 Ibid., 144, 371.

33 Ibid., 371.

and Bouyer maps this history in broad contours at various places in his writings.[34] The Word is always speaking to us, and the Spirit is always moving us, but it is very difficult to grasp their presence, as our vision has been clouded over and our will rendered chaotic by sin. It is in this context that we should place a brief word of critique that Bouyer once directed at Mircea Eliade. Although he admits to learning much from the Romanian master in the study of religion, he criticizes him in his account of hierophanies for seeing "in some way, in the timeless, things which are essentially historical and the fruit of a development that sometimes fails and falls short."[35]

In articulating the progress of the economy of salvation in the ultimately dramatic manner that he does, Bouyer insists that we must be careful not to collapse the Christian sacred into the pagan or natural sacred, whether historical or pre-historical. This perennialist variety of reductionism, as opposed to the naturalist variety that we discussed in the previous chapter, is another common temptation in the modern age, present since the time of the Renaissance. Bouyer argues that the best studies of religious symbolism, including those of Eliade, have shown that it has no grounding in actual history.[36] If progressivist naturalism reduces all manifestations of the sacred to some natural physical or psychic process, perennialist reductionism fails to distinguish sufficiently the uniqueness of the Christian sacred. Bouyer explores in *Cosmos* evidences of the wider biblical transfiguration of the sacred in royal myths.[37] His demonstrations in this regard are an essential feature of his theology of history, lending it concreteness ordered to address the persistent modern question of the meaning of religion and of Christian religion in particular. He gives attention to diachronic

34 See especially *IF*, chapters 7–9.

35 *MT*, 225.

36 Cf. *Cosmos*, 242, n. 2. Bouyer firmly asserts that there is no "uninterrupted, primitive tradition" of revelation: "This was to be the totally unrealistic notion of traditionalism, which would receive the most surprising appearances of justification during the nineteenth century in Germany from Creuzer, and which Lamennais in France would develop to the point of absurdity, claiming to justify Christian revelation, by asserting that it never revealed anything that the whole of mankind had not already known!"

37 Cf. *MT*, 97–118, 103. Bouyer speaks on page 103 of the "sacré chrétien" and the "sacré hébreu" as "une transfiguration du sacré païen."

evidences of the movement of the Word and Spirit in human cultures and civilizations. These demonstrations likewise bear on his theology and philosophy of nature.

THE ROYAL MYTH

The dialogue between the triune God and humanity in the totality of history is centered on the activities of humankind in sacred ritual and myth, as God's Word and Spirit impose a mutation on these. Bouyer seeks to clarify the precise character of this mutation with respect to ancient myths of the sacred king. In chapter three of *Cosmos*, entitled "From Myths to the Word: From Deified Kings to God the King," he begins to detail the manner in which the divine Word and Spirit, working through the patriarchs, prophets, and kings of Israel, transfigured sacred kingship and all that it purported to signify about the meaning and interrelation of divinity, humanity, and cosmos. This chapter is our focus in the present section.

Myths of the sacred king, we learn, emerged with the historical advent of the first human cities. The figure of the king around whom these myths centered replaced the shaman from earlier cultures as not only the political but the religious leader of the people.[38] The shaman, who, on Bouyer's telling of this history, has to be distinguished from the magician or sorcerer, had a sacred and cosmic role, assisting his people to live in harmony with the powers that rule the cosmos. This role was liturgical, both in the ritual and properly etymological signification of the term *leitourgia* (*λειτουργία*), which has to do with ministry or public service on behalf of the community.[39] The king, like the shaman, was understood to receive his authority from on high and to mediate divinity and cosmos, nature and culture, cosmos and polis. The royal myths thus express the link between the divine, the human microcosm, and the macrocosm of the universe.[40]

Pre-historical cultures set the pattern for early historical civilizations. Their sacred symbolism was cosmologically significant, giving evidence, from the first moment of humanity's dawning self-awareness, that the

38 *Cosmos*, 25; *Sophia*, 20–23.

39 Cf. Duchesne, *Louis Bouyer*, 29–30. See *Cosmos*, 24–25; see also *Sophia*, 18–23.

40 Cf. "Royauté Cosmiqué," *La Vie Spirituelle* 110 (1964): 387–97.

human being was quite conscious of the link between cosmos and *anthropos* in its historical embodiment. Bouyer shows that different sacred, cosmic symbols were accorded an elevated or privileged status based on the social or cultural organization of particular societies: 1) picking and gathering societies recognized the unifying presence of the sky or heavens and espied the divine presence in and through the cosmic heights; 2) hunting and fishing societies turned toward feminine cosmic powers, such as the moon and water, without forgetting the presence of the celestial realm; 3) pastoral nomads unified cult around a solar god, along with other elements vital to their needs, such as rain, wind, and thunderstorms; 4) sedentary farmers placed the divine in chthonic powers, in earth and vegetation, whose maternity presides over life and death; 5) in cities, all life was at last organized around the king, representative of gods before the people or of people before the gods.[41] In this last and culminating phase of cultural development, hierarchized and diversified functions emerged within society, and certain developments in religion occurred, such as the unfolding of a fully explicit polytheism, the emergence of a heightened tendency to absorb religion into magic, the replacement of the shaman of earlier cultures with the king as mediator between gods and human beings, and the elevation to symbolic prominence of stellar hierophanies paired with chthonian divinities.[42]

The more purely cosmic elements of earlier religion were now centered on the king, who became the "dominant hierophany."[43] Bouyer is careful to insist, in relating all that was entailed in this cultural shift, that the mediation of the king was not understood in a uniform way in the Ancient Near East. Different climactic factors led to divergent views of the link between the city and the cosmos.[44] In Egypt, marked

41 *Cosmos*, 24–25. Bouyer follows Wilhelm Schmidt in clarifying the sense of importance that differently situated societies gave to different hierophanies. See Schmidt, *Der Ursprung der Gottesidee* (Münster-in-Westphalie, 1925–1926). See also *Sophia*, 13–15.

42 Ibid., 25.

43 Ibid.

44 Ibid. 28–29. See all of Henri and H.A. Frankfort's *The Intellectual Adventure of Ancient Man*, and particularly the essay by Thorkild Jacobsen, "Mesopotamia: the Cosmos as a State," 125–28, 175–84. Bouyer's description accords especially with this essay by Jacobsen.

by the stable seasons associated with the cycles of the Nile, the king was thought to be the manifestation of the divine and the embodiment as well as guarantor of wisdom, *Maat*, or cosmic order. The succession of kings symbolized the cycle of the death of Osiris, brought back to life by his sister and consort Isis, from whom Horus was engendered. The royal myth in Egypt was linked closely with functional ontologies or pre-ontologies that saw the cosmos or cosmopolis as an endless cycle of eternal return absent real, historical differentiation of events.[45] In Canaan and Babylon, to the contrary, an unstable ecological climate fostered political perturbations that undermined the possibility of establishing a sense of stable kingship. The king was conceived as a self-made man, divinized, and not a manifestation of God from birth. He was adopted by gods and a servant set apart, given the honorific title of "son." History emerged in these lands, where the kings were thought to be responsible for the renewal of the earth each year through their ritual action in the rite of the New Year.[46] Representing in ritual battle the tutelary deity Marduk, they fought against Tiamat, the goddess of death, and their own ritual death and re-emergence brought about the healing of the land, likely, as Bouyer argues, setting the pattern for the myth of the languishing king that Frazer studied in his seminal work *The Golden Bough* (1890).[47]

Whether in Egypt, Mesopotamia, or Israel, the central function of the king was a ritual one. The king was, Bouyer says, both "the agent of sacrifice, like Saul, and its lyrically inspired commentator, like David."[48] Ritual sacrifice, centered on the person of the king, was the most important sacred event in the ancient civilizations, with, as we saw in the previous chapter, benefit (or so it was thought) to the entire cosmos. What did it mean to make a sacrifice, *sacrum facere*, to make sacred? Bouyer's answer to this question is essential to the totality of his thought, including his cosmology, and elicited controversy prior to Vatican II. He insisted, in a manner congruent with a wider phenomenological construal of sacrifice, especially drawn from the work of Gerard

45 *Cosmos*, 28-29.
46 Ibid., 29–30.
47 Ibid., 28.
48 Ibid., 25–26.

van der Leeuw, that sacrifice was not, even for pagan or natural religion, first and foremost a ritual murder carried out by human society in order to placate a bloodthirsty God. A certain type of early-modern opinion, which misleadingly passed for orthodox theology, both Protestant and Catholic, tended to encourage the view that sacrifice originates in divine bloodlust in the face of human wickedness.[49] Bouyer argued to the contrary that neither the myths of the ancient world nor the biblical narrative properly understood give support for this theology. Sacrifice, in its pure form, is not a putting-to-death. The scapegoat, doomed emissary of apotropaic rituals sent to evil spirits to placate them and repel their influence, is not the universal, primordial figure of sacrifice.[50] The primitive, phenomenological structure of sacrifice is sacred meal, a feast or banquet shared between gods and men. Sacrifice is, as Bouyer says, "a celebration recognized as the quintessential feast in which men meet the gods to enter into a profound association with them."[51] Sacrifice was thus the fundamental, constitutive, cosmic action, partaking of which individual, society, and cosmos were brought to a renewed communion by association with the gods. It was a *cosmic* feast, the meal "through which the human family constantly replenishes and recenters itself in the cosmos."[52] It was thought to unveil the basic reality, the deepest nature, of the cosmos as "a community of apparently inexhaustible life."[53] Jean Duchesne summarizes Bouyer's understanding:

49 See Grintchenko, *Une Approche Théologique du Monde*, 73–75; Duchesne, *Louis Bouyer*, 28–41.

50 *Cosmos*, 26–27, 238–29 (nn. 14 and 17). These latter notes show Bouyer's polemical fury at its most energetic, targeting René Girard's theory of sacrifice. Bouyer is certainly too polemical in his assessment, but what he says is not without some warrant, in that he is rightly concerned to stress that the divine Word of God given in history does not eliminate the sacred and sacrality but transfigures it. "Death of God" theologians, whose work encourages the view that the Mass is not a sacrifice, have taken a position on the sacred very much in line with Girard, although his work does not have to entail consequences as severe as what Bouyer targets.

51 *Cosmos*, 26. On this point, the phenomenologist with whom Bouyer most closely aligns is Gerardus van der Leeuw (1890–1950). See Duchesne, *Louis Bouyer*, 31; G. van der Leeuw, *Religion in Essence and Manifestation: A Study in Phenomenology*, trans. J.E. Turner (New York: Harper and Row, 1963), 350–60.

52 *Cosmos*, 27.

53 Ibid.

> What is sacrificed or offered is essentially food, that is, what maintains existence and (more exactly) allows us to participate in an existence no longer passive and frustrating, but free and joyous, in the reciprocity of a covenant where the initiative remains constantly divine but renders possible and even solicits a response.[54]

Sacrifice in its primitive, phenomenological structure and materiality was ordered to a recovery of the givenness of the gift of life and ultimately to discovery of the giver of the Gift. Bouyer realizes that it did not live up to this in so-called "natural religion." Christian religion alone brings true sacrifice, namely, "the adoring contemplation in which we give ourselves up to the divine will."[55] Bouyer's thought on this is aligned with that of Joseph Ratzinger, who distinguished the "replacement sacrifices" of paganism, which can go so far as to take the form of human sacrifice, from the "representative sacrifice" of Christ's Cross and Eucharist.[56] Still, the symbolism of the sacred feast, around the figure of the thaumaturgical king in the lineage of the shaman (not sorcerer), who represents the gods, reaches in unique fashion to the totality of existence in its unified meaning. Far from doing away with sacrifice or desacralizing human existence, divine revelation takes up this fundamental figure of the sacred, so naturally prone to be revelatory of cosmic meaning but distorted when it is confused with destruction or violence. Immolation and death are not the endpoint of sacrifice in its primitive meaning but the means through which, in our fallen world, the sacrifice can be rendered efficacious. They take center stage when sacrifice becomes perverted, as in the case of human sacrifice. They are placed in

54 Duchesne, *Louis Bouyer*, 31.

55 *The Spirituality of the New Testament and the Fathers*, 230. This is Bouyer's description of Saint Justin Martyr's (AD 100–165) understanding, with which he obviously agrees.

56 Joseph Ratzinger, *Collected Works*, vol. 11: *Theology of the Liturgy*, trans. John Saward, Kenneth Baker, Henry Taylor, et al. (San Francisco: Ignatius Press, 2014), 20. Cf. *Rite and Man*, 78–94, 91. In these pages, Bouyer explores the utilization and transformation of pagan sacrifice in Judaism. This includes discussion of the ambiguities of pagan sacrifice, particularly the attempt to take control of the action of the gods through ritual sacrifice.

their proper, truly representative context only in the event of the Paschal Mystery, the logical service or "sacrifice" (*λογικὴν λατρείαν*—Rom. 12:1) of Christ in which Cross and Eucharist are conjoined. Sacrifice in its primitive phenomenality does not center on the immolation of the ritual animal but on the festal communion shared by God and humanity. In its transfigured consummation in Christ, the offering on the Cross is consecrated through the sacrificial institution of the Eucharist at the Last Supper.[57] Through holocaust, Passover, expiation, and Eucharist, God communicates His marvelous deeds to us, transforms pagan sacrifice, and gives the true meaning of His transcendent yet uniquely mediatory kingship.[58] Sacrifice is increasingly recognized as God's self-offering, where He alone is offerer, priest, and victim.[59] In Christ the meaning of sacrifice is fully unveiled in a surpassing manner. Christ is perfect self-gift, the "living bread," as Vatican II teaches, "which gives life to men through his flesh."[60] There is continuity as well as discontinuity between pagan, Old Testament, and New Testament sacrifices. Through the ritual, sacrificial—ultimately Temple—matrix of Israel's shared theology, God is recognized as the transcendent source of life, glory, and moral order of the cosmos. Each of these characteristics is associated with royal dignity in the Ancient Near East, and it is with regard to each of these characteristics that Bouyer spells out, at the end of chapter three of *Cosmos*, the biblical transfiguration of the royal myth.

We see there that God is not, as in the myths, cosmic life itself but gives life to the cosmos, which is, through His creative and re-creative gifts, teeming with life as essentially personal. The biblical understanding of the king keeps the "vitalism" and "organicism" of mythic cosmology while transforming these characteristics in light of God's transcendent unity. The glory of the cosmos reflects the transcendent glory of God the One, True King. God the King is, moreover, the one and only source, in and through His Word, of the one law that rules the cosmos as well

57 Cf. *Dictionary of Theology*, 397–401.

58 *Cosmos*, 239–40, n. 17.

59 *Rite and Man*, 92.

60 *Presbyterorum Ordinis*, 5. Jean Duchesne points out that this document from Vatican II was anticipated by Bouyer's work. The document teaches, as Bouyer did well before the Council, that all Christian sacraments flow from Christ's ritual meal with his disciples. See Duchesne, *Louis Bouyer*, 34.

as the hearts of humankind.[61] Kingship is associated in Israel for the first time in human history with the one and only transcendent God's universal Wisdom and providence, sovereign freedom, perfect life, and generosity. Cosmology is given a fresh impetus, as God is distinguished from the world, and it is seen that He cannot be mediated in history through the figure of the purely human king who governs the city of man. It is on the basis of this understanding that the theological doctrine of *creatio ex nihilo* will emerge over time.[62] In chapters four and five of *Cosmos*, the concern of our next section, Bouyer clarifies further the meaning of biblical cosmology by setting it against the horizon of the mythic cosmogonies at the heart of the royal myths.

THE BIBLICAL WORD

If there is such a thing as a "primitive revelation" for Bouyer, he does not think that it should be understood with reference to a single, universal myth at the origin of history from which all religions are thought to derive and that is carried throughout the course of civilization in a continuous manner.[63] At the beginning of chapter four of *Cosmos*, entitled "From Cosmogonic Myth to Creative Word," the focus of the first part of this section, Bouyer concedes that there are recurrent religious symbols in cultures spread throughout the globe that seem, superficially, to bespeak a common revelatory source. Even pre-Colombian American myths have been found to bear similarities to biblical stories, for instance in the figures of a covetous man and woman standing next to a tree laden with fruit.[64] How can this be explained? On a theological level, it might be so by reference to an understanding of revelation and tradition that was held by Clement of Alexandria in the third century and William Wordsworth in the nineteenth, recovered in theories of primitive revelation, and summarized by Bouyer in a later chapter of *Cosmos*:

61 *Cosmos*, 30–36. We are summarizing all these pages, in which these three points are delineated.

62 Ibid., 7.

63 Cf. *Christian Initiation*, 83. Here Bouyer speaks of "primitive revelations," but the context is different.

64 Ibid., 37.

> He [Wordsworth] praised the element of truth contained in ancient myths in words which might have been penned by Clement of Alexandria: they perpetuate and revive what remains of God's initial communication through the angels with man and pave the way for the Word vouchsafed to the patriarchs and prophets, to become incarnate forever in the Son of God made man.[65]

Bouyer makes this view his own, as is clear from the final chapter of *Cosmos*.[66] There he says that the revelation of God through the angels awakened human consciousnesses to the divine law that was present in the first creation.[67] He makes clear that he holds that there is widespread prophetism, presence of the Holy Spirit, and mythopoetic inspiration in human cultures from the beginning. Prophetism emerges in various times and places in the midst of degradations of religion, which can transfigure religious traditions, but it achieved lasting success only in Judaeo-Christian tradition.[68] In the Old Testament itself the Word of God is heard by His People through the mediation of the Angel of the Lord.[69] Even the People of God first receive revelation by way of angelic mediation, defined as "a revelation of the divine plan reflected in knowledge, and refracted through the images of this carnal world, for which God's messengers, even the highest of them, can do no more than clear the way."[70] John the Baptist, "the Friend of the Bridegroom," is like an incarnate angel, and the Angel

65 Ibid., 173.

66 See especially *Cosmos*, 228–29; see also 215.

67 Ibid., 271, n. 14. Here Bouyer references his earlier works on prophetism and the widespread presence of the Holy Spirit in human culture. He points the reader to pp. 19ff. and 376ff. of *Le Consolateur* and pp. 164ff. of *IF* (see pp. 85 and 86 in English).

68 *IF*, 85–86.

69 See *Cosmos*, 229; 265, n. 35. In the footnote from the latter text, Bouyer lists the New Testament biblical passages he has in mind that imply angelic mediation of divine revelation. See Acts 7:30, 38, 53; Galatians 3:19; Hebrews 2:2. He notes that these passages, which all speak of angelic messengers mediating God's message and law, are rooted in an interpretation of Exodus 20, Joshua 5:13ff. and Deuteronomy 33:2 that was incorporated in the Septuagint.

70 *Cosmos*, 229.

Gabriel appears to Mary, but until their decisive interventions "the faithful angels could do no more than ceaselessly oppose the demons, to counter the influence of their spurious oracles, and allow flashes of the repressed truth to shine through from time to time."[71] Divine revelation is at first communicated to humanity in the midst of the theodramatic fog of warring freedoms.

This theology of angelic mediation is central to Bouyer's apocalyptic understanding of God's self-disclosure to the human race. However, in chapter four of *Cosmos*, he proceeds in a more prosaic manner, explaining similarities between geographically disparate systems of myth on the basis of more scientific considerations. He does not allude at the start to remnants of a primitive revelation given through angelic mediation. He contents himself at first to account for these similarities on the basis of psychic unity and speaks of the sacred as a perennial dimension of human consciousness "which produces similar images, everywhere and independently, in response to the same universe."[72] Bouyer holds that sacred symbols are, as Jung understood, archetypal and universally constitutive of human consciousness, giving us contact with hidden realms of the soul, but that they are also, as Eliade thought, hierophantic, signaling a transcendent source of life manifesting itself in and through cosmic realities.[73] Oftentimes, apparent similarities between systems of religious symbolism are just that: merely apparent. Common sacral elements are found throughout disparate human cultures, but Bouyer insists that we should nevertheless take care not to reduce religious symbol-systems to one another and pretend to see their origin in a common source. Rather, we should note singularities in usage and meaning and endeavor to account for the different ways in which symbols are arranged in different cultures. This, he urges, requires lexicographical analysis that goes beyond mere observation of surface appearances.[74] We might say, based on what Bouyer will suggest in later chapters, that if there is angelic mediation in the process of the constitution of the pagan sacred, it is scattered, fragmented, and mixed with demonic corruption.

71 Ibid.
72 Ibid.
73 Ibid., 36–37.
74 Ibid., 37–38.

Georges Dumézil, a comparative philologist and mythographer, is referenced by our author with respect to the important work he did in tracing multiple traditions of mythic heritage in the ancient world.[75] One tradition that Dumézil marked out extended from the earliest civilizations of India to Greece and Rome. The mythic civilizations in this particular cultural trajectory all possessed a ternary social structure, which was ordered by a sacerdotal caste concerned to maintain the sacrificial rites. This priestly caste was protected by warriors and hunters and was fed by agricultural workers. Each of these three castes or classes was representative of a different perspective on the nature of divinity.[76] A different mythic heritage arose in the Ancient Near East that became typically Semitic, although it drew its overall culture from Sumer, without adopting its language. This heritage is connected to the Indo-European tradition but transposed the sources it utilized into its own unique conditions of life in the Fertile Crescent, the so-called "Cradle of Civilization" in Western Asia.[77]

In chapter four of *Cosmos*, Bouyer focuses especially on Mesopotamian and Canaanite examples of this latter tradition, with their specific royal, cosmogonic myths, at first summarizing the cosmogonies of Sumer and Babylon, which were distinct kingdoms in Mesopotamia, the region of the ancient world that some have maintained gave birth to a primordial sense of historical consciousness. Bouyer shows in this chapter how the initial account of creation in chapter one of Genesis uses certain of the images and structures of these myths while hollowing out their agonistic content. These myths personify the powers of nature and see the world-process as a historical drama of the gods. They present all life as emerging from undifferentiated, divinized waters and the cosmos itself as an issuance of divinity, a fragmentation into a multiplicity of ranked gods. These myths relate genealogies of often sexually-differentiated gods whose productive activity is a type of sexual action. They portray the coming-into-being of the cosmos as the coming-into-being of warring deities and thus equate cosmogony with dramatic theogony. The Babylonian *Enuma Elish*, a text composed of roughly 1,000 lines

75 Ibid., 37.
76 Cf. *Sophia*, 15–18.
77 *Cosmos*, 38.

on seven tablets dating from as early as the eighteenth century BC, is exemplary in this regard.[78] It describes the annual festival of the New Year, in which Marduk's investiture as the highest, most powerful divinity was celebrated in conjunction with the royal liturgy of coronation that safeguarded the cosmos through ritual battle.[79]

These myths anthropomorphized the gods. They projected the duality of good and evil as experienced in the cosmos into the very heart of the divine being, and while the biblical narratives of creation assimilate certain of their symbolic elements, they decisively transmute their content by bringing them into a new ritual context that opens humanity to a sense of God's transcendent freedom. Bouyer argues that it is likely not, as some twentieth-century biblical scholars thought, the New Year celebration of surrounding cultures that inspired the sacred narratives in Genesis but the Passover memorial, representing Yahweh's direct deliverance of His people from slavery in Egypt.[80] For the covenantal people of Abraham, the Sabbath repose of God was understood to be the end or *telos* to which God has ordered His creation in and through humanity. This repose exists for man and not man for it. Humankind was not created to serve and offer sacrifice to gods marked by lack in their own right. There is a genealogy in the first creation account of Genesis, but it relates the history of the coming-into-being of human persons, not gods. Bouyer summarizes the biblical accounts of creation in both the first and second chapters of Genesis in terms of the *telos* of human being as mediator or *metaxu* of creation:

> On the contrary, a distinction freely effected by the transcendent God within a created but still indefinite being—always progressing in the same direction, as through an incubation of his Spirit, his own life-breath—will bring forth beings increasingly capable

78 Ibid., 41.

79 Ibid., 39.

80 Ibid., 39–40. The Norwegian biblical scholar Sigmund Mowinckel (1884–1965) thought that the New Year celebrations of Babylon and Canaan were the origin of the biblical understanding of divine kingship and were refashioned into the creation account of Genesis. See Sigmund Mowinckel, *He That Cometh*, trans. George Anderson (New York: Abingdon Press, 1968). Bouyer thus disputes this view.

> of spiritualizing this *materia prima*, i.e., the abyss or the primal chaos. Instead of a merciless struggle, we see the gradual development of an ascending ladder of life. At the top stands the one who carries God's image in himself; being able to understand the divine task and to perceive its benevolent purpose, he will assist in the completion of the undertaking, thereby perfecting himself through a cooperation which leads to friendship with the Lord.[81]

Humanity is created to share in the Sabbath repose, to be sure, but it is also imprinted in its being, as image of God, with the task to "spiritualize" matter and to enable the Word and Spirit to fulfill their mission to establish the peace and Sabbath rest of God on earth.

The second account of creation, according to this analysis, draws from while transmuting Canaanite myths, using mythic symbols such as the serpent, connected to the subterranean world and to sexuality, and the Tree of Life, symbol of the magical knowledge that gives power over death. Life, according to this mythic tradition, springs not from the sea or primal waters, as in Mesopotamian traditions, but from the barren desert through life-bestowing rains. In the first account of creation in Genesis, God transcends the primordial waters in His Spirit. In this second account, He breathes life through His divine breath (*rûah*) into the created soul (*nephesh*), who is formed not from the waters but from the slime of the earth. God does not, as in the Canaanite myths, fashion humankind with His own blood.[82] This account makes clear that God gives Adam the power to name creatures, that is, to know them in the way that God knows them, a privilege that is given to man conjointly as male and female.[83] Taking the Bouyerian exegesis of both accounts of creation together, the nuptial ontology of creation is clear: God, unlike the gods of the myths, transcends the differentiation between the sexes in His perfect being, but man and woman, by their unity-in-difference, are together the image of God, prefiguring in their nuptial personhood the eschatological God-world communion. Their task of "spiritualizing matter" is jointly given in and through their communion. The sacred

81 *Cosmos*, 41.
82 Ibid., 43.
83 Ibid., 44.

prostitution of ancient cultures, such as one finds it in the Near East, is—not unlike human sacrifice—a twisted, exaggerated desolation of the sacred. Nevertheless, it indicates inchoately that human sexuality bears a sacred, sacramental meaning connected to the human being's vocation as workshop for creation.[84]

In chapter five of *Cosmos*, entitled "The Fall in Myths and According to the Biblical Word," Bouyer clarifies the meaning of the Fall as it is presented in Genesis. The first thing to note in contrast to mythic cosmogonies, he urges, is precisely that creation is distinguished from the Fall. Humanity is not created fallen. The cosmos does not emerge from the dispersion and fragmentation of divinity from within itself. The Fall in Genesis has three successive moments: 1) that of Adam and Eve themselves, in Genesis 3, 2) that of the angels, who desired sexual union with the daughters of men, producing a race of malevolent giants, in Genesis 6, and 3) that of the Tower of Babel and the first cosmopolitan human civilization in Genesis 11.[85] There are thus three levels to the Fall, relating to the personal (Adam and Eve), the cosmic (the fall of the angels), and the social (the tower of Babel). The ordering of Bouyer's great work on the economy, which details a supernatural anthropology, a supernatural sociology, and a supernatural cosmology, correlates with this biblical ordering.[86] In the economy of salvation, the personal, social, and cosmic dimensions of created being are each drawn out of degradation and brought together into newness of life through by the divine Word and Spirit.

The summary at the end of chapter five of *Cosmos* specifies six points of distinction between the biblical Word and mythic words that helpfully reviews and clarifies Bouyer's exemplary presentation. The myths: 1) confuse divinity or the plurality of gods with the reality of nature, 2) collapse together creation or cosmogony with the Fall, 3) see evil as inherent to the being of the gods and humanity as bearing the ontological mark of evil by virtue of its very existence, 4) hold that materiality and multiplicity are the source of evil, 5) locate man at the farthest

84 Ibid., 32, 43–44.

85 Ibid., 46–47.

86 Ibid., 46–49. See Grintchenko, *Une Approche Théologique du Monde*, 104. Grintchenko's summary on these points is very helpful.

point of the falling away or degradation of the divine, and 6) maintain that the hierarchy of being in creation comes about through successive degradations of the divine.[87]

On all these points, the biblical narratives of creation and fall spell out the opposite view. The biblical Word: 1) recognizes only one transcendent God, who creates all things in and through His Word, 2) distinguishes creation from the Fall, the latter understood to come about as a result of the misuse by finite freedoms of their created gifts, 3) holds that God is not evil in His being and that creation is not ontologically evil, 4) sees that it is created spirit misusing its powers rather than matter and multiplicity that leads to evil, 5) teaches that man is created in the image of God and is thus the highest of the embodied beings of the earth, and 6) maintains that divinity exists entirely in the One or in the divine thearchy and that creation, though hierarchically ordered from the highest, angelic, pure spirits down to the level of *materia prima*, is not in itself divine.[88]

The biblical accounts of creation are thus not reducible to myths, although they make use of mythic symbolism common in the ancient world. Through this symbolism, now transformed both in structure and content, the revelation of the Word of God establishes a new relationship with the world, showing forth the radical transcendence and unity of divine being, which enables recognition of personal freedom in creation. The biblical Word promises a new creation, a re-creation established on the basis of God's beneficence. This will be a final transfiguration of the sacred in the generosity and freedom of divine being that will reveal to humankind what it is to be truly in the image and likeness of the Creator and Redeemer.

BIBLICAL WISDOM AND APOCALYPSE

History is moved by God's search for man, dimly and hauntingly perceived by man in his own search for God in and through the natural sacred. God's rescue search is mirrored in the human being's quest for God amidst the trials and travails of temporal becoming.[89] Dissatisfaction with sacred projects that reduce the divine by excessively

87 Ibid., 50.
88 Ibid.
89 Cf. *IF*, 86.

humanizing it sets in at various points in this latter search, and reforming wisdom traditions emerge to counteract these banalities. God Himself initiates a special tradition of Wisdom through and in His covenantal bond with the People of God that makes them aware at last of His true nature as well as their own. This Bouyerian account of the movement of Wisdom is highly significant to the question of cosmology, as we shall suggest in this section and the next, exploring, in turn, chapters six and seven of *Cosmos*.

The creation accounts at the beginning of the Book of Genesis reformed the mythological cosmogonies present in Sumer, Babylon, and Canaan, essentially showing forth God's transcendent goodness, freedom, and generosity, as well as the meaning of the human creature as image of God. These accounts, as we have seen, employed narrative figures and symbols that were borrowed from the myths: the Tree of Life, the Garden of Eden, or the forming of man and woman from the sea or the earth. On a deeper level, following an implied thread in Bouyer's analysis, we can say that these scriptural cosmogonies took from myth and the mythopoetic mind a sense of the unity and intelligibility of the whole of reality as well as the recognition that this unity and intelligibility can be adequately articulated only in the sort of narrative and symbolic form that is proper to mythic expression. The *logos* of Wisdom does not obliterate myth: "For the very foundation of wisdom remains that the universe in which we live makes sense. Myths express that meaning, and they alone seem to have the power to make it explicit."[90] Moreover, these creation accounts preserved the vitalism and organicism of the myths, while giving a new metaphysical, relational framework for these characteristics by locating the source of cosmic life in God's transcendent existence. Biblical revelation, as it develops, will always show forth the irreplaceable importance of myth in these just-described ways, and the canonical Wisdom literature attests to the fact, which Bouyer confirms in chapter six of his text, entitled "From Wisdom to Apocalypse."

Wisdom traditions proliferated in the ancient world. In the first civilizations of Egypt and Mesopotamia, the *logos* of wisdom developed alongside the *logos* of *mythos*, criticizing myths in order to explicate

90 *Cosmos*, 51.

their true meaning.[91] Wisdom was at first understood to be revelatory, a gift given to the king and explicated by those who surrounded him in his royal court. It was at once exemplary and practical. In Egypt, Pharaoh partook through inspiration in *Maat* or the divine wisdom and communicated it to the retinue that accompanied him in his governing functions. These latter mandarins enhanced the royal, divinely-given wisdom through their experience in day-to-day governance and purified it by a sort of proto-philosophical *logos*, by reasoned criticism, through which developed an organic tradition of practical reflection on the world and the ordering of life in the kingdom.[92] From the tenth to the seventh centuries BC, wisdom became a tradition widely passed on in Egypt in official ranks, from father to son or to functionaries who assisted them. "At this stage," Bouyer explains, "wisdom may be described as a technique for the organization of human life in the world, and more specifically in the new cities, which constituted a world in the process of humanization."[93] Though threatened by bureaucratic inertia and politicking, wisdom remained an essentially religious phenomenon, linking ritual procedure with the fundamental activities of society and its effective governance, such as war, agriculture and trade, meteorology, geometry, and astronomy.[94] The ritual governance of the city in accordance with *Maat* maintained the stable ordering of the cosmos. In Mesopotamia, by contrast, with its irregular climate, wisdom or *Me* was identified with a hero capable of overcoming adversity, and the cosmic order was understood to require periodic correction. The wise man in Egypt, land of stable climate with the periodic ebb and flow of the Nile, was silent and patient, while the wise man in Mesopotamia was heroic, more than in Egypt in the cast of a warrior.[95]

Wisdom entered Israel with the establishment of the monarchy. It was at first held under suspicion by the prophets, who thought that human kingship discouraged acceptance of God's sovereignty, and that wisdom distracts the people from attentiveness to the divine Word. Nevertheless,

91 *Cosmos*, 51–52. See *Sophia*, 19–28.
92 Ibid., 52.
93 Ibid.
94 Ibid.
95 Ibid., 52, 54.

wisdom, like myth, was eventually taken up under the guidance of divine inspiration, although given a new meaning. Its foundation was now understood to be in that "fear of the Lord" which comes from "knowledge of the Lord" given in and through the divine Word.[96] Eventually, wisdom merged with the prophetic schools for which obedience to the Word was of paramount concern. The possibility for this eventual merger, Bouyer explains, "was due to the fact that neither wisdom nor prophecy aimed at a purely objective knowledge."[97] Both the prophetic and the sapiential path "sought to learn enough of ultimate reality to model human life according to the laws governing the cosmic order, recognized as divine in its principle."[98] The wise man in Israel, such as King Solomon at the beginning of his reign—though not its end!—was characterized by humility, pious prayer, and contrition, characteristics necessary to heed the call of the divine Word.

As we see especially in Job and Ecclesiastes, Wisdom literature is oriented toward addressing the problem or mystery of evil. Wisdom in these writings is marked by hope in the effective intervention of the divine Word to overcome the seemingly permanent presence of evil in the world, which purely human wisdom could do nothing to alter. Jewish Wisdom does not make the problem of evil disappear but deepens its mystery, because it sets the mystery in the context of faith in the sovereign Goodness of the transcendent God who created all things. The more highly exalted is the divine Goodness in our understanding the more vexing becomes the question of why God would allow evil and suffering in the world. The Book of Job teaches that evil stems from sin, but those who suffer the most are not for that reason the greatest sinners.[99]

A shift from Wisdom to apocalypse, which takes place after the ordeal of the exile in Babylon, is demonstrated especially in the Book of Daniel. Apocalyptic prophecy strongly emphasized the need for an ultimate revelation at the end of history rather than within it in order effectively to unveil the lasting cure that alone can undo the reign of iniquity in the cosmos. Spurred on by the trials and travails that beset the People of God

96 Ibid., 53.
97 Ibid.
98 Ibid.
99 Ibid., 54–55.

as a result of the Babylonian Captivity, apocalyptic literature showed forth a transformed understanding of the locus of human hope: from hope in the restoration of the kingdoms of Judah and Israel within history to hope in a final denouement of history, an end to historical time brought about through catastrophe and the suffering of the just whose lasting reward will surpass all purely earthly remunerations. The message of this new prophecy, Bouyer suggests, is that God alone has the power in His own good time to overcome definitively the slavery of the world to sin and evil and to establish His everlasting Kingdom.[100]

Bouyer holds that with the advent of apocalyptic thought and literature from after the Babylonian Exile in the sixth century BC until the first century AD, biblical kingship and wisdom were definitively projected from the figure of an earthly king onto the heavenly King alone, God Himself. These writings emphasized that the establishment of wisdom on earth in the proper ordering of the human city in a cosmos fundamentally marked by sin requires apocalyptic intervention, a definitive revelation coming entirely from God.[101]

We might remark here, in this recounting of Bouyer's exegesis, that some contemporary biblical scholars hold that we must be careful not to associate the biblical Wisdom literature too closely with intertestamentary apocalyptic literature, because the latter does not have any direct, historical dependence upon the former. This argument contradicts the thesis maintained by the eminent twentieth-century Old Testament scholar, Gerhard von Rad (1901–1971).[102] Still, it is generally conceded that there are thematic connections between Wisdom literature and apocalyptic literature, for instance, that apocalyptic literature shows a concern for Wisdom and the secrets of nature, including "astronomy, meteorology, cosmology, and uranography [the mapping of stars]."[103]

100 Ibid., 55–56.

101 Ibid., 57–58.

102 Gerhard von Rad, *Old Testament Theology II: The Theology of Israel's Prophetic Tradition*, trans. D.M.G. Stalker (Edinburgh: Oliver and Boyd, 1965), 306–8. The thesis is contested by Christopher Rowland, *The Open Heaven* (London: SPCK, 1982), 204.

103 Celia Deane-Drummond, *Creation Through Wisdom: Theology and the New Biology* (Edinburgh: T&T Clarke, 2000), 155. In relating this point, Deane-Drummond draws on M.E. Stone, "Lists of Revealed Things in the

Both the Wisdom and apocalyptic genres of writing are concerned with cosmological motifs.

Whether or not apocalyptic literature has an historical dependence on the Wisdom literature, there is an evident dialectic at work in the movement of the Holy Spirit among the Jewish people, and this broad dialectic is what Bouyer wishes to describe, looking at the development of Old Testament literature as he ultimately does from the perspective of Christian faith. The progress of divine revelation, preparing for the Incarnation and the New Testament, readies God's people to receive a "supreme revelation . . . of the ultimate mystery."[104] Bouyer teaches that mystery, understood in its biblical sense, points to the inner secret of God's plan for the world to establish His definitive reign on earth. Chapter two of Daniel's Apocalypse is thus seen as precursor to the first two chapters of Saint Paul's First Letter to the Corinthians, where the Mystery of Christ's Cross is revealed as the hidden design that God had in store in the heart of His divine Wisdom from all eternity to solve the enigma of human and cosmic existence beset by evil.[105] According to the author of the Book of Daniel, only the seer inspired by the divine Word can interpret the dream of the Babylonian king—"there is a God in heaven who reveals mysteries." No purely earthly wisdom can unlock the meaning of history and the world. Only the effective apocalypse of divine Wisdom in the Word of God can do so.[106]

Apocalyptic literature has struck many in the modern age as unseemly and otiose. It is thought to be recalcitrant to the sort of demythologization that alone can render biblical symbolism acceptable to the modern mentality. It is accused of borrowing too much from mythical themes, and some have argued that it is not really essential to Judaism or to the message of Christ. The cosmology of the apocalyptic writings is thought to be particularly suspect, filled as these writings are with interventions

---

Apocalyptic Literature," in F.M. Cross, W.E. Lemke, and P.D. Miller (eds.), *Magnalia Dei: The Mighty Acts of God, Essays on the Bible and Archaeology in Memory of G. Ernst Wright* (New York: Doubleday & Co., 1976), 414–52.

104 *Cosmos*, 56.

105 Bouyer draws on the work of the biblical exegetes D. Deden, *Le "Mystère" Paulinien*, in *Ephemerides Theologicae Lovanienses* (1936) and Dom Jacques Dupont, *Gnosis* (Louvain, 1949). See *Cosmos*, 249, n. 18.

106 *Cosmos*, 55–57. See also *Sophia*, 30–35.

of angelic beings that came to the fore in this sort of literature in a way that had never been the case in the Old Testament. In much biblical scholarship in the first half of the twentieth century, apocalyptic writings were marginalized, relegated to the peripheries of studies dedicated to Old Testament theology and even held under suspicion of heresy. One particularly influential biblical scholar, Ernst Käsemann (1906–1998), initiated a change when he said in an article on the subject that "Apocalyptic was the mother of all Christian theology—since we cannot really class the preaching of Jesus as theology."[107] Käsemann, operating with the modern presumption that the Jesus of history and the Christ of faith are essentially different beings, thought that the apocalyptic view is indeed rife in Scripture but did not belong to Jesus himself. It was imposed on him by the early Church and became the foundation for the subsequent development of the Christian theological tradition. Because he made such a strong and authoritative claim regarding the influence of apocalyptic thinking on the early Church's theology, Käsemann's essay had the effect of moving apocalyptic from the periphery to the center of consideration for some biblical exegetes, although this was not in and of itself a canonization of the spiritual importance of apocalyptic literature, because, it was thought, apocalyptic motifs were central to the Church but not to Jesus of Nazareth himself.[108]

Our author, quite by contrast, takes apocalyptic to be central to Jesus's own teachings and the apocalyptic literature, whether canonical (the Book of Daniel and John's Revelation) or not (such as I and II Enoch, IV Ezra, and Baruch), to be a decisive source to get at the Galilean's own worldview. In this, his thinking aligns with a very important stream of contemporary biblical scholarship in the United States, especially centered on the Pauline writings.[109] He holds that it is necessary to clarify

107 Ernst Käsemann, "The Beginnings of Christian Theology," in *New Testament Questions Today*, trans. W. J. Montague (London: SCM Press, 1960), 82–107.

108 See D. S. Russell, *Divine Disclosure: An Introduction to Jewish Apocalyptic* (Minneapolis, MN: Fortress Press, 1992), xviii.

109 Cf. Martinus C. de Boer, "Paul's Mythologizing Program in Romans 5–8," in *The Apocalyptic Paul: Cosmos and Anthropos in Romans 5–8*, ed. Beverly Roberts Gaventa (Waco, TX: Baylor University Press, 2013), 1–20. Representative of a much wider stream of scholarship, de Boer shows Paul's theology was at once mythological, apocalyptic, and cosmological.

the way apocalyptic literature in the ambit of Judaism took up mythic views and transformed them, just as the creation literature of Genesis and the Wisdom literature did with mythic cosmogonies and the wider heritage of royal wisdom. In the final section headings of chapter six of *Cosmos*, Bouyer clarifies three basic points of distinction, which we want to expound briefly.[110]

First, in apocalyptic literature, the possibility comes to the fore of transfiguration in death or eternal life for people of justice in the presence of God as well as the resurrection of all things around them. Bouyer quotes Isaiah 26:

> Thy dead shall live, their bodies shall rise.
> O dwellers in the dust, awake and sing for joy!
> For thy dew is a dew of light,
> and on the land of the shades thou wilt let it fall.

This theme in Isaiah is, Bouyer argues, closely connected with the apocalyptic theme of the Celestial Man, the Son of Man spoken of in Daniel 7.[111] Through the Son of Man, humanity is given the possibility for exaltation onto a divine plane of existence. Daniel envisioned a Son of Man "coming with the clouds of heaven," a celestial, eschatological judge who would bring about the definitive reign of God on earth. It should be acknowledged that the proper interpretation of this passage is disputed among biblical scholars. Some argue that Daniel's Son of Man is a symbolic figure who represents the whole People of God. Others have argued that the Son of Man is an angelic being, perhaps identifiable with the Archangel Michael.[112] In other apocalyptic literature, such as IV Ezra, the Son of Man is identified with the Davidic messianic figure whose coming was anticipated in the intertestamentary period, although there is question of whether the figure maintains its celestial proportions in this text.[113] Bouyer quotes Isaiah in connection with Daniel because he

110 *Cosmos*, 59.
111 Ibid.
112 Cf. Russell, *Divine Disclosure: An Introduction to Jewish Apocalyptic*, 121–27.
113 Ibid., 123.

sees inspired coherence in New Testament writers who recognize the confluence of the Son of Man in Daniel with the Suffering Servant in Isaiah. "Around these central figures," Bouyer says, "the eschatological theme of the reign of God, replacing and abolishing all other kingdoms, was the theme in which the apocalyptic literature caused the entire revelation of the Old Testament to focus and to prepare for an even more exalted revelation."[114]

Second, apocalyptic literature promotes a strong form of dualism. Bouyer points especially to the text of the *Testament of Levi*, one of the last Jewish apocalypses, which speaks of the rule of the fallen prince Belial in our world. God's reign on earth is effectively accomplished only through an intervention, an enthronement that overcomes this Prince of Darkness. This understanding sets the context for Christ's own preaching from the beginning—"Or how can one enter a strong man's house and plunder his property, without first tying up the strong man? Then indeed the house can be plundered."[115] The apocalyptic literature should not be taken to promote metaphysical dualism. Its overarching message is consistent with the wider biblical understanding that creatures are good and reflect the perfect goodness of the Creator. The apocalypses promote instead a form of enhanced moral and historical dualism, recognizing that the order of history is marked by contravening forces because of the fall of finite freedoms.[116]

Bouyer points out that the biblical apocalyptic view takes up the dualism inherent in many ancient myths with their dramatic rendering of nature, such as the *Epic of Gilgamesh*, which, as we have seen, speaks of warring gods who bring about the cosmos as we know it. Yet the apocalyptic literature does demythologize, in that it projects the drama of history and cosmos especially toward the future and definitive Sabbath that God wills to establish in and through catastrophic rupture at the end of history. The apocalyptic writings do not, like mythic cosmogonies, speak of conflict and rupture as the source of created being but instead project it into the future as the crucible through which all must pass in order to enter the heavenly banquet of God. Moreover, this literature associates

114 *Cosmos*, 58.
115 Ibid. See Matthew 12:29.
116 *Cosmos*, 59–60.

the final, redeemed, transfigured state of the world with the garden of the gods that myths from the Ancient Near East located at the beginning of time.[117] Jewish apocalyptic points us to the Golden Age of future glory rather than to a lost, primordial garden in communion with the gods. For Bouyer, this sets the model for the meaning of truly Christian nostalgia or memory, an affective accompaniment of the virtue of hope which he described in his first published study of Cardinal Newman.[118]

Third, the apocalypses provide the basis for all future angelology.[119] Angelologies, as we indicated, proliferated in the intertestamentary period as never before. In the biblical Word, the gods of the myths were no longer accorded divine status but were recognized as creatures. The apocalyptic literature oftentimes describes in detail the role and effect of the angels throughout history and the cosmos. Bouyer argues that it is likely true that the influence of Iranian or Persian religion on the Israelites led to this new emphasis, yet the angelology that develops in the ambit of Jewish apocalyptic has its own specific character in continuity with the entire history of the divine Word's enlightening intervention in salvation history:

> For Judaic angelology is but an extension of Israel's primary assertion that the cosmic powers worshipped by their neighbors are at the root of the manifestations of sacrality and divine glory in the cosmos, far exceeding anything man can find on his own level, although these powers are no more divine than himself. To treat them as divine would therefore be to share in their downfall, brought on by their own pride and lust for domination.[120]

In recent years, some advocates of Temple theology or, more generally, of the recovery of "lost forms of Christianity," have argued that we should take *I Enoch*—whose full text, in Ethiopic translation, was rediscovered for the West only in the nineteenth century—with the utmost seriousness

117 Ibid., 59.

118 "Newman et le Platonisme de l'âme Anglaise," *Revue de Philosophie* 45 (1936): 285–305.

119 *Cosmos*, 59–60.

120 Ibid., 60.

in order to get at the heart of Jesus of Nazareth's teachings.[121] Bouyer, in *The Eternal Son*, recognizes the importance of this text and expounds it as an important source for linking together, as Christ and the Church do, the figures of the Son of Man and Suffering Servant.[122] He does not expound its angelology in his writings, but the apocalyptic, angelic cosmology that he developed has some striking parallels to it, which merits a mention at this point.

I Enoch was thought to be an inspired document by some early Christians up to the second century and maintains this status for some Eastern Christian traditions into our own day. The text narrates a history of the Fall and the prospect of future salvation for the righteous elect. In clear reference to Genesis 6, the text tells us that two-hundred "Watchers" or angels "lusted after the daughters of men."[123] This is the ultimate source of the race of giants who corrupt the earth in this text from Genesis. In their midst, the angel Asael brought knowledge and power to the earth: metallurgy for warfare and ornaments or adornments for "beautifying" women — a rather strange priority, it seems![124] This heavenly knowledge was given to humanity without wisdom, in a Promethean fashion. Enoch, the biblical figure who "walked with God, and he was not, for God took him,"[125] is given a vision of the secrets of the cosmos that points to its intimate connection with the angelic beings. He pronounces judgment on the fallen angels, and he is assimilated to the Son of Man. The angel Uriel reveals to him the very secrets of the laws of nature and the heavenly bodies, which are overseen by the ministry of the angels.[126] Bouyer's recognition that human culture originated through the gift of a mysterious spiritual inspiration, which can be developed from the standpoint of Wisdom or, alternatively, in a spirit of usurping power, is congruent with the apocalyptic warning

121 Cf. Dominic White, *The Lost Knowledge of Christ: Contemporary Spiritualities, Christian Cosmology, and the Arts* (Collegeville, MN: Liturgical Press, 2015), 44–45.

122 See *ES*, 120–21.

123 *Cosmos*, 47–48.

124 *I Enoch*, chapter 8. For a full text in translation, see *The Ethiopic Book of Enoch*, trans. M.A. Knibb (Oxford University Press, 1978).

125 Genesis 5:24.

126 Cf. *I Enoch*, chapters 72–82.

of *I Enoch*. This apocalyptic text points to the fundamental choice that human societies have to make with regard to their relation to the cosmos: either contemplate and develop the cosmos in a spirit of humility and Wisdom, obedient to the call of the Word of God, or use the cosmos in a spirit of malevolence, seeking to bend it to the utilitarian aspirations of our fallen nature. We have to make a choice between religion and magic. Bouyer's clear distinguishing of these in *Cosmos* as elsewhere shows his understanding of and sympathy for the overall weight of the apocalyptic message.

THE ROYAL CITY

The myths of the first civilizations centered on the person of the king as mediatory figure linking heaven with earth. The biblical Word took up this theme but transformed it, asserting that God alone is true King, the King of all nations and people on earth as well as of the cosmic powers themselves. God is King of the cosmos. The story of Babel in Genesis (11:1–9) gives an image of the prideful city that tries to achieve deification through magical technology. Bouyer argues in chapter seven of *Cosmos*—entitled "The Heavenly City of Wisdom," and our focus in this brief section—that Babel represents the cities of the ancient world, awash in magic and idolatry, while the biblical Word proposes a new city built by God Himself in His Wisdom. This City is the Celestial Jerusalem in which God will gather all of His dispersed children throughout the earth, purified through eschatological trials. The new and definitive mediator between God and man at the head of this city, the City of God, is not an earthly king but the Celestial Man, "the new Adam, the ultimate and definitive man" as identified by Saint Paul.[127]

Creation and salvation, though distinguishable, are inseparable for Bouyer, as we have seen. No part of creation can therefore stand under ontological condemnation in the form of annihilation. As Grintchenko explains Bouyer's hermeneutic of revelation, the divine Word's "cosmic perspective suspends the world between the original divine blessing of all human work collaborating in the blossoming of creation and the curse aroused by the refusal of man to recognize the giver in accepting

127 *Cosmos*, 62.

the gift, enslaving himself to the cosmic powers."[128] The Fall described in Genesis, in its successive phases, has necessitated a reformation of human action. God has responded by initiating the sacrificial economy of divine revelation. Proper human reception of this revelation requires ascesis, a dying to self, losing oneself in order to find oneself.[129]

Bouyer explains in chapter seven of his text that the prophetic Word of Israel established contrasting yet complementary views of justice and divine blessing. If the just man in the time of the patriarchs was blessed with many earthly goods such as numerous children, a fruitful wife, land, and flocks, a new figure of the just one emerged over time. The prophet Jeremiah, for instance, was called to celibacy: "You shall not take a wife, nor shall you have sons or daughters in this place" (Jer. 16:2). The figure of the Suffering Servant (Is. 53–54) became a paradoxical paradigm for justice and an inspiration for the apocalyptic writers. This was an outcast figure, rejected by men. Yet he emerged as the very one who would redeem the sins of the people as well as of all humanity and show thereby that the suffering of the just is necessary to consummate creation in the Eschaton.[130]

The New Testament, Bouyer explains, will confirm the need for ascesis in the lives of all those who desire to follow the will of God. God calls His people to radical poverty. The beatitude of the poor in the last biblical psalms prepares for the coming of the Gospel. The poor man "defeated by life and who lacks everything . . . whose inability to rely on anything but pure and invincible faith" is a true figure of creaturely wisdom.[131] A fundamental reversal takes place that was confirmed by the Gospel but already well established by the time of Christ's birth in a manger. Bouyer says that whereas the statement "'blessed are the rich' seemed the logical consequence of the fundamental goodness of creation, there had been a shift to 'blessed are the poor,' especially the poor by choice, since it is only by freeing oneself beforehand from the world as it is that one may prepare to greet the renewed world of the

128 Grintchenko, *Une Approche Théologique du Monde*, 110.
129 *Cosmos*, 62–65.
130 Ibid., 65.
131 Ibid., 65–66.

Kingdom of God."[132] The biblical Word thus admonishes us to embrace radical poverty and to leave the earthly city in expectation of the new City that God wills to give to His people. Our author argues that this is demonstrated in the Book of Exodus and given a new spiritual impetus in the intertestamentary period, in the time of the apocalyptic writings, by the spiritual community at Qumran and in the figure of John the Baptist. Christian monasticism, East and West, inherits this spiritual cosmology. The correct form of life is one of eschatological expectation, of confident hope in the destruction of the malevolent cosmic powers and in the transfiguration of all cosmic life.[133]

Bouyer insists that the apocalyptic theology of divine nuptials that we see in John's Revelation has to be understood in this light. The divine Word rejected from the first the sexed gods of the myths, but this was in order to give a properly sacramental understanding of sexual relationality among creatures. God wills to espouse His people to Himself in the Celestial City, forming a one-flesh union with them.[134] The establishment of this "nuptial City" on earth can alone fulfill our deepest human desires. Significantly, Hosea denounced idolatry as a form of adultery and prostitution. He saw it as deeply akin to sexual perversion. God continues to love His people in spite of their adulterous idolatry, and He intends them for purification and regeneration, so that they become worthy of being the eschatological Bride of the Lamb. Isaiah 54 applies nuptial images to the Celestial City built by God Himself in Wisdom.[135] Bouyer argues that while the myths speak of the bejeweled abodes of the gods, Isaiah speaks of the heavenly cosmopolis into which God wills to draw His people as "set . . . in stones of antimony . . . foundations with sapphires . . . pinnacles of agate . . . gates of carbuncles . . . [and walls] of precious stones."[136]

Our author associates the Celestial City of John's Apocalypse, the Bride of the Lamb, with Wisdom as personified in Proverbs 8, Sirach 24, Baruch 3:9–4:4, and Wisdom 6–9. Wisdom, he insists, is truly personal in these passages and not simply metaphorical personification.

132 Ibid., 66.

133 Ibid., 66–67. See Grintchenko, *Une Approche Théologique du Monde*, 110.

134 Ibid., 67–68.

135 Ibid., 68–69.

136 Ibid., 68. We take parts from Bouyer's direct quotation of Isaiah 54.

Modern exegetes dismiss the idea that it is so by pointing out that in Proverbs 8 Wisdom is contrasted with the person of Folly, which is clearly, they assert, only a metaphor. Bouyer argues to the contrary that Folly should also be taken in a truly personal sense, because it represents the spiritual figure of evil from an early tradition. Both the personification of Folly and that of Wisdom intend more than metaphor.[137] Wisdom, Bouyer says in chapter seven of *Cosmos*, is the eternal thought of God on His work, the plenitude of content of the divine Word ordered to the perfection of the cosmos in the economy and thus inseparable from the apocalyptic city.[138] The People of God are led by the divine Word on a progressive journey away from visions of the world that view it in idolatrous and magical ways, beset by evil, surrounded by or infused with powers that humanity delusively thinks it can ritually manipulate according to its whims. God's revelation in the Old Testament progressively points the way to the coming in the flesh of the definitive King who will effect, in word and deed, the final unveiling of the Mystery of creation, hidden in his Wisdom from the foundation of the world.

SHEKINAH AND *MERKABAH*

To this point in the present chapter, we have expounded the main themes of chapters three through seven of *Cosmos*. In these chapters, Bouyer gives a richly detailed, historically-based account of biblical theology of Wisdom and apocalypse, showing how the developing disclosure of the divine Word transfigures ancient mythic themes of royalty and the cosmic city. Bouyer was fascinated by Jewish mysticism, and he frequently returned in his writings to the mystical themes of *Shekinah* and *Merkabah*, associated with Moses and Elijah respectively, the two Old Testament figures who stood with Christ on the Mount of Transfiguration.[139] These themes are not spelled out in *Cosmos* in the

137 Ibid., 69–71.

138 Ibid., 70.

139 *Memoirs*, 71. On this page from his *Memoirs*, Bouyer says that the "monk of the East," Lev Gillet, introduced him to "the labyrinth of Merkabah mysticism" which is connected to Saint Antony's theology of uncreated light. Recall that Bouyer did his doctoral thesis on Saint Antony.

way that they are in other writings of his, but if we turn to a couple of these writings in conclusion of this chapter, especially *The Meaning of Sacred Scripture*, we can better understand an implicit tension in biblical thought of which Bouyer is well aware and that helps us to get under the surface of his chapters dedicated to the Old Testament in his book on cosmology.[140]

One of the most important questions having to do with divine revelation in relation to the natural sacred and to myths pertains to the status of the image in Old Testament thought. By virtue of being called out and set apart, Israel recognized Yahweh as the one and only true God. The images of the gods found among the surrounding cultures were thereby castigated as idols. Did this lead the People of God to reject all images of the divine? Bouyer emphasizes in *The Meaning of Sacred Scripture* that the gods worshipped by surrounding peoples were not for this reason denied real existence, except relative to the one and only life-giving God who made of Israel His chosen People. In reacting to the sacred cults of the surrounding peoples, and in recognition of the uniqueness of the God who had graciously revealed Himself to them, Israel developed a suspicion of the use of images and representations of God in worship, but this hardly meant that images as such were entirely prohibited from the life of praise. Bouyer points out that the angels could still be taken to image the divine, as in the case of the three men or angels who visited Abraham.[141] The most telling attestation, in our author's view, of the continued place of the image in Israel is found in Ezekiel's vision, recounted in the biblical book to which his name is attached. The prophet saw the invisible Presence of God in the Holy of Holies as "in the likeness of a man."[142] Bouyer argues that the theological meaning of Genesis 1:27 is elucidated in this: man and woman are indeed made in the image of God, but this must mean that there is eternally in God "above all, in His creative Word in the expression of His breath of life, something 'human.'"[143]

140 *The Meaning of Sacred Scripture*, trans. Mary Perkins Ryan (South Bend, IN: University of Notre Dame Press, 1958), 140–55. See also *IF*, 138–42.

141 Ibid., 140. Genesis 18.

142 Ibid., 142–43. Ezekiel 1. See *IF*, 139.

143 Ibid., 143.

This exegesis is coherent with the theme of divine-human correlativity in Bulgakov, and, like Bulgakov, Bouyer develops the idea in his Christology around the theme of the "eternal humanity" of the Son, a point that we shall explore in our fifth chapter. The French Oratorian sees this as a crucial dimension of Jewish mysticism centered on the *Shekinah* and *Merkabah*. Herein resides the tension of Old Testament thought to which we refer. The human image shows forth the inner life of God, but God is also the Wholly Other who appears in fire, lightning flashes, and unbearable brightness. The *Shekinah* refers to the divine presence which comes down into our midst: in the tent of the covenant on Sinai or in Solomon's Temple. Yet if the divine presence dwells with His people, in their sacred sites, it cannot be contained by these. God can withdraw Himself suddenly. The *Shekinah* has a Face—God has a Face!—but Moses cannot see it directly without dying. He can grasp God's passing glory only "from behind."[144] The *Merkabah* intervenes as the chariot of fire beside the four Cherubim with human faces that lifts Ezekiel beyond the burning walls of the world. God comes down to His chosen people and dwells among them in the *Shekinah*, while he "ravishes them out of the world and of themselves" in the *Merkabah*, to join the divine presence at the throne of glory.[145]

We can learn from this that God's presence in the world is not static and immobile. He is intimately present to His people. He has made us in His image and to His likeness, but He is free of earthly bonds and can "pass over" His people. He is not bound to a place but is pure liberty. Bouyer emphasizes, contra what he takes to be the view of Greek philosophical mysticism, that one does not encounter God's presence simply by learning to open one's eyes.[146] God is everywhere present as a personal presence, unlimited in the glory of divine power and Wisdom, but He is also, as *Shekinah* and *Merkabah*, self-revealing. God gives Himself to His people only through the boundless grace of divine *hesed* or mercy. As important as the sacred is to God's ability to manifest Himself, God cannot be contained by the natural sacred. Even the supernatural sacred does not give us the fullness of His eschatological

144 Ibid., 149–51. See Ex. 33:18–34:8.

145 *IF*, 141.

146 *The Meaning of Sacred Scripture*, 157.

presence—a lesson that Bouyer long understood but that struck him with particular vividness in his encounter with the Hagia Sophia, the most glorious site of the supernatural sacred in Christendom, where the divine liturgy is no longer celebrated. This has implications for how we understand Christian life lived *in concreto* as well as the task of those called to be teachers in the Church. Bouyer insists that attentiveness to the manner of God's presence in the cosmos requires attentiveness to the "creative event," to the historically unique instantiations of God's gracious self-disclosure understood not only in terms of the destiny of individuals but of the entire course of history.[147] It should compel recognition that existence in the sight of God calls forth eschatological reorientation of self. Bouyer urges that theologians should be especially careful to maintain a standpoint of eschatological attentiveness, of awareness that we can envision God's glory only "from behind." The Kingdom of Heaven has already come to us in Christ, yet it remains hidden in the present age. This understanding should inculcate an overall awareness of history, its particularities, its contingencies and developmental path, and this, we think, is manifested in our author's way of showing forth the progressive transfiguration of the sacred by the divine Word, ever orienting us in hope toward the future. Obviously, this does not mean that he thinks we should forsake what has come before us and that has been given to us through Tradition. Far from it, as we saw in the last chapter: he thinks delusory the idea that there can be a pure, creative futurity. Yet he always insists that Christian existence requires an eschatological standpoint, and that this must be fully embraced by those who carry out the theological vocation. Like our father in faith, Abraham, we are called to follow the Lord in exile, as nomads. Like the Son of God himself, the disciple has no place to rest his head. This brings us back again to the figure of the

147 See especially *IF*, 302. Here Bouyer recommends the need to correct Hegel along the lines of Kierkegaard: "The only way to achieve something of value would be to elicit, *a posteriori* from human history as a whole viewed in the light of faith, some essentially historical theology in the truest sense of the word. In other words, we need that respect for the creative event so well expressed by Kierkegaard, as long as one avoids his mistake of restricting it simply to each individual history but rather finds it in that human history, the inseparable warp and woof of which consists of what, on the one hand, is essentially personal, and, on the other, essentially collective, in a cloth which is wholly interpersonal."

monk, who is, even in the way of life of Benedictine stability, a disciple in exile.[148] Our author thinks that this eschatological standpoint should move the theologian to recognize the provisional character of his or her achievements.[149] The theologian must look back on the whole of history in order to show the character and meaning of the works of God as God gives Himself from Himself in His mysterious Presence. On this basis, he or she must point the way to the future coming of His Kingdom in its as yet undisclosed plenitude.[150]

148 See *Introduction to the Spiritual Life*, 244: "Hence, even if the monk is in fact fixed by the vow of stability in a monastery chosen once for all (as in the Benedictine Rule), his is to be a nomadic existence, and one of eagerness for the desert—which makes it almost the exact opposite of such a life as found in a world that tends with all its weight toward settling down, becoming established."

149 See *IF*, 302, 308–11.

150 Ibid., 310. It is in this way that theology must be a "phenomenology of the Spirit." We shall develop this idea in our final chapter.

CHAPTER 4

From Myth to Wisdom (II): Philosophy

BIBLICAL REVELATION TRANSFIGURES MYTH. It ultimately moves beyond the confusions of myth through the historical intervention of the divine Word in his apocalyptic Wisdom, but it is truer to say that it consecrates mythopoetic thinking, correcting and elevating its religio-cultural productions by showing forth in a definitive manner God's transcendent personality, than that it overcomes it. Concurrent with the development of biblical Wisdom out of the seedbed of mythopoetic imagination, philosophical *logos* emerged, itself only seemingly in radical opposition to *mythos*. Bouyer shows that philosophical wisdom in its Greek form, as much as biblical Wisdom, drew inspiration from the wisdom traditions centered on the royal myths in the Ancient Near East. The Greco-Roman philosophical tradition entails its own forms of theoanthropocosmic synthesis in mythopoetic mode. Bouyer summarizes these in chapters eight and nine of *Cosmos*, our focus in this chapter. In his wider writings, we see that our author's genealogy of cosmological wisdom embraces a global and even pre-historical perspective. In this chapter, we want to trace this genealogy in its broadest outline. First, we shall briefly explore the manner in which Bouyer treats of the roots of philosophical cosmology in religious pre-history and the global scope of these roots as they flowered into the philosophical systems of China and India. Second, following directly the aforementioned chapters of *Cosmos*, we shall narrow in on the Greco-Roman philosophical tradition that arose in dialogue or dispute with biblical revelation, exploring with Bouyer three main dimensions of its development: 1) the move from the pre-Socratic philosophers to Plato, 2) the relation of Aristotle and the Stoics to the earlier philosophical traditions of cosmic wisdom, and 3) the movement from the philosophy of Philo of Judaea to that of the Neoplatonists. We shall ultimately zero in on the theme of cosmic exemplarism, which is deeply rooted in mythopoetic experience and has a transformative corollary in biblical revelation. We shall summarize and conclude the chapter by highlighting some implications of Bouyer's treatment of philosophy and emphasize especially his view that the Christian monk is the symbolic figure of the true philosopher.

THE RELIGIOUS ROOTS OF PHILOSOPHY

Human cultures, though irreducible to one another, are universally interrelated. This is an evident principle in Bouyer's work and essential to his understanding of Tradition. On the basis of this principle, he recognizes that one cannot grasp the development of Greek philosophy and the subsequent tradition of Western cosmology if one limits one's focus to the immediate cultural and historical circumstances of Greece itself. There is a universal dialectic at work in the concrete forms in which philosophy relates to religion and myth, even though one has to take care not to reduce particular philosophical systems to one another, any more than one can dissolve the different frameworks of myth in which religious symbolism is deployed into an undifferentiated unity.[1] His sapiential reading of history centers all of human and cosmic history on Christ and his eschatological promise to transfigure the world. The history of philosophy, crucial to humanity's search for wisdom, must be told as an essential part of the story of God's meeting with humanity, but this history is not limited to the particularities of Greco-Roman antiquity.

If Bouyer mostly limits himself in *Cosmos* to the exploration of the Western tradition of philosophy, this is not true of his wider writings. He does point out in *Cosmos* that while Greek philosophy seemed to pit dialectical *logos* against *mythos*, this sort of antagonistic tension does not

1 Cf. *IF*, 3–82. These pages constitute the whole of part one of the book. Bouyer places the development of philosophy in the wider context of the "dialectical process of human knowledge and with the religious evidence where its development admits of observation" (14). He explores in this book arguments for the existence of God in the Christian theological tradition. He roots these arguments in mythopoetic experience. If philosophical arguments are critical of the mythic construal of divinity, they nevertheless do not leave mythopoetic experience behind. This is especially evident when the different approaches to proof for the existence of God—cosmological, moral, and transcendental—are taken together and the aesthetic dimension of human experience is taken into account.

represent the only avenue of approach to the concept through which we know the cosmos.[2] There are alternative traditions of philosophy that are incontestably sympathetic to mythopoetic experience. He mentions these other traditions in *Cosmos*, but it is not there that we find his most developed consideration of them. Instead, these considerations are most fully articulated in *Le Consolateur* and *The Invisible Father*. In these books, the Oratorian shows that Eastern religious cosmologies, especially in China and India, are explicitly rooted in the mythopoetic experience that subtends the earliest flowering of human religion.[3]

Bouyer's theology of history teaches us that the Word and Spirit of God are universally present in human existence throughout the ages from the very beginning, even though their self-communication is obscured by the persistent tendency of humanity to withdraw into a condition of self-involution. Yet their presence does not leave the development of philosophy unaffected. Philosophy, as all cultural achievements, originates in consciousness or experience of the spirit, which is first given, by a gift of mysterious inspiration, to a particular figure of immense cultural importance whom we have mentioned, the shaman, who is the first inspired pioneer of human society. This primordial religious figure shares his inspiration with the tribe or clan to which he belongs through "communicative enthusiasm" (*enthousiasme communicatif*). Shamanic experience is, Bouyer says, "the first witness of the divine spirit who surpasses us, at the common sources of culture and religion, but which can equally, across the development of these, complete and transfigure us."[4]

Shamans are the original sages in (pre-) history, but if there are proto-philosophical stirrings in shamanic cultures, these are doubtlessly, Bouyer argues, "too primitive" (*trop primitive*) to have left the trace of a definitive testimony that would give access to an articulated cosmological

2 *Cosmos*, 20.

3 *LC*, 19–35; *IF*, 108–22. Bouyer links mysticism and the development of philosophy. See *Mystery*, 187–205, 260–77. See also "Mystique cosmiques et mystique interpersonnelles," in *Des Bords du Gange aux Rives du Jourdain*, 2nd ed., eds. Hans Urs von Balthasar, Louis Bouyer, Olivier Clément (Paris: Saint Paul, 1983), 150–53.

4 *LC*, 22.

vision.[5] In *Le Consolateur*, he turns to historical civilization in the Chinese tradition of religious thought in order to locate a properly philosophical culture that can provide access to the pre-historical roots of human understanding of the world. In China, he argues, philosophy developed and progressively refined, over the ages, the original experience of the shamans. He finds Taoism to be especially significant in this regard, as he takes this form of Chinese religion to be more deeply rooted in the depths of the Chinese religious soul than the other significant religious traditions in China, Confucianism and Buddhism. Taoism gives us, in living form, a practically-oriented cosmological wisdom that was able to "systematize the primitive experience of cultures awakened by Shamanism."[6] Exemplified preeminently in the *Tao te Ching* attributed to Lao Tzu (c. 604–531 BC), this most native form of Chinese religion proposes that the Tao or "the Way" is a continuous flow of cosmic life that is profoundly one yet varied. The Tao is analogous to the "nature" of modern ecology, to the cosmic "subtle fire" of Stoicism that "insinuates itself in all things," and to the continuous current of fire in Heraclitus's cosmology. It promotes a subtle materialism that is, perhaps not so paradoxically, a pan-spiritualism of cosmic being.[7] The Tao confers proper form on all things but also abolishes formal singularities in the death of the individual or in a periodic *ekpyrôsis* (*ἐκπύρωσις*) or cyclical conflagration. Bouyer describes it as at once the reality most transcendent to and most immanent in all things. It is bi-polar in its metaphysical structure, at once Yang and Yin, heaven and earth, or masculine and feminine.[8] It is both the constituting "stuff" of all things that relates them in unity and continuity and a sovereign, quasi-personal reality that reconciles beyond all human conception the One and the Many. When Buddhism was exported onto Chinese soil, the *Tao* was associated with the *chi* of Buddhism, which is the breath or air that fills all things through ever-extending *yang*, *chi* in its active phase, and contracting *yin*, *chi* in its passive phase.[9]

5 Ibid., 23.

6 Ibid., 25.

7 Ibid., 24. Bouyer thinks that "Tao" is translatable as Pascal's *les chemins qui marchent* ("the walking paths").

8 Ibid., 23–24.

9 Ibid.

Taoist philosophy articulates ancient microcosmism in a way that clearly is of great interest to Bouyer. In both *Le Consolateur* and *Cosmos*, he references Chinese religious painting in the Sung period (AD 960–1279).[10] Does Bouyer see himself in the miniscule human figures in landscapes from this period upon which he comments? These humble spectators seem to him to be annihilated in contemplation of the "soft beauty" and "serene grandeur" that first captures one's attention in viewing them. The entirety of the vision portrayed in the Sung landscapes is self-consciously the outcome of an act of contemplation. The contemplator portrayed is identifiable, Bouyer says, with "a humble, all-enduring openness to the integral reality: not to the empirical world, but to a superhuman world, which he discovers beyond all sensible experiences, even more all reasonings."[11] These landscapes represent a flowering of the human spirit in a "rediscovered cosmos," merging together both Taoism, "with its sense that man is rooted in the cosmos that, although transcendent, reveals itself only to him and through him," and Buddhism, "the greatest pre-Christian doctrine of liberation from the all-invading ego."[12]

If Neoconfucian humanism widened "humanity to cosmic or even hypercosmic dimensions and fullness" and met with the Buddhist critique of ritual and myth, the outcome for Chinese philosophy, deeply rooted in pre-historical experience of the Tao, was not a-cosmism but attestation of the existence of "the hyper-cosmic super-humanity of a mysterious and wholly divine compassion."[13] This vision of hyper-cosmic human expansion is a kind of veiled precursor to the celestial, eschatological humanism of the Christian tradition that we shall begin to explore directly in our next chapter. Bouyer suggests that this Chinese religious

10 In *Le Consolateur*, Bouyer notes in this regard Peter C. Swann, *Chinese Painting* (New York: Universe Books, 1958); Victor Fennellosa, *A History of Chinese Painting* (New York, 1969). We have not been able to find more precise bibliographical information for this latter book. Perhaps Bouyer means to reference the famous scholar of Oriental art Ernest Fenollosa (1853–1908), who taught in Japan. In *IF*, he references Nicole Vandier Nicolas, *Art et Sagesse en Chine* (Paris: PUF, 1963). It should be pointed out that one of the most important influences on Bouyer in his youth was an uncle by the name of Francis who was a dealer in Oriental art. See *Memoirs*, 18, 27.

11 *LC*, 25–26.

12 *Cosmos*, 160.

13 Ibid., 21.

philosophy accords with mythopoetic rationality and expresses it with such profundity that Christian thought should be obliged to engage the Chinese tradition more deeply. Within the limitations of what we have called in the modern age "natural religion," it tends toward the accomplishment of the vocation of human *logos* in poetic *nous* to rediscover the unity of the cosmos through the differentiation of an endless presence of harmonies that are the self-multiplication of the original unity of being.[14] Bouyer sees the sacral instinct native to Chinese soil, even in our day, as possessing a religious potentiality which could provide inspiration for evangelical expansion and even for the development and renewed expression of Christian theology.[15]

As regards India, our author seems a little more ambivalent about the tradition of religious philosophy on offer, although he does recognize that Indian wisdom attempted to harmonize *logos* and *mythos* in a way that the Western philosophical tradition seems at first glance not to have done. It developed all the dialectical possibilities inherent to the experience of duality that Taoism articulated by virtue of the principles of Yin and Yang. Indian philosophy is thus deeply religious in origin. It reflects both the Earth-Mother cultic vision present in the *Upanishads*, which hearkens to a pre-Vedic form of religion, and the solar and astral cultic theology of the *Vedas*, in which Brahma emerges in human consciousness through the practice of ritual sacrifice as the source of regularity and vitality in the cosmos.[16]

Brahma is, in origin, "the divine virtue of the sacrifice, in which all existence is suspended."[17] For the writers of the *Upanishads*, "Brahma

14 Ibid., 20.

15 *IF*, 109. With regard to the continuing interaction of Taoism, Confucianism, and Buddhism, Bouyer remarks: "It may even be that we stand unawares on the eve of a new victory for their assimilative power, this time over Marxist Communism (which seems incapable of developing in the West). Christians in the know as regards the Far East think that such a victory could prelude a wholly new beginning for the Christian mission and, above all, provide twentieth-century Christianity, which so often seems moribund, with an unexpected instrument for the vigorous expression of eternal truths."

16 *LC*, 26–27. Bouyer shows that the *Upanishads*, although written later than the *Vedas*, are in fact a reflection of the most ancient religion of India, which the Aryan invaders who gave us the *Vedas* attempted to supplant.

17 Ibid., 27.

becomes identical with the universal vitalism which animates the autochthonous, pre-Vedic religion."[18] The texts of the *Upanishads* maintain at the same time that Brahma is the only substantial reality underlying phenomena in the world, to the point of being identical with Atman, which is the very breath of human life. Subsequent Indian philosophy will, Bouyer explains, attempt to reconcile various positions that arise from reflection on these religious texts in the context of sacrificial praxis with the formation of a vast array of metaphysical positions. These range from solutions to the problem of existence that are akin to the second account of creation in Genesis, for which Atman would be in the image of Brahma, to those that affirm an absolute identity between Brahma and Atman.[19] Indian philosophy does not arise with the express purpose to abolish mythopoetic thinking. It works within myth, but its vast array of metaphysical positions ends up emptying myth of its real power to give witness to the living unity of being. Bouyer says in *Cosmos* that in Indian philosophy "eventually myth became a mere poetic gloss, and a beguiling but superficial unity covered a schizoid acceptance of the most bewildering constellation of contradictions in terms."[20] He does not see in Indian philosophy the same sort of organic, harmonious unity between philosophy and mythopoetic experience as in Taoism, as its dialectic veered in his view too far in a rationalist direction.

Yet there is another dimension of Indian religious thought to which Bouyer accords great importance and which is a central consideration in his overall account of the history of religious philosophy, with its continuities and discontinuities. This is the tradition of the "Primordial Man," which he mentions in *The Eternal Son*. Bouyer understands the apocalyptic figure of the Son of Man in the Old Testament and the Gospels to be remotely connected with the myth of Purusha that is found in the early *Vedas* and in myths akin to it. He holds that the biblical understanding transforms the meaning of these myths in light of the theology of creation in the Book of Genesis and the Wisdom literature. Purusha in India, in the early *Vedas*, is the "cosmic man" from whose sacrifice all life comes into being. The cosmos derives from the fragmentation or

18 Ibid.

19 Ibid., 28.

20 *Cosmos*, 20.

dismemberment of Purusha, who then reappears anew at the beginning of each new age of the cosmos.[21] Unlike Purusha, who is the cosmic figure from whom all things originate as the explanatory principle of creation, the figure of the Son of Man in the apocalyptic texts has eschatological significance. The Son of Man is soteriological and eschatological rather than ktisiological.[22] He comes at the end of history, not at its beginning. Still, the myth of Purusha is, on this interpretation, an important foreshadowing of the Celestial Man of biblical revelation inasmuch as it points to the existence of cosmic or supracosmic humanity. Does this mythic figure not give expression to an inkling of the existence of a universal human, maintained in the midst of the frenzy of philosophical rationalizations in India? This cosmic figure is, our author maintains, the sign of a profound instinct for the unity of God, humanity, and the world, which the Christian tradition will consecrate in a new form, through the revelation of God's universal Fatherhood in the Incarnate Word, who bears uniquely the humanity of the Son of Man.

GREEK PHILOSOPHY: FROM THE IONIANS TO PLATO

Does the Greek tradition of philosophy, typically interpreted in the modern age to be the enemy of the mythopoetic mentality, ultimately encourage the evisceration of this inchoate, mythic sense of theoanthropocosmic unity? The development of the *logos* of philosophy in Greece seems to present a very different form of relationship to the *logos* of myth than in China or India. From the time of the development of Ionian rationalism, classical Greek mythology in Homer and Hesiod fell under withering critique. Mircea Eliade once said: "If in every European language the word 'myth' denotes a 'fiction,' it is because the Greeks [Ionians] proclaimed it to be such twenty-five centuries ago."[23] The most eminent successor of Eliade in the study of world religion, the Belgian scholar Cardinal Julien Ries (1920–2013), identified three classical schools of rationalist thought concerning myth in the Western philosophical tradition. The first, that of the Ionians (7th–6th century BC), saw the myths as covering over scientific concepts and proposed theories of physical nature to replace

21 *ES*, 91.
22 Ibid.
23 Mircea Eliade, *Myth and Reality*, 148.

the myths of Homer and Hesiod as the source of cosmological understanding. The second was that of Plato (428–347 BC), for whom myths retain importance inasmuch as they provide an image of truth. The third, that of Euhemerus of Sicily (c. 330–250 BC), understood myths to be deifying stories of ancient kings or heroes.[24]

Bouyer recognizes the distinctive character of the Greek philosophical tradition with regard to myth. It is in this tradition alone, he says in chapter two of *Cosmos*, that "the development of our knowledge was to go through the three phases recognized and described by Auguste Comte [1798–1857]: the theological, metaphysical, and positivist stages."[25] Yet this development was delayed, for it was only after the Renaissance that the tripartite Comtean schema was instantiated in full.[26] The ancient Greek philosophers themselves provide a more ambivalent testimony. Targeting the stories of Homer and Hesiod, these philosophers aimed to abolish decadent mythology rather than living and vital religious myths.[27] Bouyer argues in *The Invisible Father* that Plato's desire was not to do away with the myths as such but to purify them of distorted, anthropocentric images of the gods. Plato wanted to associate the gods with the Good, with self-diffusive being itself, and to show that they were free of envy and therefore worthy of worship.[28] Plato's philosophical quest was not, then, anti-religious or dialectically opposed to myth such that it sought its destruction, for he knew that the exercise of philosophical reason could not in and of itself bring salvation, and that one must return to the myths to have any hope for liberation from injustice.[29] In this return to the myths, the philosopher's understanding of them would be transformed, but the counsel to go back to them signals that Plato did not sanction the idea that philosophy can replace myth *tout court*. It cannot be said of Plato that he held, in the manner of Hegel, that religion has primacy of place in human culture only until the scientific, philosophical concept

24 Julien Ries, *Symbole, Mythe et Rite: Constants du Sacré* (Paris: Les Éditions du Cerf, 2012), 280. Euhemerus, Ries shows, drew his interpretation of myths from the Stoic philosophers.

25 *Cosmos*, 21.

26 Ibid. We shall return to this issue in greater depth in the next chapter.

27 Cf. Eliade, *Myth and Reality*, 147–54.

28 *IF*, 54.

29 Ibid.

has been established by enlightened philosophers.[30] It is on the model of Plato's approach that Greek philosophy can be understood to carry out a needed demythologization that is itself a kind of *praeparatio evangelica* connected with the biblical Word's own manner of correcting the myths.[31]

Nor should it be forgotten, as Bouyer points out in *Sophia*, that the advent and progress of Greek philosophy was connected to political developments in the ancient Greek city-state with regard to the role of kingship in society:

> The power of the kings, in Greece, was only ephemeral, quickly desacralized, if not destitute purely and simply, by turbulent oligarchs, ceding in their turn, especially in Athens—in this opposed to Sparta—to the democratic wave. This itself, gliding to demagogy, will end in dictatorial tyrannies, before opening paradoxically from one of them, the Macedonian, to the first sketch of realization, even the first idea, of an ecumenical empire.[32]

Greek philosophical rationalism is often tied to the promotion of a new kind of individualism, but Bouyer points out that the Greek philosophers maintained a vision of the whole that always connected the individual to a social and cosmic view of human flourishing. Greek philosophy also has connection to the royal myth. It voices "the permanent tension

30 Ibid. Bouyer references A. J. Festugière, *De L'Essence de la Tragédie* (Paris: Aubier-Montaigne, 1969). We point to the similarity between Bouyer's understanding of Plato and that of Josef Pieper, one of the greatest scholars of Plato in the 20th century. See Pieper, *The Platonic Myths*, trans. Dan Farrelly (South Bend, IN: Saint Augustine's Press, 2011), 41–62. Pieper argues that Plato believed the myths are true. The "ancients" are the authorities in the transmission of the myths. It is they who received the sacred tradition from a divine source and passed it on. Yet, Plato recognizes, there are degrees of truth in the myths and philosophy performs its critical function based on this point. One can distinguish true, eschatological myth from myths that lie, and the former has salvific bearing, such as at the end of Plato's *Republic*, with the soldier who returns to life on the funeral pyre with the true story of the afterlife which "also saves us when we believe in it" (Plato, *Republic*, 621c1).

31 *IF*, 54–57. Bouyer sees the development of rational arguments for the existence of God as important especially inasmuch as they clarify our understanding of God, freeing it from idolatrous, mythic images.

32 *Sophia*, 38.

between this individualism [of the Greek city-state] and a cosmism, more exactly a physicalism, where the development of the properly Greek city will bring the very individualism which characterizes it paradoxically to nourish itself from a reinsertion of the individual into an evolution which is that no longer of the city, even enlarged, but of the whole Universe."[33] This vision parallels the development of Wisdom in the ambit of divine revelation. Divine revelation in the Old Testament and Greek wisdom both recast the royal myth by extending its scope, and both foreshadow the supernatural cosmopolitanism of what will become the Catholic Church's ecclesial cosmology, that is, its understanding of the City of God in its final, cosmic breadth at last realized in the plenitude of the eschatological Church.

Confining ourselves with Bouyer in chapter eight of *Cosmos*—entitled "Development of Greek Cosmologies"—specifically to the cosmology or physical theory of the Greeks, which was the chief concern of the Ionians, it seems indeed that mythopoetic experience was in fact consecrated in important ways within a conceptual framework.[34] In this sense, Greek wisdom, no less than Chinese or Indian wisdom, "never really replaced myth, but simply incorporated it under another form."[35] The "vitalistic" and "organic" qualities which the Ionians projected onto the *ούρανός*, heaven and earth (the totality of being, which Greek philosophers only later described as *κόσμος*), gives evidence of this.[36] Bouyer demonstrates the point in chapter eight by briefly summarizing the positions of Thales of Miletus (624–546 BC), Anaximander (610–546 BC), and Anaximenes (585–528 BC), highlighting the typically mythic characteristics present in their cosmologies. For Thales, who gave us the first philosophical usage of the word *φύσις* or "nature," the *ούρανός* is "ensouled," *εμψυχον*. It is an

33 Ibid., 39.

34 *LC*, 26. Bouyer references John Burnet, *Early Greek Philosophy* (London: Adam and Charles Black, 1892), and R.G. Collingwood, *The Idea of Nature* (Oxford, 1970). Bouyer's monograph on cosmology is methodologically akin to this book by Collingwood—a dear friend of Tolkien—although Bouyer's standpoint is of course theological rather than philosophical. We shall have recourse here to this book by Collingwood to fill in a bit Bouyer's exposition in a way that is concordant with the Oratorian's understanding.

35 *Cosmos*, 72.

36 Ibid.; see Collingwood, *The Idea of Nature*, 31.

animal or living organism. Bouyer suggests that his understanding that water is the source of all being puts him in line with the worldview of Mesopotamian mythic cosmogonies and the stories of Homer and Hesiod. Nature is, for him, "the mysterious process through which everything derives from the divine power infused into the basic substance."[37]

Anaximander detected a fundamental metaphysical contradiction in Thales's attempt to ground all being, in its vast diversity, in a single, undifferentiated substance. He proposed instead as explanatory principle the existence of an undifferentiated infinite, the *ἄπειροη*, which he identified with God. Innumerable worlds, of which ours is one, arise in the medium of this boundless infinite, like eddies condensing the original and amorphous condition of the world.[38] Anaximenes, for his part, returned to a reductionist view akin to that of Thales. He thought Anaximander's explanation of the process of cosmic evolution to be absurd and resorted instead to the view that one of the formed elements of the world is the source of all being, although, instead of identifying this source with water, as Thales did, he identified it with air, from which, he thought, various degrees of "condensation or rarefaction" give rise to the multiplicity of beings that we experience.[39] Later, Heraclitus (c. 535–475 BC) developed his metaphysics along the lines of the earlier Ionians, rooting all cosmic process in the element of fire and stressing that the cosmos is fundamentally characterized by perpetual change, which is nevertheless governed by an underlying cosmic law or *logos*. His chief rival, as histories of philosophy so often present it, Parmenides (b. c. 515 BC), while not disputing this latter point, understood change to be mere appearance or illusion, because, he thought, becoming could have no coherent metaphysical explanation, as being cannot become without thereby ceasing to be.[40]

The Ionians, and the philosophers they inspired, thus consistently consecrated vitalism and organicism in their cosmologies, but there is another dimension of the mythopoetic mind that a different direction of Greek philosophy sanctioned through dialectic. We refer to the cosmic exemplarism that Eliade thought to be the essential hallmark of myth

37 *Cosmos*, 72.
38 Ibid.; see also Collingwood, *The Idea of Nature*, 33.
39 *Cosmos*, 73.
40 Ibid., 74.

in the philosophy developed by the Pythagorean tradition.[41] Bouyer has a brief word to say about Pythagoras (570–495 BC) in chapter eight of *Cosmos*. The Orphic philosopher, he explains, introduced dual metaphysical principles for explaining the process by which nature is formed. The first, geometrical, mathematical, archetypal forms, gave coherent structure to the second, amorphous matter. Out of this distinction arose philosophical hylomorphism in its various permutations, which are often radically different from one another. Pythagoras, unlike the Ionians, held that the process of world-formation cannot be understood only by comprehending the inherent structure of matter. He thought that the pivotal insight needed to grasp the intelligibility of the world is to recognize the capacity of matter to be shaped or in-formed geometrically.[42] He understood numbers to be the explanatory principle of the "unified yet differentiated structure of the universe" and the key "to the arrangement and interrelation of figures in space."[43]

Our author emphasizes in the final sections of chapter eight that Plato and Socrates (c. 470–399 BC) were not Pythagoreans *simpliciter*, for their respective philosophies were shaped by the pragmatic, ethical concerns of the Sophists (at least initially) and the critiques these latter figures leveled at the speculative cosmologists in a period of Athenian history marked by moral malaise and religious skepticism. Bouyer urges that these Sophists were the first philosophers to seek the perfection of democratic governance in accordance with a cosmic natural law that supersedes the particularities and divergences of local laws. Two Sophists in particular, he holds, Protagoras and Gorgias, if they did indeed succumb to extreme skepticism, nevertheless sharpened Socrates's dialectical skills in his search for the Good.[44]

Socrates, according to this genealogy, is best understood along the lines of Aristotle's interpretation of him, not, that is, as a cosmologist or

41 Mircea Eliade, *The Myth of Eternal Return*, trans. Willard R. Trask (Princeton, NJ: Princeton University Press, 2005), 23.

42 See Collingwood, *The Idea of Nature*, 52. We turn to Collingwood on this point to supplement Bouyer's overly succinct presentation. What Collingwood says is perfectly in line with Bouyer's understanding, which is largely drawn from Collingwood.

43 *Cosmos*, 73.

44 Ibid., 75.

metaphysician but as "a demanding moralist, intent on developing accurate definitions through a rigorous critical process."[45] Socrates understood himself to be wisdom's friend: *φιλόσοφος*. Bouyer takes Socrates's cryptic utterances regarding his *δαίμων* or *daimôn* with the utmost seriousness. These indicate, Bouyer holds, the religious character of his thought, that he understood himself to be the subject of inspiration by a mysterious being to whose prodding he owed obedience.[46] Our author is quite sympathetic to the ancient Greek philosophical view that all knowledge and wisdom comes from a divine source, a view that was revived by nineteenth-century Traditionalists. He relates Socrates to the Prophets of Israel and sees him as the surpassing summit of Greek thought in his period. Socrates was driven by the "overpowering attraction of a never-satisfied demand for purity and authenticity," which impelled him "to pursue a still unknown good, but one sufficiently foreshadowed so that any inferior substitute or caricature would be unhesitatingly rejected by critical reasoning." "This twofold impulse," Bouyer explains, "involving a return to the myths and an effort to reach beyond them in a spirit of faithfulness to the essence of the mythic instinct, is characteristic of the entire period, but reaches its purest expression in the doctrine of Socrates."[47]

Plato, on this telling, inherited the Orphic myth from the Pythagoreans, according to which humans were born from the ashes of the Titans struck down by Zeus. The Pythagoreans interpreted this myth to mean that human beings are divine spirits covered over by earthen shells, and Bouyer sees this understanding as the basis of Plato's anthropology. At the same time, he holds that Plato inherited the Ionian critique of the anthropomorphic character of the mythic gods related in Homer and Hesiod, and that he scorned, with Heraclitus, the religious mysteries of his day for their pretense to bear salvific efficacy. In line with Socrates, Plato carried forth a critique of the Sophists that has a prophetic character, moved by an impulse which gave birth in Socrates to "humanism beyond the merely human."[48]

45 Ibid., 76. See Aristotle, *Metaphysics*, 987a.

46 Cf. *Sophia*, 40.

47 *Cosmos*, 76.

48 Ibid., 77. Without indicating a page number, Bouyer points to A.J. Festugière, *Socrates* (Paris: Éditions du Fuseau, 1932).

Bouyer explains that the Platonic theory of forms, which is of obvious cosmological interest, was developed in the dialogues in four stages.[49] It was inspired both by the Pythagorean notion of geometrical forms and the Socratic quest for the Good. In the ultimate stage, as found in *The Sophist*, *The Statesman*, *The Timaeus*, and *The Laws*, Plato projected the forms entirely into a noetic world, of which the material world is only a "feeble and imperfect copy."[50] Because *The Timaeus* was the only work of Plato's known in the Latin Middle Ages, it is of immense significance in the history of Western cosmology. In this writing, which recapitulates Pythagorean cosmology, Plato presents the material world as a perpetual flux. It is given to us only through the senses, which tend to deceive us, but nevertheless participates in or imitates the transcendent forms in the noetic world. The world is "a changeable image of unchanging eternity" through which the forms are copied eternally but successively onto the material plane.[51] For Plato, the *Demiurge*, "which seems only to be the *daimôn* of Socrates extended to the dimensions of the universe,"[52] contemplates the forms and infuses a world soul into the indefinable material supposit, the receptacle that the Pythagoreans had already identified. The world soul connects the noetic realm with the material or physical, and it is through it that all the forms are copied. But how does all of this relate to the Good that Socrates identified in his quest for the natural law and which Plato accorded a metaphysical and cosmological status as self-diffusive, the ultimate source of all that exists? Bouyer argues that Plato never successfully integrated his conceptions of the Good, the ideas, and the *Demiurge*. His successors were left to piece together the puzzle. It was through biblical revelation of the triune God alone that these elements could at last be integrated coherently, in a way that surpassed the possibilities evident to Greek mythographers or philosophers who did not have biblical revelation to guide their steps.

49 See Wincenty Lutoslawski, *The Origin and Growth of Plato's Logic: With an Account of Plato's Style and of the Chronology of His Writings* (London: Longmans, Green, and Co., 1897). This is Bouyer's reference in marking out these four stages.

50 *Cosmos*, 78.

51 Ibid.

52 *Sophia*, 40.

ARISTOTLE AND THE STOICS

The cosmology of Aristotle (384–322 BC) likewise does not, in the story as it is told in *Cosmos*, provide justification for absolute demythologization of the world. Bouyer turns to Aristotle, the Stoics, and Philo of Judaea (20 BC to AD 50) in chapter nine of *Cosmos*, which he entitles "From the Cosmos of Aristotle to that of Philo Judaeus." In this section, we shall limit our attention to Aristotle and the Stoics and turn to Philo in the next section, where the question of biblical influence on later Greco-Roman philosophy will be considered. The French Oratorian follows the scholarly opinion according to which there are three phases in the development of the Stagirite's work. In the first, he remained a disciple of Plato, while, in the second, he began to criticize the Platonic doctrine of the forms. In the third and final phase of his thought, Aristotle directed attention exclusively to the study of "physics," that is, to scientific scrutiny of living beings and nature. This last phase includes the twelfth book of his *Metaphysics* (*Lambda*), which is his major essay on cosmology.[53]

Bouyer reads Aristotle with the sort of interpretive lens that Saint Thomas Aquinas provided, rejecting thereby Nominalist and modern materialist interpretations of his thought.[54] He understands Aristotle to be the sort of "moderate realist" that he himself clearly aspired to be. He summarizes in chapter nine five dimensions of Aristotle's philosophy that are essential for grasping the significance of his cosmology. First, Aristotle held nature or *physis* to be an immanent principle in things, in contrast with a principle of activity coming from the outside, as in the work of a craftsman or in acts of violence. Second, in contrast with Plato, he held that nature is an immanent efficient cause, the principle

53 *Cosmos*, 79–81. See Werner Jaeger, *Aristotle: Fundamentals of the History of His Development*, 2nd ed., trans. Richard Robinson (Oxford: The Clarendon Press, 1948); W.K.C. Guthrie, *History of Greek Philosophy*, vol. 6: *Aristotle: An Encounter* (Cambridge University Press, 1981). Bouyer references both these scholars on this point. Guthrie's volume was published just prior to the publishing of Bouyer's *Cosmos*.

54 See Anton-Hermann Chroust, *Aristotle: New Light on His Life and on Some of His Lost Works*, vol. 1 (London: Routledge, 1973). Bouyer thus follows Chroust's interpretation of Aristotle. He rejects that of William David Ross's *Aristotle* (London: Methuen & Co., 1923).

which self-actuates the unrealized potencies that it bears within itself, in its eternal, circular movement. Third, he postulates the real existence of forms in order to make sense of the reality of change, but he holds that these cannot pre-exist in the potency of matter, otherwise there would be no change or becoming, as the end that inspires movement would already be possessed by things. These forms exist instead in a cosmic intelligence, in the "thinking thought" (*νόησις νοήσεως*) of the Prime Mover, God. Fourth, the Prime Mover, without knowing anything outside its own self-thinking thought, exerts an attraction on all things which inspires in them an *eros* (*ἔρως*) or desire to move toward it. This attraction is mediated through lesser cosmic intelligences in the upper heavens. The cosmos is essentially noetic. Fifth, the matter of the cosmos is only a limit of unfulfilled potencies.[55]

Bouyer confines his discussion of Aristotle in this chapter to the just-summarized points, although he will return to the Stagirite in summary fashion once again in his discussion of matter and spirit in chapter twenty-one of *Cosmos*, and we shall expound him on this in our own tenth chapter.[56] In chapter nine, he next moves quickly through Democritus and Epicurus to the Stoics, who may be, to his mind, the most religiously-minded of all the Greco-Roman philosophers. He stresses in his writings the mythopoetic vitalism and organicism of Stoic thought and the similarity between their cosmic vision and that of ancient Eastern religious philosophies. He explores the development of various Stoic philosophies in some detail in *Le Consolateur* and summarizes briefly their most important contributions to cosmology in chapter nine of *Cosmos*.[57] The Stoics have a particularly significant place in his book on the Holy Spirit because their thinking centers on the notion of *pneuma* (*πνεῦμα*) or spirit. In the various places in his trilogies where he writes about the Stoics, he emphasizes the importance of their thought in the development of biblical Wisdom writings and commentaries. He notes

55 *Cosmos*, 80–81.

56 Ibid., 219–20. See *IF*, 187. Here Bouyer summarizes Aristotle's position in an even more succinct fashion. He stresses once again the essentially religious or quasi-theistic character of Aristotle's thought: Aristotle's "own universe exists only in dependence upon the transcendent existence of a 'first unmoved mover.'"

57 See *LC*, 28–34.

that the Stoics were a decisive influence on Philo of Judea. With regard to the earlier Greek tradition, their pneumatology represented to his mind a kind of synthesis of earlier schools of thought, while also going beyond the earlier philosophers:

> The Stoic *pneuma* . . . will not be without incorporating to the old physicalism [of the pre-Socratics] all the intellectuality of Plato himself: this vital "spirit" which constitutes, which moves all things, reveals itself as endowed with an interiority which extends to the dimensions of the cosmos the Platonic and Aristotelian *logos*, the true thought that thinks itself, but this time in thinking all things.[58]

The Stoics anticipated many of the future developments of Neoplatonism. This is so true for Bouyer that he gives a nod of assent to those scholars who understand Neoplatonism to be, in a way, a form of Neostoicism.[59]

The author of *Cosmos* focuses in this text on the three most significant early Stoics, whose philosophy was, at once, cosmological and anthropological. Without distinguishing differences in their respective philosophical positions, as he does in *Le Consolateur*, he synthesizes their shared cosmological, religious philosophy. The three major figures are Zeno of Citium (c. 334–262 BC), Cleanthes of Assos (c. 330–230 BC) and Chrysippus of Soli (c. 279–206 BC), who all developed a cosmology based on anthropology. They drew on the tradition of Sicilian medicine according to which, as Bouyer says in *Le Consolateur*, "*Pneuma* is a hot air which is loosened from the blood itself and that the heart propagates throughout the organism."[60] This is in contrast to the view of Hippocrates (c. 460–370 BC), according to which the *pneuma* or breath of man is drawn from the breath of the cosmos. The cosmos, for the Stoics, is suffused with a common life. Everything is a condensation, in various degrees, of *pneuma*, the fiery air or "material breath" (*soufflé*

58 *Sophia*, 41.

59 Ibid. See Edwyn Bevan, *Stoics and Sceptics* (Oxford: Clarendon Press, 1913). This is the scholar Bouyer references on this point.

60 *LC*, 29.

materiel) at the innermost depths of all bodies.[61] This *pneuma* is, at the same time, *logos* or immanent, divine reason.

Distinctive, unified structures of being and vitality are owed to the "seminal reasons" (*λόγοι σπερματικοί*) through which the universal *logos* is spread abroad ubiquitously yet without being divided in itself. There is an ascending series of cosmological realizations of the *pneuma-logos*. Chrysippus distinguished four: *hexis* (*ἕξις*) or development in the inorganic realm; *physis* or development in plants; *psyche* (*ψυχή*) or the soul of non-rational animals; *hegemonikon* (*ἡγεμονικόν*) or rationality and freedom in the human being.[62] The human soul is thus a special sort of fragment of the divine *logos* and *pneuma*, the center of reason given the task of recognizing the divine necessity in all things. The development of this idea was particularly important to Stoic ethics and later influenced Jewish and Christian asceticism. Bouyer explains that the Stoics held a "distinction between two constantly opposed aspects or tendencies of human and cosmic life: the *poioun*, or positive activity working to fulfill the *logos*, i.e., the divine reason in all things; and the *paschon*, a mere passivity tending to reabsorb all things into a diffuse and tensionless materiality and leading to absolute nothingness, nonexistence, or unconsciousness."[63] The cosmos is an eternal return of conflagration or *ekpyrôsis* and rebirth, and biblical thinkers would transform this eschatological vision in light of the biblical, apocalyptic understanding of the unified *telos* of cosmos and history with one ultimate origin and end, a historico-cosmic process characterized by unrepeatable occurrences moved by the drama of uncreated and created freedoms and whose overall guidance from within creation is under the aegis of the human being, if he or she should live in accordance with divine Wisdom.[64]

From this vantage point, our author briefly explores the connection between biblical and Stoic wisdom traditions. Biblical writers, as we saw in the previous chapter, employed myth and wisdom themes from the royal traditions of the Ancient Near East, critically reinterpreting them in light of God's intervention in history in the divine Word and Spirit.

61 Ibid.
62 *Cosmos*, 83.
63 Ibid.
64 Ibid., 83–84.

In chapter nine of *Cosmos*, Bouyer argues that this process of critical employment and re-interpretation of myth and wisdom traditions continued throughout Hellenistic times. Jewish traditions of commentary on the meaning of divine Wisdom made use of the Platonic theory of ideas, the Aristotelian understanding of God as self-thinking thought, and the Stoic cosmology of *pneuma-logos*. Jewish commentators came to conceive of divine Wisdom as "the ultimate and total object of God's thought on creation and the history of salvation, but without ever immersing God into the world or divinizing the world."[65] Indeed, the old cyclical vision of eternal return was overcome, as the cosmos was recognized to be finite both in space and time and only thus able to fulfill "definitively through the freedoms divinely created and sustained, the immutable plan of a wholly benevolent Wisdom."[66] The human soul was given an elevated status beyond that which it attained in the Greek philosophies. It was now seen as irrevocably connected in its individuality and sociality to the divine plan for creation as well as to its physical body, foreordained to be transfigured in the Body of Christ with the general resurrection of the dead.

DIVINE EXEMPLARISM

The development of human wisdom in ancient Greece and Rome is in fact inexplicable without reference to divine revelation. Not unlike many Church Fathers, Bouyer argues that as the Word of God took up and transfigured myth and wisdom it did not leave human wisdom unaffected in the latter's intrinsic development in history. In order to affirm this, one does not have to take things as far as those Church Fathers did who thought that Plato received his philosophy from the Torah. We may not be able to speak of a biblical influence on the earliest Greek philosophers, but its influence on Greco-Roman philosophy by the time of the emergence of Neoplatonism is, in Bouyer's estimation, little deniable. The Oratorian points in his wider writings to three philosophers in the Platonist trajectory of thought for whom a certain level of biblical understanding is verifiably present. The first is Numenius, a Platonist from the second century, who, although not a Christian, acknowledged

65 Ibid.
66 Ibid., 84.

his indebtedness to the biblical tradition and even to Jesus himself.[67] This philosopher drew from Greek, Jewish, Egyptian, Persian, and Indian sources.[68] The second is the Jewish sage Philo of Judea (c. 25 BC–c. AD 50), and the third is the one-time Christian philosopher Ammonius Saccas (c. 175–c. 245). In chapter nine of *Cosmos*, Bouyer gives a brief treatment of Philo.[69] In *Sophia*, he briefly expounds the latter two figures, and we want to summarize these expositions—that of Philo in *Cosmos* and Ammonius in *Sophia*—which, together, help us to develop a better understanding of how he interprets the theme of exemplarism in philosophical thought and its biblical transformation.[70] These expositions serve, moreover, as a point of departure for understanding his wider reading of pagan and Christian Neoplatonism.

Philo, Bouyer insists, has been misunderstood by commentators who see him primarily as a man of Hellenistic philosophical convictions for whom the Bible and Jewish tradition is of secondary concern. It is truer to say of him that his thinking was deeply steeped in these sources: "Philo was motivated above all by the justification of Judaism—in the eyes of the most religious and thoughtful pagans—as a form of life: the *basilike hodos* [οδός βασιλική] or royal road leading to the actual reign of God over all human existence, to match His reign over the cosmos."[71] The whole of Philo's work takes on a new significance when seen in this light. He is not first and foremost an eclectic philosopher whose primary allegiance is to Hellenistic categories of thought but a thinker rooted in biblical revelation who uses Hellenistic categories to articulate biblical truth in a new context. This is true of his doctrine of "powers," the angels of Judaism in Hellenistic guise, and of his doctrine of the *Logos*, which is, for Philo, the creative Word of the Book of Genesis, the

67 *Sophia*, 44. *The Spirituality of the New Testament and the Fathers: History of Christian Spirituality*, vol. I, trans. Mary P. Ryan (New York: Desclée and Co.), 256–302. These pages discuss Clement and Origen of Alexandria and their philosophical context. They provide a decisive reference for Bouyer's own interpretation of the Church Fathers in relation to the Greek philosophers and Bouyer refers to these pages in his subsequent volumes in the trilogies.

68 *IF*, 203.

69 *Cosmos*, 84–85.

70 *Sophia*, 46–49. *Cosmos*, 84.

71 *Cosmos*, 84.

figure he uses to take over the Platonic understanding of the intelligible world.[72] He does, Bouyer holds, lack a sense of eschatological tension in his work, of the final judgment understood on a unified, cosmic scale, and instead reduces the plan of divine Wisdom to the destinies of individuals taken separately with respect to their faithfulness or not to the *Logos* of God, but this is not inimical to certain threads of biblical thought, such as in the writings of the prophets or in the biblical Wisdom literature. "Philo," Bouyer concludes, "focuses strongly on the identification, by the believer turned philosopher, of the entire biblical history with the history of his own meeting with God and the perfecting of his resemblance to the Most High."[73]

According to Porphyry (c. AD 234–c. 305), who edited and published the *Enneads* of his teacher Plotinus (204–270), Ammonius Saccas—the teacher of both Plotinus and the Alexandrian Church Father, Origen (185–254)—was the son of Christian parents but rejected their religion when he learned Greek philosophy. This is contrary to the views of some Church Fathers, such as Eusebius of Caesarea (263–339) and Saint Jerome (347–420), who considered Porphyry to have lied about the point. Bouyer holds that there are reasons to doubt Porphyry's account, which he thinks was based on the unfounded prejudice that one could not be both a philosopher and a Christian.[74] Following René Cadiou's famous study of Origen, *La Jeunesse d'Origen* (1936), Bouyer highlights a fundamental distinction that Ammonius recognized between the aims of biblical Wisdom and those of Hellenist philosophy: only in biblical Wisdom is God recognized as transcendent to the world and the One who brings it into being from His will alone.[75] This means that biblical Wisdom eliminates all emanationism, if this expression is taken to mean the confounding of creation with the Fall, as well as dualisms that take matter to exist in total independence from the divine. The cosmos is creation, in the Christian understanding, constituted by finite personal freedoms. Origen, Bouyer argues, adopted Ammonius's view that the existence of the qualitative diversity of finite, free beings in the

72 Ibid.
73 Ibid., 85.
74 *IF*, 203, 211–13.
75 *Sophia*, 48–49.

dispensation of creation is the result of creatures falling away from the divine in acts of misused freedom.[76]

Philo and Ammonius are decisive figures in the development of Neoplatonism.[77] Bouyer suggests that it is their biblically-influenced doctrines that moved the Neoplatonist school to seek, however ineffectively in the end, for greater acknowledgement of the transcendence of the divine than was present in the earlier forms of Greek philosophy. Plotinus is the most important philosopher of all the Neoplatonists, and Bouyer expounded his doctrine of the divine triad in several writings.[78] The dialectic of divine transcendence and immanence in Plotinus is, to Bouyer's mind, the most subtle of all in the history of ancient Greek thought, or even in global pagan philosophy. Bouyer explains that the One in Plotinus's doctrine is God but not really transcendent to the world. It is, instead, the world seen in terms of its concealed, underlying unity. The divine transcendence is within the world, yet excluded from all worldly existence in what we mean by the word.[79] The Plotinian Mind or *Nous* (*νοῦς*) is a mediating reality, emanated from the One as the One is opened to multiplicity, enabling the One to encompass the Many without losing its unity, and it is from the Mind that the World Soul—the immediate principle of cosmic life that triumphs over the evil of non-being—is emanated.[80] Plotinus's ideas are not mere "fleshless generalizations" or "concrete individuals." The One is "boiling with life."[81] The ideas in the divine Mind through which the One is mediated to the world were understood by him to be individual consciousnesses in communicative integration with one another "within that transcendent thinking which thinks itself by thinking all of them."[82]

Plotinus's cosmic philosophy is inherently religious in that it teaches a way of "conversion" (*epistrophe, ἐπιστροφή*), of salvation and liberation.

76 Ibid., 49.

77 Ibid., 46–50.

78 *The Spirituality of the New Testament and the Fathers*, 300–2; *IF*, 208–11; *Cosmos*, 182–83; *Mystery*, 188–205; *Sophia*, 49–50. We take these to be the most important reference points on Plotinus in Bouyer's writings.

79 *Sophia*, 43.

80 Ibid.

81 See *Enneads*, VI, 12 (in the Brêhier edition: vol. 6, pt. 2, p. 83, lines 22 and 23). We follow here Bouyer's reference in *Women Mystics*, 26, n. 11. Plotinus prefigured Eckhart's expression, discussed in the previous chapter.

82 *IF*, 209.

In conversion, the World Soul returns, with the matter it has informed, through the ideas of the Mind to the Mind itself, and the Mind returns in parallel fashion to the One. The individual mind must be intellectually converted to the truth of its non-being as distinct existent, and this carries with it, because the divine One is also the perfect Good, the need for rigorous, ascetical, moral purification.[83] We might recognize, as some commentators have, that Plotinus understood the way of conversion to be grace-infused, and that he defended the goodness of the physical world against the Gnostics. Surely he does not wish to abolish the world? On both these fronts, his thinking seems to come very close to that of the Gospel. Bouyer emphasizes that there is another side of the coin. His doctrine of matter did not in fact wholly escape Gnostic distortions, in that he thought it to have an existence independently from God. His dialectic of transcendence and immanence or the One and the Many maintained a deceptive harmony between them:

> God and the world, according to Plotinus, are both one and distinct. To the extent that they appear distinct, they become alien to one another. Either God makes himself known to us and his identity changes, or we return to him and are no longer ourselves. The alternative is inescapable.[84]

Conversion, in the Plotinian scheme, does not issue in liberation through eternal beatitude for the individual soul in eternal communion with God and the elect. He did not in fact consistently and in the depth of his insights effectively demythologize the confusion according to which salvation is de-creation or the elimination of finitude by its return into the undifferentiated origin of being.

The Church Father and convert to Christianity Marius Victorinus (c. 290–c. 364), himself at one time a propagator of Neoplatonism, showed the congenital weakness of this philosophy vis-à-vis the biblical understanding of God and creation, and Bouyer, perhaps most especially in his final writing, *Sophia*, takes his Christian triadology, developed only after his late conversion to Christianity (c. 355—decades after the Council of

83 Ibid., 210.
84 *Cosmos*, 183.

Nicaea in 325), to set the exemplary pattern for future Christian Trinitarian reflection. Marius saw that Greek philosophy, even after it has been elevated by biblical creationism in the thought of the Neoplatonists, could not escape the reduction of love to an *eros* or desire present only in creatures, whose object is God.[85] Only in the plenitude of God's revelation in Christ was God revealed as *agape* (ἀγάπη), pure, eternal self-gift or generosity in Himself. God is, Victorinus taught, essentially paternity projecting Himself eternally in His only Son, giving the very Spirit of paternal love to him, enabling him to give himself actively in self-abandoning return to the Father in the Spirit, perfecting eternally the triadic consummation of *agape*.[86] "For Victorinus," Bouyer summarizes, "the only possible escape from the religious and speculative dilemma of Hellenism will be in the discovery of the biblical God, attaining only in orthodox Christianity to its complete expression, such as the Council of Nicaea formulates it in qualifying the eternal Son of the Father *homoousios*, that is, consubstantial with Him."[87]

Our author follows the direction of Victorinus's thought in recovering the properly Christian understanding of creation as having come forth from the Wisdom and will of God in His eternal, perfect, triune generosity, through a divine action that is indeed Trinitarian.[88] This gets at the heart of the difference between biblical and philosophical processes through which mythic cosmogonies are surpassed. Still, he sees the development of Neoplatonism alongside the biblical commentaries of the Church Fathers as having been necessary in order to bring to light the transformative Trinitarian message of the divine Word given in and through Scripture. Moreover, for all of his failure to escape from the mythological residue present in Greek philosophy, Bouyer thinks that

85 *Sophia*, 50.

86 Ibid.

87 Ibid., 45.

88 Cf. Vincent Holzer, "Karl Rahner, Hans Urs von Balthasar, and Twentieth Century Catholic Currents on the Trinity," in *The Oxford Handbook of the Trinity*, ed. Gilles Emery and Matthew Levering (Oxford University Press, 2014), 314–27, at 324. In an article largely dedicated to the two most influential 20th-century Catholic theologians, Rahner and Balthasar, Holzer points out the importance of Bouyer's Trinitarian theology of creation among the list of theologians he commends.

"there was plenty in Plotinus moving him away from the orbit of Greek thought and toward that of biblical or Christian revelation."[89]

One of the crucial developments in Neoplatonism had to do with its increasing clarification of the Platonic doctrine of the forms. The confluence of biblical and philosophical wisdom traditions had perhaps its happiest outcome with respect to the theme of forms, ideas, paradigms, or exemplary models. These were perhaps the most essential characteristic of myths and point to the most important area in which philosophical and biblical Wisdom traditions take up the mythic stratum of human pre-ontology. Bouyer argues that biblical revelation goes beyond philosophy and captures the heart of living myth by personalizing the divine ideas or exemplars that give meaning to the cosmos. As we saw above, Plotinus construed the divine ideas as personal, and this may well reflect biblical influence on his thinking. In biblical revelation, the assimilation of Wisdom to the divine Word is, in its broad development, a progressive personification.[90]

The *telos* implicit in the royal myths and the wisdom associated with them had to do with the perfection of the human city, and this is a great inspiration for Greek philosophy as well as the biblical Wisdom literature. The growing personalism of the Bible, present in both Wisdom and apocalyptic writings, was connected to a growing understanding of the true nature of the City of God. Bouyer traces a line of biblical instantiations of the theology of divine love realized in the perfect City from Isaiah 62, to Hosea with his nuptial theme, to Psalm 45, and finally to the Canticle of Canticles. The consummating, surpassing perfection of the revelation of this love of God realized in the eschatological City is related only in the New Testament, for instance, in the Johannine Apocalypse, which tells us of the Eternal Bridegroom, the Word and Son himself, descending from heaven onto the earth (21:2), and in Hebrews (12:23), which speaks of "the Church of the predestined whose names are inscribed in the heavens."[91] The final unveiling of the "great Mystery" that Saint Paul speaks of in Ephesians (5:32) is the full transfiguration of what is most fruitful in the ancient wisdom traditions.

89 *IF*, 211.

90 *Sophia*, 51.

91 Ibid., 53. I follow Bouyer's translation here.

Biblical Wisdom in the Old Testament is a feminine, personal figure (cf. Proverbs 8:22—discussed in the previous chapter). She transfigures the exemplarism of the myths and of philosophy by imposing herself on the development of the whole of creation:

> She is, perhaps one can say, like an equivalent—but much more alive!—of the Idea of the Good according to which the Plotinian demiurge sculpts all things. However, this Good is the very divine source from whom everything proceeds, who, projecting onto the emptiness of nothingness His plan for all things, will realize by the power of His Word, and, under the influence of His Spirit (Gen. 1), awaken in this image of God the living resemblance which could be nothing other than the very love which has loved in advance His creature: *berakah*, a substantial Eucharist in which what the word has emitted, penetrated by the very Spirit which inspires it, returns to its source.[92]

Bouyer projects back onto the Old Testament feminine figure of Wisdom a theological understanding that is derived from New Testament revelation. Wisdom, he holds, exists in God eternally, in the procession of the Son on whom rests the Gift of the Holy Spirit and is the plan for the world that includes its eschatological completion. The Word and Spirit can be detected already in the Book of Genesis. The *berakah*, the thanksgiving of creation, is a Jewish ritual theme that prefigures and prepares the way for the Christian Eucharist.[93] The theme of cosmic liturgy does not emerge for the first time as if *ex nihilo* with specifically Christian apocalypse.[94] God the Father has no eternal feminine companion, yet Wisdom can be

92 Ibid., 55.

93 *Eucharist: Theology and Spirituality of the Eucharistic Prayer*, trans. Charles Underhill Quinn (South Bend, IN: University of Notre Dame Press, 1989). This is one of Bouyer's most famous books. It traces the roots of Christian liturgy in Jewish synagogue prayers. See Jean Duchesne, "L'enracinement dans le Judaïsme du mystère Chrétien," *La Théologie de Louis Bouyer* (Paris: Parole et Silence, 2014), 179–90. See also Nicolas-Jean Séd, "Tradition Juive et Christianisme," *La Théologie de Louis Bouyer* (Paris: Parole et Silence, 2014), 191–95.

94 See especially, *Cosmos*, 62–71.

said to pre-exist in Him as exemplar of the cosmic role that she will play in her actualization in history.

At the end of chapter four of *Sophia*, Bouyer summarizes the similarities between Greek wisdom and biblical Wisdom. There are, in fact, things which the progress of biblical Wisdom, in its growing personalism, owes to Stoic and Platonic wisdom. For one thing, biblical Wisdom, in the Book of Wisdom (for instance), consecrates the philosophical idea, so in accordance with the myths, that the whole cosmos is alive and in some manner one. This lends itself well "to the emergence of a person in the process of the opening . . . in the whole creation tending toward its terminus, to its end willed by God."[95] At the same time, the Book of Ecclesiastes is not without some indebtedness to Platonic philosophy in its vision of the hierarchy of the cosmos: "There is something, indubitably, which recalls the vision of the physical world reflecting multiple 'ideas,' coordinated by their source, who is also their good end: that of the Good, clearly here the divine generosity."[96] Moreover, Proverbs and Job, as well as these other just-mentioned Wisdom books, might be said to reflect a Pythagorean view according to which everything has been created "according to measure and number."[97]

Biblical Wisdom brings a new sense of the personal character of the noetic cosmos that exists not only beyond but within the world of sense-perception. Biblical Wisdom shares in common with Greek wisdom and with myth a sense that the material world is an echo of a primordial, spiritual world, but it clarifies the participatory hierarchy that is presumed in this ancient understanding of the world. Biblical revelation gives us to know, ultimately, God as eternal, transcendent, triune goodness who, in His transcendence, is able to be more intimate to us than we are to ourself. It distinguishes between angels and demons, on the one hand, and angels and gods, on the other. It teaches us that creation is given being, freely, in and through the loving, generous Wisdom of God, in order to return that love in the cosmic Eucharist or *berakah* of praise. In this context, Bouyer invokes the celestial sanctuary in Exodus (25:40) that is reproduced in this world by the Mosaic tabernacle. The

95 *Sophia*, 59.

96 Ibid.

97 Ibid. Book of Wisdom 11:20.

spiritual universe of rational and free beings constitutes an immense hierarchy, unified in Wisdom, whose glory is to resound in chant and praise, glorifying the One who gave them life.[98]

FURTHER IMPLICATIONS

Before moving on to the next chapter, where we shall look at Bouyer's Christological cosmology, we want to pause for moment to explore briefly some of the implications of Bouyer's genealogy of ancient philosophical cosmology as recounted in this chapter. There are five points that we wish to highlight. The philosophical dimension of the Oratorian's thought is underdeveloped, if taken as a separable component of it, and his discussion of the history of philosophy in the chapters of *Cosmos* recounted here has an encyclopedic tonality, which we find it difficult to evade in seeking accurately to expound it. Yet his overview of ancient philosophy in relation to *mythos*, on the one hand, and biblical *Logos* and Wisdom, on the other, makes some important contributions. The first has to do with the way he relates philosophy with faith. Bouyer is not on board with modern theologians who pit Christian revelation against Hellenist philosophy, as if these two paths of wisdom were purely contradictory avenues to approach the truth of the world. Clearly, he maintains an emphasis on the point that Christian revelation is irreducible to a pre-existent, pagan cosmological religious framework.[99] He constantly emphasizes the uniqueness of the biblical vision of the God-world relationship. Still, his reading of the interpenetrating influence of Jewish and Greek thought, with mythopoetic experience underlying both of them, enables him to avoid falling into the

98 *Sophia*, 61.

99 See Rodney Howsare, *Balthasar: A Guide for the Perplexed* (New York, NY: T & T Clarke, 2009), 35–37. In these pages, Howsare helpfully summarizes what Balthasar means in speaking of a "cosmological reduction" in the Church Fathers. The High Scholastic theologians, on Balthasar's reading, overcome this tendency present in the earlier tradition. Bouyer's thought is not opposed to this, but there may be a subtle differences between the two. See Grintchenko, *Une Approche Théologique du Monde*, 85–86. She explains that whereas Balthasar begins from the heart of faith, with the mystery of Christ as figure of revelation, Bouyer's approach "addresses the mystery under diverse aspects." She suggests that the place that Bouyer gives to mythic thought allows him to unify without confounding "cosmic beauty" and the "glory of God" at the origin of all human thought.

trap of certain radical programs of de-Hellenization.[100] He recognizes that mythopoetic experience unites theology with philosophy. His manner of connecting both Jewish and Greek wisdom traditions with the global heritage of living myth and religious philosophies carries with it the possibility of utilizing his insights in the growing dialogue between Catholic theologians and philosophers with representatives from other religious traditions. His liturgical or ritual focus, with its amenability to mythopoetic exemplarism, gives a unique emphasis that could be particularly fruitful in this dialogue. Moreover, he helps us to see that liturgy, metaphysics, physics, poetry, and biblical hermeneutics give us approaches to the truth of the world that, if distinguishable, are ultimately inseparable.[101]

Second, he shows us, as we have stressed, that biblical revelation is not inimical to organicist and vitalist conceptions of the cosmos. If biblical revelation corrects immanentist or pantheist forms of vitalism and organicism, it does not thereby consecrate, as some ecologist critics of Christianity maintain, those objectivist and atomist ways of construing the world common in the modern West since at least the time of Neoscholasticism in the sixteenth and seventeenth centuries that led us to think of the world as a machine.[102] Bouyer's biblical, sophiological rendering of the cosmos does not succumb to the view that a Christian *apologia* can be adequately formed on the presupposition that the material world is lifeless and disconnected from the vast hierarchy of uncreated and created intelligences and freedoms. Biblical revelation, when understood as centered on the Wisdom tradition at its heart, does not sanction human violence with respect to cosmic processes. Without deifying the world, as if the world and these processes were in themselves God or gods, it does not give a justification for the mechanistic materialism that prods us to relate to the world as if it were a vast series of objects to be manipulated according to our untransformed desires. It does not sanction the splitting

100 Cf. Pope Benedict XVI, "Faith, Reason and University Memories and Reflections," University of Regensburg, September 12, 2006. This is the famous Regensburg address, which was roundly attacked, especially by segments of the Church's intellectual community who oppose "Hellenism."

101 See Grintchenko, *Une Approche Théologique du Monde*, 151–53. The title of this section of her work is "Interdependence of metaphysics, physics, and biblical hermeneutics."

102 Cf. Howsare, *Balthasar: A Guide for the Perplexed*, 10–11.

apart of reality into two disconnected spheres, *res cogitans* and *res extensa*. Biblical revelation, while giving us access to the God who is transcendent to the cosmos, is not absolutely inimical to ancient philosophies that maintained the view that the world is in some manner a living whole or at least unified by a common purpose, even if it does not support the position that it is, in its depth dimension, a single, divine substance.

Third, by the same token, the French Oratorian's analysis shows that biblical revelation should not be taken as a justification for the elimination of the noetic plane of reality or the ideas. The invocation of this realm is not a form of unbiblical Gnosticism, as is commonly asserted by certain theologians. The Church Fathers did not falsely Hellenize the faith with their various doctrines of divine ideas or *logoi*. Biblical revelation itself, in the Wisdom tradition become apocalyptic, showed forth the supreme reality of the world of spirits, of God Himself first of all, and then of the angels and demons. Beyond what we have already said, we wish to point out that Bouyer emphasizes a difference between biblical revelation and Greek philosophy on this score in that he ultimately recognizes that the ideal plane and the world of sensory experience constitute together a single cosmos in the biblical view. This point foreshadows much of what we shall discuss in the coming chapters: the final nuptials of the eschatological Church with the Lamb of God at the end of history will be a wedding in perfect communion of the noetic and sensory domains, of the triune Creator with *esse creatum* in its entirety. In this vein, it should be recognized that the apocalyptic eschatology of biblical revelation both acknowledges the real existence of the spiritual plane and overturns the violent sundering of spirit and matter that is common to Gnostic and Manichaean religious philosophies. The exemplarism of apocalyptic Wisdom encourages attentive expectation for the coming of the Parousia, of cosmic realization in the divine ideas by the comprehensive development of finite being moved by the Holy Spirit. It includes history, time, and individuality in its encompassing purview. Bouyer can be seen thus to develop Newman's theology in a sapiential or sophianic direction. In his theology of revelation and salvation history, the English Oratorian stressed that the unified Idea of revelation develops in time and history, remaining ever within the sensible realm as a living, fecund, and moving reality that makes possible the development of doctrine. Bouyer applies this understanding to the entirety of creation, centering all ideas on the Idea of ideas, the Mystery of God from the heart of divine Wisdom made present in the flesh of Christ. For both Oratorians, humanity does not unite to the ideal world by fleeing from the sensible plane of existence. They hold that the world of ideas is ultimately united to humanity by God Himself in the Incarnate Word of God, although Newman does not develop this understanding in its vastest cosmic scope as systematically as Bouyer does.[103]

Fourth, in all of this discussion we should keep in mind the centrality in Bouyer's thought of the mystery and uniqueness of Trinitarian love communicated through the nuptial communion of divine and created Wisdom, even if he does not himself make this explicit in the early parts of *Cosmos*. Wisdom, as a general search for the perfect city, bridges Greco-Roman *eros* and biblical *agape*. The stark opposition that the Swiss Lutheran theologian Anders Nygren (1890–1978) famously held to exist between these two figures of love cannot be fully maintained if one's interpretation of biblical revelation is sufficiently sophiological.[104] With Marius Victorinus, Bouyer holds that the world is indeed moved by desire for God, but that this movement is rooted in the Trinitarian gift of creation. For Bouyer, the potentialities of creation are actualized only through the incessant and ubiquitous work of the Father in and through His two hands. In a biblical view, the cosmos and all of the vast hierarchy of individual personal beings who constitute it are known and loved perfectly, individually and all together, in the Father's eternal self-knowledge and love. God is not self-thinking thought unaware of what is not God, obliviously luring the potentialities of the world to their actualization. In His divine Wisdom, in the Eternal Son, the Father knows all that is or ever will be, more perfectly than it can ever know itself by its own created power. In the eternal Gift of the Holy Spirit, the Father wills the perfect Good for all things. He desires to establish communion with creatures. This speaks to the existence of a kind of divine *eros* in the triune life, a

103 See Solari, *Le Temps de Découvert*, 50. Solari discusses Newman here in relation to Bergson. But we think what he says applies to the sophiological doctrine that Bouyer will develop as well.

104 See Benedict XVI, *Deus Caritas Est* (2006), 1:3–8. In these passages, Pope Benedict XVI brilliantly proposes the path of reconciliation—respecting their unity in difference—of *eros* and *agape*. We shall explore Bouyer's Trinitarian discussion of this in chapter 8 of our study.

point that we shall explore further in chapter eight.

Fifth, and lastly, it is important to recognize once again the monastic dimension of Bouyer's cosmology, this time in relation to his discussion of philosophy, although he does not spell this out explicitly in *Cosmos*. The French Oratorian numbers among the French scholars of antiquity in the twentieth century who recognized that, for the ancients, philosophy was not just an exercise in speculation for the sake of satisfying one's curiosity but a way of life.[105] He argued in 1950, in *The Meaning of the Monastic Life*, that early Christians understood this quite well, and that they recognized that it was the monks who were the authentic heirs to the ancient philosophers, that they were the "true philosophers." In fact, philosophy had become increasingly religious by the time monasticism took hold in the Empire. It had become fully a search for salvation through rational mysticism. This is how Bouyer ultimately understands the philosophy of Plotinus.[106] These ancient philosophers sought salvation through gnosis, by which the mind of the human being could become like that of the gods. To this form of gnosis, two centuries before Plotinus, Saint Paul opposed an *epignosis* (ἐπίγνωσις), which Bouyer translates as "super-gnosis."[107] Human beings could not attain this supernatural gnosis by virtue of their own efforts. It was a free gift of God, "a grace granted to prayer and purity of heart."[108] It could be acquired only through the gift of faith. Faith is not opposed to gnosis but brings *epignosis*, a knowledge which is inseparable from inner transformation through the gift of charity. Christian gnosis is the outcome of the discovery of the *agape* of God in the Mystery of Christ, whose earthly mission is the key of the divine Wisdom. This *epignosis* "effects the divinization that gnosis had dreamed of."[109] The creature is thus metamorphosed into its divine model, not by way of a magical enterprise through which it wrests control of the divine but by "the fruit of the gift freely given by Him to whom the creature abandons itself in return."[110]

105 See especially Pierre Hadot, *Philosophy as a Way of Life: Spiritual Exercises from Socrates to Foucault* (Malden, MA: Blackwell 1995).

106 *The Meaning of the Monastic Life*, 206.

107 Ibid., 207. See Romans 3:20.

108 *The Meaning of the Monastic Life*, 207.

109 Ibid.

110 Ibid.

The monk, through the ascetical detachment and contemplation that is the heart of his or her vocation, living the Christian life "at its maximum purity and intensity," is the exemplary bearer of *epignosis*:

> Gnosis, an off-shoot from Wisdom, sought not directly to live, but to know. Yet its knowledge, of man, of the universe and of their common mystery, irresistibly tends towards life under its highest form, by the fact that it tends to what we call to-day mysticism: to a knowledge which shall be identification, identification with the supreme intelligence who is also the supreme Being. In the mystery of the bleeding yet radiant heart of divine Wisdom, the epignosis of the monk discovers, in an experimental knowledge of the Cross to which it brings him, the secret of salvation, the secret of immortality.[111]

The monk is thus the symbolic, transfiguring inheritor of wisdom and *logos* along its two paths, the course of the Old Testament as well as the philosophy of antiquity. The mysticism practiced by the Christian monk is the surpassing culmination of the entire heritage of religious humanity globally. The monk, as all Christians who live their Christian vocation to the fullest, finds his or her true self in the "unity in divine love which is the goal of all authentic mysticism."[112] This implies the reunification of creation in a communion that is at once collective and personal through filiation in the Incarnate Son of God. Our author understands the theological rediscovery or recovery of the cosmos in the light of this mystical end, this communion with the triune God in Christ, and he sees it as giving thereby a surpassing validation to the religious *eros* of any philosophy that is a genuine search for truth. Theological cosmology, as practiced by Bouyer, is a mystical theology, centered on the Mystery of Christ, which is not indifferent to the quest for wisdom undertaken by the ancient philosophers. It is to the uniquely Christian understanding of Mystery, of the Son of Man and Son of God incarnate for our salvation, that we now directly turn our attention.

111 Ibid., 208.
112 *Mystery*, 292.

CHAPTER 5

Christology and Cosmology

THEOLOGICAL COSMOLOGY EXPLORES THE ORIgin, ontology, and end of the cosmos in the light of God's salvific revelation in Jesus Christ. Cosmology and Christology should be intimately related endeavors for any theologian who aspires to put forth a comprehensive theology of creation. Modern theologians were not always inclined to understand this or to link the cosmos with its triune Creator in Christ in as intimate a way as some theologians began to do after the work of Teilhard de Chardin and the proponents of Patristic *ressourcement* in the twentieth century. Louis Bouyer, in *Cosmos*, situates the chapter on Christ as revealed in the New Testament at a central location, chapter ten, which he entitles "The Evangelical Vision of the Cosmos." It does not appear until he first explores the joint movement of Word and Wisdom in the history of the world that prepared the ground for the Son of God's Incarnation in the flesh of historical humanity. As we have seen, he gives in the first nine chapters of *Cosmos* a survey of the history of the divine Word's encounter with human wisdom in its diverse manifestations until the time of God's definitive self-revelation through the Son's assumption of human nature. Chapter ten, a relatively succinct chapter, describes the decisive new fact that entered into our history in Christ, shedding decisive illumination on the past, on the entire history of salvation that prepared the way for his coming. Knowledge of the unfolding of biblical revelation in the Old Covenant helps us to know Christ, but only in light of the Mystery of the Incarnate Word does the meaning of antecedent history take firm shape. Like the biblical theologian Oscar Cullmann (1902–1999), the teacher he revered most on the Protestant faculty of theology in Strasbourg where he studied for ministry, Bouyer understands Christ to be the center of time.[1] We shall, in the

1 Cf. *Gnôsis*, 60. Bouyer interprets this differently than Cullmann. For Cullmann, Christ was the center of a straight line of continuous intervention that crosses our time from end to end. For Bouyer, "the straight line of divine

present chapter, examine this Christic centering of Bouyer's cosmology by exploring in a first section his exegesis in chapter ten of Jesus's own cosmic view as articulated in the New Testament. In a second section, we shall turn to his exposition of Patristic cosmology in chapter eleven of *Cosmos*. This section will conclude with an exploration of the theology of divine and created Wisdom in the writings of Saint Athanasius and Saint Augustine that Bouyer sums up near the end of the chapter. In a third section, going entirely outside of the text of *Cosmos*, we shall explore a theme that Bouyer does not explicate in his book on cosmology but that is an implicit, underlying factor in it, that is, his theology of the "eternal humanity" of the Son of Man. Finally, we shall consider a Christological theme of major importance in all his work and especially his cosmology: the predestination of all things in the Son of Man, the eternal, immolated Lamb who is the very reason and purpose for creation. Bouyer's text on Christology, *The Eternal Son*, will be consulted throughout this chapter, particularly in the final sections, in order to understand better the chapters of *Cosmos* that are our consideration here.

THE ESCHATOLOGICAL MAN

Chapter ten of *Cosmos* is the only chapter of the book in which Bouyer provides sustained exegesis of New Testament cosmology. He argues in it that the cosmology implicit in the New Testament texts is steeped in that of the intertestamentary apocalyptic literature, which sums up and surpasses the goal and ideal of Wisdom in the Old Testament. Once these writings were set forth, he seems to suggest, there remained only to await the appearance of the "new fact" of Christ in our historical horizon, allowing us to grasp them in their true, existential meaning.[2] For Bouyer, this new fact, the direct, personal intervention of God into

---

eternity comes across the semp-eternal cycles of a universe turning in the round. It enlarges them into a progressively opening spiral toward the eschatological tangent of accomplished redemption, in order to project us finally from our time into divine eternity."

2 Joseph Ratzinger, *Eschatology*, trans. Michael Waldstein (Washington, DC: CUA Press, 1988), 92. Ratzinger uses this expression to describe the transformation that the New Testament brings to Old Testament ideas regarding death and eternal life. It does not bring new ideas but a "new fact." We think that this understanding coheres with Bouyer's treatment of the history of cosmology.

our world in the flesh, draws all the ideas, symbols, and images of the vast history of revelation that preceded it into its resplendent light. The New Testament thus carries forward the old mythic patterns of human thought that inspired humanity to see the world in personal terms, as "I-Thou" relationality, and as an organic unity, while also embracing the demythologization carried out by the divine Word in the progress of the Old Testament traditions that we have already expounded. This is not to say that the Old Testament authors laid a Procrustean bed onto which the Savior of the World could be fitted readily. Bouyer recognizes that there is a decisive metamorphosis of both mythical and earlier biblical traditions in the New Testament theologies. The already metamorphosed themes of the royal myth in the Old Testament are now infused with a new and unexpected content in the figure of Jesus of Nazareth, who is at once Son of Man, Messiah, and Suffering Servant. Christ now becomes the all-determining antitype for pre-existing Old Testament themes and typologies, which have no center of unifying intelligibility without him. There is surpassing newness in Christ and discontinuity between Old and New. All of this is duly emphasized by Bouyer, but he also highlights the continuity between the Old and New. As he explained in his *Dictionary of Theology*, divine revelation, from the time of the inspiration of the prophets of the Old Covenant to the death of the last apostle, is but "the progressive communication of *one* truth with manifold aspects."[3] Given the nature of Tradition as Bouyer has explained it, we can say that Jesus of Nazareth is the ultimate figure of Tradition. He assimilates and bears forth what was given in the totality of human experience, especially in the direct lineage of his chosen People before his coming, but he opens this experience to a new understanding of God's kingship and paternity.

In his Christology text, *The Eternal Son*, Bouyer traced in detail the development of the earliest Church's Christology in general accordance with the chronological ordering of historical events as deciphered by the prevalent historical-critical scholarship of his time. His discussion of the New Testament tradition in *Cosmos* presumes the work he had already done in this earlier book, and this is doubtlessly a major reason why his

3 *Dictionary of Theology*, 209. This quotation is in Bouyer's entry on "Revelation." The parallels with Newman's thought are obvious.

exegesis of New Testament cosmology in *Cosmos* is truncated.[4] We must ourselves turn to this earlier writing in order to grasp better what Bouyer says much more succinctly in chapter ten of *Cosmos*, where he naturally focuses his attention strictly on the cosmological question.[5]

In this earlier monograph, Bouyer explores the development of the mysterious figure of the "Celestial Man" in the Old Testament, a hero with many faces, whether that of King, Messiah, Suffering Servant, or the Son of Man of intertestamentary writings. The biblical Celestial Man is, on Bouyer's reading, the surpassing, concrete exemplar of the ancient mythic figure of a Primordial Man, which is itself a personified confluence of two pre-existing mythic figures.[6] The first of these is the primordial cosmic being from whose dismemberment the whole world emerges and in whom all things will be reunited, such as Ymir in ancient Nordic mythology, Gayomart in Iran, and, as we noted in the previous chapter, Purusha in India. The second figure, Oannes in Babylonia or Adapa in Sumer, a figure akin to the biblical First Adam, is a man found within cosmic history, the king of paradise, who is not the cause of the world's coming-into-being.[7]

The myth of the Primordial Man is related to the story of the First Adam in Genesis and to the apocalyptic figure of the Son of Man who makes his appearance in the intertestamentary literature, but Bouyer

4 This cannot be the only reason that the discussion of the New Testament in *Cosmos* is cut short. This is so because in this book, and others besides, a great deal of space is given to exegesis of the Old Testament, and this is repeated in *Cosmos*, while the New Testament exegesis is not repeated at the same length. Bouyer emphasizes the concordance of Jesus's cosmovision with that of the apocalyptic literature.

5 See Weill, *L'Humanisme Eschatologique de Louis Bouyer*, 53–93. In these pages, Weill gives a masterful summary of Bouyer's exegesis of the New Testament centered on the theme of the Eschatological and Perfect Man of the Gospels and Pauline writings. She surveys the widest reach of Bouyer's writings on the topic. We are indebted to her for some of what follows in the present chapter. See also Bouyer, "La Notion Christologique du Fils de l'Homme a-t-elle Disparu dans la Patristique Grecque?" in W.F. Albright, F. Amiot, P. Auvray, eds., *Mélanges Biblique Rédigés en l'Honneur de André Robert* (Paris: Bloud et Gay, coll. "Travaux de l'Institut Catholique de Paris," n. 4, 1957), 519–30.

6 *ES*, 91–92. See Mowinckel, *He That Cometh*, 423–25. Bouyer follows Mowinckel in distinguishing these two forms of the myth.

7 *ES*, 91–92.

insists that the mythic figures and their biblical counterpart do not bear the same theological significance. The Primordial Man of the myths is, he explains, enchained within the endlessly recurring cycles of cosmic being.[8] He is ultimately not so much an eschatological figure as is the Son of Man of biblical writings, who is explicatory of humanity's end, but ktisiological or protological, explicatory of its beginning, an expression of humanity's nostalgia for the lost, Golden Age of paradise. Bouyer interprets Saint Paul's words in First Corinthians regarding the Second Adam to be a definitive refutation of the backwards-orientation of the perennial myths:

> But as it is, Christ is now raised from the dead, the first fruits of those who have fallen asleep. Death came through a man; hence the resurrection of the dead comes through a man also. Just as in Adam all die, so in Christ all will come to life again, but each one in proper order: Christ the first fruits and then, at his coming, all those who belong to him.[9]

Christ is not the "First Man" but indeed is the "Final Man" or "Eschatological Man" (*ὁ ἔσχατος Ἀδὰμ*).[10]

Bouyer stresses that Christ as Son of Man has a heavenly origin, but that he will be manifested only at the end of time "as the eschatological redeemer of humanity and the universe, as universal judge, as triumphant over the powers of evil and as cause and source of universal resurrection."[11] Paul connects in Jesus Christ the cosmic or celestial human person with the king of the paradise of the future, and this was not achieved by the earlier, intertestamental literature. It is, Bouyer holds, in this cosmic and ultimately supracosmic light that Paul understands Christ to be "the ultimate human being."[12] He argues that this understanding is made

8 Ibid., 91.

9 Ibid., 229. 1 Cor, 15:20–22. The myth of the Primordial Man would have actually been known to Saint Paul through the writings of Philo of Alexandria, whose speculations on this front Bouyer associates with the development of later, heretical forms of Gnosticism.

10 Ibid. 1 Cor. 15:45.

11 *ES*, 229.

12 Ibid., 230.

especially evident in the Letter to the Ephesians and, once recognized, sheds conclusive illumination on the whole of the Pauline writings.

Bultmann insisted that Saint Paul demythologized Jewish apocalyptic by evacuating it of its cosmic significance, emptying out the cosmic powers from influence over human decision, thereby getting to the purely anthropological core of Christ's earthly message.[13] In line with Cullmann, Bouyer insists to the contrary that Paul embraced and developed the cosmic dimension of apocalyptic theology, and that what he did in this regard flowed directly from the teachings of Christ himself as transmitted in the earliest days of the Church by the apostolic community. Modern biblical criticism raises the question of whether the teachings found in the texts of the New Testament go back to the Gospel that Christ himself preached. Bouyer argues at the beginning of chapter ten of *Cosmos* that in the cosmological domain the answer to this question is assured: "Jesus' thinking has come down to us precisely as it was expressed to his disciples."[14] Jesus was himself a resolutely cosmological and apocalyptic teacher. The Synoptic Gospels, Bouyer argues in *Cosmos*, evince a unified agreement on this point, which signals the derivation of their teaching from a single source in Christ himself. Each of them makes clear "that Jesus adopted the vision of the created universe that emerges from the intertestamentary apocalypses."[15]

Our brief excursion into Pauline theology sets a context for chapter ten of *Cosmos*, which begins with consideration of the Synoptic Gospels taken together and includes pithy sections on the teachings they contain concerning God's fatherhood, the knowledge that is given to us through the Eucharist, and on Jesus's conflict with the Devil. We learn that Matthew, Mark, and Luke themselves presume the teachings found in intertestamentary literature regarding creation, the Fall, and the eschatological completion. But they also communicate the way that "Jesus transfigured these elements in his assertion of the divine fatherhood,

13 See Rudolf Bultmann, "New Testament and Mythology: The Problem of Demythologizing the New Testament Proclamation," in *New Testament and Mythology and Other Basic Writings*, ed. S.M. Ogden (Philadelphia: Fortress, 1984), 1–43.

14 *Cosmos*, 86.

15 Ibid.

which would extend to all men."[16] Bouyer emphasizes that in the Synoptic texts Jesus the Son displays absolute filial reverence to the Father, fulfilling in a surpassing way the tradition of Jewish piety for which all of creation and the whole of history in all of its unique and individual occurrences speaks of God's benevolence. "Jesus' attitude," Bouyer says, "became as it were the last word of this address, a perfect reflection of the Father's *agape*, the creative, saving and paternal love in which the Lord's justice and mercy join and are absorbed into a hitherto unimaginable and inconceivable generosity."[17]

Every created reality, the whole of creation, comes from God's transcendent goodness and liberality. The parables with which Christ teaches his disciples educated them to see the cosmos as a sign of the Father's unfathomable goodness.[18] God's gifts should move us to response, to enter into communion with the Father in and through the Son. This requires total, though joyous, surrender on our part, a giving over of ourselves to the Rabbi whose yoke is easy and burden light.[19] In following the way of Jesus, we learn to love everything as he loves it. The cosmos is meaningful, precisely because it is a sign of the loving presence of God, whose glory resides in the heart of all things. God the Father is transcendent, but He desires to make all of us His children. Bouyer says that the Gospel teaches us to see everyone in this light, to regain the innocence of the child's way of seeing, "which is possible only if Christ's essentially filial experience is shared by us."[20] This requires ascetical detachment, especially voluntary poverty. The properly eschatological standpoint or emplacement of the human person can be instantiated in no other way. There is no alternative path

16 Ibid.

17 Ibid.

18 Ibid., 86–87.

19 Cf. Mt. 11:30. We take Bouyer's *Memoirs* to exemplify the point that both his polemical edge and ascetical theology were tied to a persona that was ultimately playful, impish, even childlike, all the while remaining shrewd as a serpent. It was not always the case, though, that his criticism maintained a playful air. His assessment of Annibale Bugnini (219, 221, 224–25, 229) demonstrates the point. However, even in postconciliar works like *Decomposition of Catholicism* and *RC*, his underlying perspective is one of a certain bemused frustration more than angry condemnation. Bouyer's *Memoirs* show that his own ascetical practice was joyously embraced and was not a form of absolute world-denial.

20 *Cosmos*, 87.

through which to recover or rediscover the cosmos. Bouyer urges that the beatitudes in Luke and Matthew need to be understood in this light: only in following the beatitudes can we regain possession of the entire cosmos as gift in which the glory of the Creator radiates in and from its depths.[21]

The creature's proper response to the Creator constitutes a Eucharist or "thanksgiving" in itself. Following especially the Pauline and Johannine writings, Bouyer emphasizes in chapter ten that the Son of Man communicates his own eternal Eucharist or thanksgiving in the Trinitarian life to us in time and history in the economy of salvation.[22] This Eucharistic message, truly cosmic in scope, is the culmination of the Synoptic Gospels as well, and the French Oratorian points to the importance in these Gospels of the so-called (by some modern biblical scholars) "Johannine Meteorite," according to which it is in and through Christ's communication to us of his own filial status in relation to the Father that we can know, with the knowledge of love, the Father's *agape* and His plan for all creation.[23] This is a saving knowledge, which liberates us from the hostile, fallen powers of the world that set it against its Creator. In this light, the temptation of Jesus in the desert, where he was led by the Spirit, takes on its full significance. This is, Bouyer argues, the most crucial experience of the Son of Man, where he chooses obedience to the Father over the Devil's delusory offer to him of worldly power in order to rescue humanity and the cosmos from the Devil's snares.[24] Christ's mission is at once to preach the Kingdom and to cast out Satan.

21 Ibid. See Mt. 5:3–12; Luke 6:20–22. Bouyer references here the Cistercian Saint Isaac of Stella (1100–1169). Without providing specific bibliographical information, he points to the monk's commentaries on the Beatitudes. See Bouyer, *The Cistercian Heritage*, trans. Elizabeth Livingstone (Westminster, MD: The Newman Press, 1958), 161–89. Here Bouyer discusses this mysterious Cistercian in depth. All of Isaac's writings are published in *PL* 194:1689–1896. Saint Isaac was a 12th-century Englishman, the Abbot of Stella, in the diocese of Poitiers.

22 *Cosmos*, 87–88.

23 Ibid.; cf. Mt. 11:27; Luke 10:22.

24 *Cosmos*, 88. Cf. Mark 1:12. Bouyer points the reader to Harald Riesenfeld's *Studies in the Gospel Tradition* (Philadelphia, 1968). This reference is in *Cosmos*, 248. It is likely that Bouyer is in fact referring to Riesenfeld's *The Gospel Tradition* (Philadelphia: Fortress Press, 1970). It is Riesenfeld, he says, who has conclusively shown that the temptation in the desert was such a decisive experience for Christ.

Bouyer does not engage in depth in *Cosmos* as he does in his Christology text Saint Paul's theology of the Eschatological Adam that we have briefly described, but he does show in *Cosmos* that Paul's cosmology, like that of the Synoptics, flows from the tradition of apocalyptic teachings, with their sophiological resonances. He dedicates four ensuing sections in chapter ten to Paul's cosmology (entitled "Pauline Cosmology," "The Enemies of God," "Two Contrasting Dispensations," and "The Pauline Eucharist"). In these sections, he follows those scriptural exegetes running counter to Bultmann according to whom Paul's Christology is directly informed by the account of creation contained in the Wisdom literature of the Old Testament and by the biblical transition from Wisdom to apocalypse.[25] The second chapter of First Corinthians, Bouyer explains, evokes the second chapter of the Book of Daniel, where the transition from Wisdom to apocalypse is most evident in the Old Testament texts. In First Corinthians, in parallel with Daniel, the wisdom of the world is confounded by the Wisdom of the Cross, God's own eternal Mystery made manifest, appearing as folly to the world.[26] God's plan for history is true Wisdom, and its content is the Mystery or divine secret that God alone knows and anyone to whom He freely chooses to reveal it.[27] Mystery, Wisdom, and apocalypse are united in the figure of Christ. Christ's Cross is the final key that opens the Mystery of divine Wisdom, and it is communicated to all those little ones who surrender themselves in him by the Spirit to the will of the Father.[28]

In this unified vision of the cosmos and history, which transfigures with a definitive but non-annihilating demythologization the mythic heritage of the human race, Paul includes an understanding of the material world itself that draws (as the last Wisdom books of Scripture had done) on themes in Stoic philosophy, such as "the consonant notions of creation as *cosmos* and of man as *soma*."[29] Saint Paul "was particularly

25 Cf. D. Deden, "Le 'Mystère' Paulinien," in *Ephemerides Theologicae Lovanienses* (1936); Dom J. Dupont, *Gnosis* (Louvain, 1949); André Feuillet, *Le Christ Sagesse de Dieu d'après les Épîtres Pauliniennes* (Paris, 1966).

26 Cf. *ES*, 232.

27 Cf. *Mystery*, 5–18.

28 Cf. Mt. 19:14.

29 *Cosmos*, 89.

conscious of the organic nature of the entire creation and of human life inserted into creation, so that only human life can express the ultimate meaning of the created world in its entirety and its unity."[30] This quotation, already given in our introduction, is central to the whole of *Cosmos* and is especially evocative of this famous passage from Romans: "We know that the whole creation has been groaning in labor pains until now; and not only the creation, but we ourselves, who have the first fruits of the Spirit, groan inwardly while we await for adoption, the redemption of our bodies."[31] In the overarching Pauline interpretation that Bouyer promotes, everything is created in the Son and is meant to share in the glorious life of his Resurrection, extending the glory of the Spirit of God to the farthest reaches of the cosmos.[32] Yet all of creation, in a condition of slavery and enmity in the current age, groans in the pangs of childbirth awaiting the completion of the Body of Christ that will be given as the final action of divine grace in the Parousia.[33]

Bouyer highlights in *Cosmos* Paul's development of the apocalyptic distinction between the present age and the age to come with reference to the angelic powers. The Pauline cosmos is ruled by the elements of the world.[34] The "god of the world" referenced in Second Corinthians, Job's accuser, the apocalyptic Belial, has headship over these spiritual forces.[35] Man, through his own sinfulness, has allowed himself to become entrapped by the snares of these beings, but God overcomes their wiles and their tyranny through the Cross and Resurrection. Bouyer emphasizes that for Saint Paul the old law is associated with the rule of the angels over the cosmos. We receive the law through their ministry, and the physical world was subjected to their power from its origin. Paul's

30 Ibid. We drew in our introduction on this quotation, which we think is decisive in the Bouyerian understanding.

31 Romans 8:22–23.

32 *Cosmos*, 90.

33 See Weill, *L'Humanisme Eschatologique de Louis Bouyer*, 72–76. In these pages, Weill shows the importance to Bouyer's thought of the Pauline theme of the completion of the Body of Christ through and in all of those who are joined to him awaiting their perfect union with him at the end of history. We shall return to this theme especially in our penultimate chapter.

34 See Galatians 4:9, Colossians 2:9 and 20, 1 Corinthians 2:6 and 8, and Ephesians 2:2 and 3:10.

35 2 Corinthians 4:4.

enigmatic statements with respect to the law (such as in Romans 5:20–21) can appear to be antinomian. Bouyer insists that his teachings in this regard have to be understood in connection with his angelology. It is with the Incarnation of the Eschatological Adam that the authority originally given to the angels has passed over "into the hands of the new man and the renewed mankind which derives from him."[36] There are two ages corresponding to two dispensations and ultimately to the two Adams. The meaning of the law in the reign of the Second and Final Adam is inscribed in the liberality of mercy that constitutes the new and definitive dispensation of divine grace (*χάρις*) effectively communicated through the ministry of Christ and the Spirit.[37]

The Church's Eucharist, the thanksgiving that we give to the Father in and through the Son, so prevalent a feature in the Pauline epistles, becomes all-pervasive in the Letter to the Ephesians.[38] We glorify in the Eucharist the Cross and Resurrection of the New Adam, the glorious work of redemption through which God has elevated His children to a condition of freedom and glory. All of creation is "recapitulated" in its source around humanity renewed in the crucified flesh of the Son, "the firstborn of every creature, who thus becomes the firstborn of the dead."[39]

The chapter from *Cosmos* that we are considering devotes two brief sections to John's cosmology, but in *The Eternal Son*, Bouyer explores the Gospel of John in depth in connection with the Gospel of Mark. There he accepts F.C. Baur's (1792–1860) demonstration of the close affinity of these two gospels.[40] The Johannine theology of the Word is highly pertinent to cosmology, and Bouyer emphasizes that John does not derive his theology of the Word from Greek philosophical sources but from the earlier tradition of the biblical *Dabar*. The "Word of God,"

36 *Cosmos*, 81.

37 Ibid., 92.

38 Ibid., 92–93.

39 Ibid., 93. See Ephesians 1:10 for the language of "recapitulation." Paul in fact uses the verb *ἀνακεφαλαιώσασθαι*. Saint Irenaeus will convert this into the noun form. See also Colossians 1:15 and 18: "He is the image of the invisible God, the firstborn of all creation . . . the firstborn from the dead."

40 Cf. Friederich Christian Baur, *Die Epochen der Kirchlichen Geschichtsschreibung* (Tübingen: Drud und Verlag Ludwich Friederich Fues, 1852). Bouyer argues that Baur deserves credit for being the first genuine New Testament critical scholar. See *ES*, 139–40.

for John, is the "Word of the Lord," Christ himself. This identification is made in biblical sources outside of the Gospel of John, but John extends it to the Word of creation found in the first chapter of Genesis.[41] John, in line with the earlier Wisdom literature, identifies the Word with Wisdom, "the divine plan inscribed in the very structure of all things," through and in which they find their meaning.[42] The Word creates all things in Wisdom from the very beginning, and it is through Wisdom that the Word is present in all things.

When John says "The Word became flesh and made his dwelling among us," he identifies the Word of the Lord with the *Shekinah-Yahweh* of Israel. This mystical figure, as we saw in chapter three, is the very glory or radiance of God made present in the tabernacle with which Israel journeyed through the desert. The *Shekinah-Yahweh* is, Bouyer insists, a sign of the divine Word's active presence in the cosmos through salvation history. God in this mysterious figure of glory "pitches His tent" in the supernaturally sacred sites of the People of Israel and finally, in plenitude, in the humanity of Christ. In Sirach (Ecclesiasticus) 24, the Word and Wisdom of God are identified with the *Shekinah-Yahweh*. John says: "The Word became flesh and made his dwelling [literally, set up his tent] among us."[43] Jesus reveals himself to us as the Word and Wisdom of God who has a unique filial relationship to the Father. He communicates to us by giving us—through pitching his tent in us—a share in his unique knowledge of the Father.[44]

41 *ES*, 268–29.

42 Ibid. See Weill, *L'Humanisme Eschatologique de Louis Bouyer*, 48, n. 3. Weill points out that the term "wisdom" appears in John's writings only two times, both of them in the Apocalypse, where it is counted among the divine attributes: 1) "Worthy is the Lamb that was slaughtered to receive power and wealth and wisdom and might and honor and glory and blessing" (Rev. 5:12); 2) "Amen! Blessing and glory and wisdom and thanksgiving and honor and power and might be to our God forever and ever! Amen." Bouyer takes the conjunction of the two—Word and Wisdom—to be implicit in the Johannine theology, especially, as we see in the next note, in connection with the theme of the *Shekinah-Yahweh*.

43 *ES*, 269–70. The quotation is from John 1:14. Bouyer himself provides the bracketed statement regarding the literal meaning of the word ἐσκήνωσεν, which means "to have one's tent," or "to dwell."

44 Ibid., 273–76.

In chapter ten of *Cosmos*, Bouyer focuses on two other themes in John's writings, which both pertain to apocalyptic dualism. The first concerns contrasting dispensations between light and darkness, especially in the Book of Revelation, which presents "the history of the Church . . . as the completion of cosmic history."[45] The theology of the Cross and the theology of glory are inextricably linked together. The immolation of the Lamb brings victory over the serpent to all of those who follow him wherever he leads. Key here is the figure of the celestial Jerusalem, the cosmic or supracosmic Bride of the Lamb, adorned for intimate communion with the Bridegroom, who comes to earth at the end of time.[46]

The second theme has to do with conflicting ways of perceiving the goodness of the world, which we touched on in our first chapter. On the one hand, John says that the world should not be the locus for our love, because it is under the power of the Evil One. On the other hand, he says that God so loved the world that He sent His only Son for our salvation.[47] These seemingly contradictory assessments of the cosmos can be reconciled, Bouyer suggests, by recognizing that the world is fundamentally good in its creation but rendered corrupt by finite freedoms who have misused the gifts that God gave to them. Corruption in the world is historical and ethical in scope. It is not intrinsic to the world's deepest ontological formation. This carries forward the development of the Old Testament theology of creation taken up by the apocalyptic books. The world has been corrupted by the one who has usurped God's glory for himself and become the "prince of the world." If the Father's love in the world is to be recognized anew, it can only be, Bouyer says, by "contemplating in a spirit of faith the cross of Christ, which is the manifestation and triumph of divine love."[48]

This entails that Christian love for the world must be placed in the horizon of eschatological hope for the transfiguration of all things that will be completed only on the Last Day.[49] Bouyer warns us to sail a course between the Scylla and Charybdis of two popular but misguided

45 *Cosmos*, 93.
46 Ibid.
47 Ibid., 94. See again 1 John 2:15, 5:19; John 3:16.
48 *Cosmos*, 94.
49 Ibid., 94–95.

interpretations of New Testament eschatology. Christ did not preach, on the one hand, an imminent, catastrophic end. On the other hand, his teaching did not imply that the Paschal event brought absolute, eschatological presence to the cosmos without need to wait for anything substantially new. In their existence in the world, Bouyer insists, Christians cannot escape Christ's own paradoxical relation to the cosmos in his own "already-not-yet" relation to the End. Bouyer concludes chapter ten of *Cosmos* by advising the reader that the Kingdom, the center of Christ's teaching connected to his person, is with us and in us in the Spirit of Christ, but that it suffers violence, hidden as it is "by the still persisting externals of Belial's reign."[50]

DIVINE AND CREATED WISDOM

Christ himself taught an apocalyptic cosmology, and those who were given the mantle of discipleship to him in the early Church advanced in understanding of the world along these lines. Bouyer argues this point in chapter eleven of *Cosmos*, entitled "The Struggle Against Gnosticism and Arianism: From St. Irenaeus to St. Athanasius and St. Augustine," where he surveys the responses made by the Church Fathers to the challenges of these two initial forms of Christian heresy. He highlights, as we shall see, the distinction made by both Saint Athanasius and Saint Augustine of divine from created Wisdom.

The Church Fathers have often been accused by modern scholars of projecting a Platonizing metaphysical schema onto the historicity of the message of the Gospel preached by Christ. This schema is presumed to be a residue of a pagan, mythic mentality, harboring an idealist and exemplarist view of the cosmos. Bouyer endeavors to show that the theology of creation in the Church Fathers, in its best attestations, flows from the divine Word's own self-communication to the Church. The first opponents to the orthodox faith in the early Church, on the other hand, the Gnostics, are the ones who failed sufficiently to carry out an effective demythologization.[51]

There are many theories of the origin of second-century Gnosticism. In chapter eleven of *Cosmos*, Bouyer lists four of them, which followed

50 Ibid., 95.
51 Ibid., 99–100.

one another in a succession of modern theorizing. Either Gnosticism: 1) developed from within Greek philosophy (Adolf von Harnack and Eugène de Faye), or 2) emerged out of a syncretistic religious milieu in the Hellenistic period (Wilhelm Bousset and the early Richard Reitzenstein), or 3) derived from a developed Iranian Mazdaism (the later Reitzenstein), or 4) was the result of the decomposition of Jewish apocalyptic (Gilles Quispel, Robert M. Grant, and Cullmann).[52] Bouyer favors this fourth explanation. Jewish apocalyptic promotes, as we have seen, ethical and historical dualism, although not dualism of a metaphysical kind. Bouyer contends, along with proponents of this fourth theory, that the shattering of eschatological hope following the failure of the Jewish wars in the first century AD led some of the Jewish faithful to lose hope in the prospect that God would ever again intervene in history to save His people. The hope for a historical and eschatological salvation, with its concomitant historical dualism, was replaced by a soteriology of escape from the world. Apocalyptic dualism was thereby hardened into the metaphysical dualism of unregenerate myth. This was the first stage in the long development of heretical Gnosticism in the early Church. Bouyer argues that Gnosticism developed over time and integrated other elements into it.[53] In the end, it brought back the confusions of the myths for which cosmogony is equated with theogony, creation with the Fall, and salvation with escape from creation.[54]

Our author vehemently opposes the thesis of the Bultmannian School that the Son of Man in the Gospel is derivative of the Gnostic figure of the Celestial Man.[55] It is, he argues, the other way around: the Gnostic Celestial Man is a plagiarized version of the truly Celestial, Eschatological Man of Jewish and Christian apocalypses. Bouyer notes that

52 Ibid., 98; 251, nn. 1–12. See Zordan, *Connaissance et Mystère*, 594, n. 1. In this lengthy footnote, Zordan summarizes the context of Bouyer's assessment of Gnosticism with respect to other approaches to understanding it. Bouyer distances himself from the *Religionsgeschictliche Schule* headed by Bousset, according to which the resurrection of Christ and his celestial humanity was a Gnostic myth imported into Christianity. Instead, Bouyer is convinced of and argues for the central role of Jewish thought in the entirety of the intellectual milieu of the first Christian centuries.

53 See Grintchenko, *Une Approche Théologique du Monde*, 134.

54 *Cosmos*, 98–99.

55 *ES*, 117–21.

Gnosticism did not emerge in full, poisonous blossom until the second century, whereas the Christian messianic figure was already firmly implanted in the consciousness of the Church by the time of the Pauline writings as a transfigured inheritance of intertestamentary Judaism.[56] The Celestial Man of Gnosticism was a docetic figure, representative of the metaphysical dualism that Gnosticism embraced. The Pauline Eschatological Adam, on the other hand, is a truly apocalyptic figure who brings personal salvation in and through the recreated cosmos.[57]

This story of the origin of Gnosticism provides proper context for understanding the cosmology of the Church Fathers, beginning with Saint Irenaeus.[58] Bouyer appropriately focuses in chapter eleven of *Cosmos* on Irenaeus's Pauline theology of "recapitulation."[59] For Saint Irenaeus, recapitulation means, Bouyer explains, "both a re-establishment from its principle of the divine work, thwarted by the Fall, and a summation of that work in the adoption—meant literally and not merely in a legalistic sense—of the human creature in the Son of God made man, 'to make the children of men into children of God.'"[60] Saint Irenaeus understood that the salvation offered to us in the flesh of Christ gives hope for the renewal of the entire cosmos in bringing about the renewal of the totality of our humanity. God communicates His glory cosmically in and through humanity. The divine glory is infused "into the whole of our life, that we can truly live only through the living knowledge of God."[61] While not accepting Irenaeus's millenarianism, our author commends the link between theology of history and creation in the first Bishop of Lyons. He embraces the way that Saint Irenaeus sees all of creation in the history of salvation: as "a flawless continuity in the Creator's plan."[62] The Creator's mind is reflected in the unfolding drama of relationship

56 *Cosmos*, 99.

57 Ibid., 29–30.

58 See *Gnôsis*, 155–68. This text provides a fuller account of the Church Fathers on gnosis.

59 *Cosmos*, 100. The language of recapitulation is, again, drawn from Ephesians 1:10.

60 *Cosmos*, 100. Saint Irenaeus, *Adversus Haeresies*, 5, 16, 1–2. See *Cosmos*, 251, n. 17.

61 *ES*, 307.

62 *Cosmos*, 100.

between infinite Spirit and finite spirits from the first moment of creation and the Fall to redemption in Christ. This in no way diminishes the radical discontinuity between God's transcendent, infinite life and the life of the cosmos.[63] The cosmos can be given life as a gratuitous gift only because God is transcendent to it. Bouyer's own way of relating theology with economy in his wider works reflects this Irenaean understanding of continuity in the *telos* of creation within the radical discontinuity that obtains between the perfections of eternal being and finite being.

An essential dimension of the theme of recapitulation is that the divine plan is ordered as a unity, which the struggles between orthodox and heterodox theologians engaged in combat during the Arian crisis in the fourth century brought to a deeper level of reflection. This deepening reflection was a matter of urgency in the wake of Arian challenging of the divinity of Jesus Christ. One of the most significant dimensions of this reflection, which has not been emphasized in most standard histories of the development of Christian doctrine, was the establishment of a distinction between divine and created Wisdom by the greatest champions of Nicaea. Doubtlessly inspired by his reading of Cardinal Newman on the development of doctrine, Bouyer noticed at the very beginning of his career in his thesis on Saint Athanasius that this monastic defender of Nicene orthodoxy laid the groundwork for a sophiological understanding of God's relationship to creation by suggesting the distinction (in unity) between Wisdom in its uncreated and created manifestations.[64]

63 Ibid., 101.

64 See *MSW*, 46. Newman says in his *Essay on the Development of Christian Doctrine* (4.2.8): Jesus "indeed was really the 'Wisdom in whom the Father eternally delighted,' yet it would be but natural, if, under the circumstances of Arian misbelief, theologians looked out for other than the Eternal Son to be the immediate object of such descriptions. And thus the controversy opened a question which it did not settle. It discovered a new sphere, if we may so speak, in the realms of light, to which the Church had not yet assigned its inhabitant.... Thus there was 'a wonder in heaven': a throne was seen, far above all other created powers, mediatorial, intercessory; a title archetypal; a crown bright as the morning star; a glory issuing from the Eternal Throne; robes pure as the heavens; and a sceptre over all; and who was the predestined heir of that Majesty? Since it was not high enough for the Highest, who was that Wisdom, and what was her name, 'the Mother of fair love, and fear, and holy hope,' 'exalted like a palm-tree in Engaddi, and a rose-plant in Jericho,' 'created from the beginning before the world' in God's everlasting counsels, and 'in Jerusalem her power'?

The Arians were able to challenge the divinity of Christ on apparently sophiological grounds, and Saint Athanasius gave a response in his *Second Discourse Against the Arians* that brought much needed nuance to the problematic. Chapter eleven of *Cosmos* mentions this point, but it is important to show Bouyer's invocation of this theme in other writings, where he develops it a little more fully.[65]

The Arians followed the exegesis of Scripture admitted by the Church in their time for which the figure of Wisdom in certain texts of the Old Testament should be applied directly to Christ himself. As the Christological controversy developed in the fourth century, this became a troubling issue. In Proverbs 8:22, Wisdom is clearly a created being, not eternal, the first of all the divine works and not one in being or *homoousios* (*ὁμοούσιος*) with God. If Wisdom is identified with Christ, it would seem to follow that Christ is a creature and not divine. He would not possess the plenitude of the divine being that belongs to the Father alone. In his early thesis on Saint Athanasius, Bouyer showed that the great saint responded to this challenge not by denying the personal and created character of Wisdom in Proverbs and its application to Christ. He did so instead by distinguishing divine from created Wisdom in their intimate connection with the Second Person of the Most Holy Trinity.

In the second book of his *Discourse*, Saint Athanasius attempts to clarify the relation between the Wisdom of Proverbs 8:22 and Christ.[66] He shows that there is a divine Wisdom of creation that we can call "created":

> This figure . . . of Wisdom being created in us and in all his works, it is natural that the true and creative Wisdom, assuming in it what belongs to its figure, says: "The Lord has created me for his works," because what the Wisdom says that is in us, the

---

The vision is found in the Apocalypse, a Woman clothed with the sun, and the moon under her feet, and upon her head a crown of twelve stars. The votaries of Mary do not exceed the true faith, unless the blasphemers of her Son came up to it. The Church of Rome is not idolatrous, unless Arianism is orthodoxy."

65 *Cosmos*, 100. See also Weill, *L'Humanisme Eschatologique de Louis Bouyer*, 38–42. Saint Athanasius, *Second Discourse Against the Arians*, section 76 (*PG* 26:307ff.).

66 See Weill's discussion, *L'Humanisme Eschatologique de Louis Bouyer*, 38–42. See also *Sophia*, 89–102; *Cosmos*, 104–5.

> Lord himself says as belonging to him, not that the Creator is himself created but he says that of the image of himself that he has created in his works, as if it were of himself.[67]

For Athanasius, Christ, who is indeed *homoousios* with the Father, can be called created, but only inasmuch as he has assumed our humanity in the womb of Mary. The very possibility for this to occur exists from before the creation of the cosmos. The possibility for life and salvation has been prepared for us eternally in the Word of God. It is in the Eternal Word that God possesses the plenitude of life, and the Word does not need to become incarnate among us in order that God may increase in divinity. It is we who have need of the Incarnate Lord in order to live. All things have been created in the Word out of God's inexhaustible, divine liberality, and thus there is an abiding, intimate connection between the production of creatures and God's own Word. The Wisdom of God exists in the eternally engendered Word, in which all of creation "pre-exists" eternally "before" creation "in time."[68]

Plato postulated that the eternal, immutable, perfect models of all the mixed, changing, partial realities of the cosmos exist in some separated, eternal domain of being, and it might seem that Saint Athanasius follows a Platonizing direction of thought in construing the cosmos and the history of salvation in terms of an eternal, sophiological exemplar in the Son and Word. Bouyer, in line with what we suggested at the end of the previous chapter, adamantly maintains that there is a difference between Athanasius's Christ-centered, sophiological exemplarism and Platonic idealism. For the latter, he insists, temporal being can only ever

67 Saint Athanasius, *Second Discourse Against the Arians*, 78. Quoted in *IÉ*, 144.

68 See Weill, *L'Humanisme Eschatologique de Louis Bouyer*, 39–42. See especially p. 41, nn. 2 and 3. Quoting Saint Athanasius's *Second Discourse Against the Arians*, 76, Weill explains that Saint Athanasius maintains a profoundly unified understanding of Christology, ecclesiology, and sophiology. "The Church," she explains, "is literally *in* the incarnate Word, *Body* of the incarnate Word, predestined in it from all eternity, in such a way that Christology and ecclesiology are only one. Sophiology, in its turn, rejoins Christology and ecclesiology, since it consists in scrutinizing the unique mystery of the Word, living will and wisdom of the Father, in its eternal engendering as in its economy in our favor."

be a shadow or reflection of the immutable ideas or archetypal realities that alone account for the possibility of form and meaningful existence, while for Athanasius what is anticipated eternally in the Word of God is the final, perfected, subsistent destiny of all finite, temporal beings in Christ, who is the Incarnation of the eternal Word.[69] God, according to this understanding, wills to put in the human being created Wisdom, the very image of divine Wisdom "in whom all things are seen eternally as God wills them."[70] Bouyer says:

> God not only creates us in order to unite us all to the only Son, but, in eternally contemplating him, He contemplates us in this union consummated in time, so much so that He no longer sees Himself separately from us. In the integral vision that the Father has of things, divine and created Wisdom are not confounded, as the Arians wanted it, but espoused.[71]

The Athanasian texts concerning Christ and Wisdom are founded on a theology of the eternal, divine vision for all of creation that includes the Incarnation of Christ and his Cross as the center of the world and of the historical destiny of all personal beings predestined by God for glory in His eternal thought and will. God's vision of creation includes all of its becoming and the end to which He has willed it, which is to join creation to Himself in perfect communion at the climax of history. For God, in His infinite being, the plan for creation actualized in history is an eternal reality, which belongs to Him in His divine heart. We are created and saved through the Incarnate Word in time, but, as far as God is concerned, we are eternally present in the totality of our existence to Him in His Wisdom. Saint Athanasius gives a new, metaphysical lens of interpretation in which to regard the Old Testament theology of the

69 *Sophia*, 95–96.
70 Ibid., 96.
71 *IÉ*, 145. Quoted by Weill, *L'Humanisme Eschatologique de Louis Bouyer*, 42. Weill rightly points out that all 2,800 pages (in the original French editions) of Bouyer's triple trilogy can be understood in some way in the light of this sophiological attempt to understand the relationship between God and creation. This was the point of our own dissertation in 2007, which seemed to us, at the time, a rather risky interpretation.

apocalyptic man as well as the celestial city of Jerusalem that is forever in God in His eternal thought and love.[72]

After the dust settled on the Christological controversy, and the Arian heresy no longer posed the same level of threat to the Church as it did in the time of Saint Athanasius, Saint Augustine took up and developed Athanasius's sophiological affirmations. He did so especially in book twelve of *The Confessions*. Bouyer makes brief mention of this in *Cosmos*.[73] In *Sophia*, he points to the importance of Balthasar's book *Das Ganze im Fragment* for bringing Saint Augustine's sophiology as found in *The Confessions* to light.[74] In *Sophia*, Bouyer includes a lengthy passage from this writing that takes up more than a page of his own text.[75] The crucial point he highlights is that the Bishop of Hippo, like his Alexandrian predecessor, understood that from all eternity the creation of finite reality is expressed within the Father's self-expression or eternal Word, in order to return the love of the Father back to Him as it is given to creation by the Son in the Spirit of filiation.[76] Creation, redemption, and the salvific Incarnation of Christ proceed from this eternal expression of Wisdom in the Son: God has planned to come to us in order to lead us to Him. This is not to absorb us into the divine being but instead to consummate our life in His, in filial communion. Inasmuch as it is associated with the eternal life of the Trinity, Wisdom is divine, yet, in itself, as finite and distinct from God, Wisdom is created, bearing the uncreated within. It is through Wisdom that the entirety of the cosmos is associated with God. As created, Wisdom is not the divine Word but "the whole humanized

72 *Sophia*, 98.

73 *Cosmos*, 105.

74 *Sophia*, 98. See Hans Urs von Balthasar, *Man in History: A Theological Study* (London: Sheed and Ward, 1968), 1–42. On p. 11, Balthasar quotes a decisive passage from Book XII, 15 that we can reproduce in part here: ". . . created wisdom, that intellectual nature which by the contemplation of light is itself light: for this, though created, is likewise called wisdom: there is as much difference between the Wisdom which creates and the wisdom which is created as between the Light which gives light and the light that is so only by reflection, or between the Justice which justifies and the justice which results from being justified."

75 Ibid., 99–100. Augustine, *Confessions*, XII.15 (*PL* 32:832). This passage is too lengthy for us to quote, although Balthasar's chosen quotation, from our previous footnote, is contained within.

76 Ibid., 101.

cosmos through the divinization bestowed on it as a free gift by the Son (the eternal *Logos*) when he became man, and therefore one of us, a creature, so that the Spirit of his own sonship might be ours."[77]

It is within a sophiological framework that Bouyer recovers the doctrine of divine ideas first enunciated by the Church Fathers, and he ends chapter eleven of *Cosmos* with this theme.[78] In God's Wisdom, the ideas or *logoi* are unified in their living reality, ordered to the perfection of historical becoming in the Incarnate Word of God. Our author reads Saint Maximus the Confessor's theology of the *logoi* in a sophiological light, and in *Sophia* he connects Aquinas's theology to this sophiology as found in select Athanasian and Augustinian texts.[79] The theology of Wisdom in this untapped tradition requires us to recognize that the physical world, the world of matter, is dependent upon a spiritual world of guiding spirits, God and the angels. We shall explore the point in greater detail in chapters nine and ten, but it is important to mention here that Bouyer thinks of the divine ideas in connection with the angelic intelligences who are the first created rectors of the physical cosmos. This is a controversial point in his writings, but he holds that if God's exemplary plan for creation includes all historical becoming in His eternal Wisdom, the divine ideas are not perpetually ordered to some static, lifeless being, and simply to be equated with the divine essence. The divine ideas have a primordial manifestation in or even as the angelic persons themselves.[80]

When, in chapter eleven of *Cosmos*, Bouyer describes the Patristic, particularly Cappadocian, vision of the noetic world in relation to matter, he is not merely engaging in an exercise in historical reporting. He is instead operating a recovery of positions that he himself sanctions, although not without critical correction. The Cappadocians are known to hold that matter is limit, and that it exists only as a proviso for sin, the means through which the Atonement of Christ can be made effective. Bouyer agrees with their view that matter exists only in connection with created spirit in order to be itself spiritualized. He explains:

77 *Cosmos*, 104.
78 Ibid., 104–7.
79 Ibid., 104–5. See also *Sophia*, 123–32.
80 *Cosmos*, 106–7.

> Gathering us to his own resurrected being, Jesus thus prepares the spiritualization of the entire universe. This universal liberation will be the extension to all creation of the glory bestowed upon the children of God.[81]

For the Cappadocians, matter safeguards humanity from falling into absolute corruption in the attempt "to be self-sufficient" and in the failure to heed "the inspired *eros* which answers within . . . the call of the divine *agape*."[82] The New Adam reveals the meaning of matter in his reversal of human sin and slavery to the Devil. This brings us to our next topic. Bouyer does not follow the Cappadocians in holding that God creates matter only as a safeguard for the consequences of sin, but he does emphasize that the materiality of the world has meaning and can be fathomed ontologically in the concrete order in which we live only in light of a Trinitarian and Christological understanding of the whole creation centered on the atoning Cross and Resurrection.

ETERNAL HUMANITY

Bouyer thinks that the concept of "eternal humanity," found in Cardinal Pierre de Bérulle (1575–1629) as in Bulgakov, is a development of aspects of this Patristic, sapiential or sophianic Christology.[83] He does not utter the expression "eternal humanity" in *Cosmos*, yet this book is indecipherable without reference to it. The concept is found especially in *The Eternal Son*. Because it is essential to all that we shall explore of Bouyer's thought moving forward in our book, and because it is an essential dimension of his Christology, it is appropriate to introduce the idea at this juncture. The Oratorian explores the idea of eternal humanity in God in the final chapter of his book on Christology in order to help address the central Christological riddle of how we can

81 Ibid., 105.

82 Ibid.

83 *MT*, 216. Bouyer actually roots this concept in the Church Fathers and says of Bérulle: "For the Fathers, and more especially still for a certain number of theologians in the XVIIth century, in particular Cardinal Bérulle, Christ is the eternal man, the celestial man, but he is more than a man since he is God."

understand the universal and cosmic importance of the Incarnation as the singular event in history that makes human salvation possible.[84]

An adequate response to this question requires investigation into the ontology of human nature and development of a proper sense of the capacities of the humanity that Christ has assumed. How could Christ, in his particular humanity, bear us all—and the entire cosmos with us—in his Body? Bouyer argues in *The Eternal Son* that we must affirm, with Leontius of Jerusalem (485–543), that the humanity of Christ is "enhypostatized" in the divine person of the eternal *Logos*. He quotes Leontius:

> The *Logos*, in these last times, having clothed with flesh his hypostasis and his nature, preexisting in relation to his human nature and before the ages being without flesh, has enhypostatized that human nature into his own hypostasis.... Christ does not possess a human hypostasis that like ours is particularized and distinct in relation to all beings of the same species or to different species, but He possesses the hypostasis of the *Logos*, common and inseparable in regard to his human nature and to the divine nature that surpasses it.... That is why Scripture calls the humanity of Christ "flesh," since it is a generic term that implies human nature in its totality, and it is not only an individual in the human race but is all the humanity that has been united to the divinity in Christ.[85]

The ontological subjectivity of Christ is not that of a human hypostasis that renders his humanity intrinsically divided from other humans and the cosmos. His hypostasis is that of the eternal *Logos*, which is common both to God and to humanity. He thus possesses "human nature in its totality," not only a fragmented individuality but the entirety of the species. Bouyer explains that because human nature is enhypostatized in the Son of God, all its potentialities or capacities are able to be unlocked. Christ's personality is, he says, a "common personality... a humanity that

84 *ES*, 393.

85 Ibid., 337. Bouyer quotes Leontius of Jerusalem, *Adversus Nest.* 5.28, *PG* 86:1748d [before the first ellipsis]; 5.29, 86:1749bc [between the first and second ellipses]; 5.29, 86:1749d to 1752a [after the second ellipsis]. See *ES*, 343.

is common to all humanities recapitulated."[86] Christ is one among us, with an individual human soul, but he is nevertheless uniquely capable of encompassing all of us and all of creation in the personal achievement of supernatural actualization that he operates in his life and mission. It is in both his divine and human natures, perfectly united, without confusion, in the one subject of the Son of God, that Christ embraces the entire experience of creation in his body.

How is it possible to attribute to Christ a humanity that is projected into some eternal, celestial plane, without seeing it as having little to do with the concrete, individuated humanity that defines each particular human being? Would this not dehumanize Christ? Bouyer proposes that it is necessary to understand Christ anew as the concrete universal of humanity incarnate in time and space in the womb of Mary from the seed of Abraham.[87] If Christ is fully ensouled with an individual human soul, he is so by virtue of his unique ontological subjectivity, which is capable of actualizing the potentialities that each individual human soul bears within.

There are two suggestions that Bouyer makes, contained in dense and compact passages in *The Eternal Son*, in order to elucidate what he means on this point. The first is that it is a proper characteristic of individual human beings to possess a potential universality. This is what makes human thinking possible. The human soul is in some manner all things (*quodammodo omnia*). If it is proper to us in our individualized humanity to be particularized by our body, which is a limited share of cosmic materiality belonging to and defining us, it should be recognized that there is nevertheless a potential universality even in this. Each of us occupies, Bouyer says, a corporeally-limited "particular situation in time and space."[88] Our spirit is subject to bodily conditioning, a point which modern science, as much as biblical revelation and Thomistic metaphysics, affirms. According to Bouyer, advances in modern physics have also made possible a renewed recovery of ancient microcosmism precisely in its Christian form. Dumitru Staniloae's language of the "macrocosm" or "macro-*anthropos*" in Christ seems to express well the

86 *ES*, 396.
87 Ibid., 398.
88 Ibid., 397.

central idea in this quotation from Bouyer describing what modern physics enables us to see regarding the cosmic potentiality of our nature in its individual instantiations:

> We do not mean by this what many ancient philosophies meant, i.e., that the person is a microcosm of the world or a little universe, but, rather, and much more interestingly perhaps, that our bodies have no limit except those of their own sensations and impressions. And these seem to have no frontiers except those of the universe itself. Contrary to ancient physics, which imagined bodies as being mutually defined by their reciprocal exteriority, modern physics sees them as mutually permeable systems that are defined, not according to an apparent exteriority, but according to a variety of perspectives, a set of coordinates or rather a formula of unique coordination containing the same elements that make up what is common to all that exists—something like the Leibnizian monads which, each in its own way, encompass the whole universe but have form, are a single universe together.[89]

This passage gets at the heart of the metaphysical aspect of Bouyer's way of linking cosmology with Christology. He encourages us to think that modern physical science can help us to contemplate the mystery of human universality in its concrete particularity. The concrete, individual human being is potentially linked to the total universe of being even in his or her bodily existence precisely through the mutual permeability of material beings. Yet it is not just any hypostasis or concrete subjectivity that can actualize or has actualized this potentiality. God alone has done so, in and through His divine, personal intervention in history in the Incarnation of His *Logos*, the absolute concrete universal. We shall further explore the implications of this view in our chapter ten, particularly with respect to the philosophy of materiality in the Cappadocians that Bouyer recovers.

The second possibility our author raises regarding how we might better grasp Christ's concrete universality in his individual humanity concerns

89 Ibid.

linking the cosmic universalization of human nature not with exteriority and quantity but with interiorization, quality, and empathy, an idea which calls to mind the phenomenologies of affect found in the works of Max Scheler (1874–1928) and Edith Stein. Human and cosmic achievement, Bouyer suggests, is ordered by way of affect, consummated in charity, to the common *telos* of creation realized in and through salvation history. He puts forward the idea that human nature is brought to fulfillment socially and historically precisely through the endeavors of concrete individuals, guides of humanity, prophets, priests, sages, saints—and ultimately Christ himself. These "guides" and "accomplishers" bring "the entire race to its final destiny."[90] Indeed, each individual has a singular mission with implications for the "common history" and "entire race" of humankind, but clearly some individuals, remaining steadfast to the call of God, are uniquely significant in this regard.[91] Jesus Christ, in the particular humanity that he has assumed, recapitulates the universal mission of the People of God with respect to the creation in its entirety. Christ as Son of God is the supereminent, transfiguring concrete universal who actualizes the vocation intrinsic and latent to human existence through the mercy and empathic event of his Cross, by which all of history is reintegrated into the eternal plan of divine Wisdom in his suffering, redemptive flesh. The union of cosmic suffering and love are elevated onto a supernatural plane. God's own life, in the humanity of the Son, becomes the subject of the world's becoming in love through suffering, without losing the plenitude of its eternal being.[92]

Christ's humanity, Bouyer argues, though uniquely capable of unlocking the potentiality of human nature, is at the same time uniquely his own. It is in this connection that we can rightly speak of an "eternal humanity" of the Word of God. It does not have to be taken as a threat to our affirmation of the individual humanity of Christ or to the recognition that the Word becomes incarnate and suffers only in time to say

90 Ibid., 398.

91 Ibid.

92 Sergius Bulgakov, *The Lamb of God*, trans. Boris Jakim (Grand Rapids, MI: Eerdmans Press, 2008), 197. The way we have described matters in this sentence follows closely Bulgakov's way of putting it on this page. Just as stated, the thought of the two coincides.

that this humanity is enhypostatized in the Second Person of the Holy Trinity. Bouyer suggests that reactions to the contrary evince a failure to understand what it means to say that we are created in the divine image (Genesis 1:27). In God the Son, the Word of the Father exists eternally as His living image, "the icon of the invisible God" (Col. 1:15). It is according to this image that we have all been created in time. Thus we cannot grasp "the possibilities for our nature, as made so by God, until we consent to allow our history to be fashioned, by the internal force of the Holy Spirit, into definitive form as an encounter with the divine Word, in whom the Son, the Father's unique beloved, comes to us."[93] This entails that in God there exists eternally the model of perfect humanity. If the principle of our existence is divine, there must be in God something that we can call human: there is divine-human correlativity.

Our author takes the idea of eternal humanity in God to be a development of the theology of the Son of Man in the Old Testament and of the Pauline Second and Final Adam in the New Testament. Human being reflects divine being, particularly in the Word or living image of God who possesses a celestial humanity. Just as God is eternally Creator, even though, as far as we are concerned, He creates us in time, so God assumes humanity eternally, even though He does so, as far as we are concerned, from within a particular moment in time. He both creates and saves us from His own being, even though the creative act is "refracted, so to speak, through nothingness."[94] Bouyer argues that the Father eternally generates the Word as Creator and as made flesh. This does not mean that God needs to create or to redeem us in the flesh, but that God's supereminent, generous willingness to do so is necessarily tethered to His Wisdom, which is one with His eternal, necessary essence in perfect freedom.

THE IMMOLATED LAMB

In line to some extent with the kenotic theology of Bulgakov, Bouyer understood the humanity of the Word of God to be constituted by the particularities of the Son's eternal predestination to his mission in time. The Incarnation and the Mystery of the Cross are not divided from one

93 *ES*, 399.
94 Ibid., 401.

another in Bouyer's thought, nor are they considered to be fortuitous additions to the act of creation. Instead, they are understood to manifest and realize God's eternal Wisdom. Already in his thesis on Saint Athanasius and in his early work *The Paschal Mystery*, Bouyer emphasized that the adoration of the Church is directed to the Lamb whose sacrifice on the Cross "was designated in advance before the foundation of the world" (1 Pet. 1:20), and that Christ is "the Lamb which was slain from the foundation of the world" (Rev. 13:8).[95]

The vocation of the Word of God with respect to our salvation and deification was eternal. Bouyer said in his early book *The Paschal Mystery* that the Word of God was eternally seen by the Father, united to his mission to become incarnate and as entering into the Way of the Cross: "in uttering Him (the Word), stupendous as this thought may be, the Father destines Him for us."[96] God sees us eternally in the unity of the Son, as distinct, personal ideas, on whom rests the Gift of the Holy Spirit. He sees the Word and all creation in the unity of divine Wisdom, whose key is the Cross. The Father eternally knows the Word as "uniting Himself to us to the extent of assuming in Himself our suffering and death."[97] It is God's eternal, divine design to perfect in us the living image of His eternal Love. The Word is eternally ordained to unite "us to Himself to the extent of placing in us the reflection of that one and perfect image that He forms and of carrying us on, through the torture of the cross, even to the living union that the Spirit of love seals between Him and the Father, between the Father and Him."[98]

Marie-David Weill has highlighted the controversy that has surrounded the sort of theology of "predestination to the Cross" that Bouyer develops in his work.[99] For one thing, the translations of 1 Peter 1:20 and Revelation 13:8 that he favors are disputable, a fact that Bouyer

95 1 Peter 1:20: "προεγνωσμένου μὲν πρὸ καταβολῆς κόσμου, φανερωθέντος δὲ ἐπ᾽ ἐσχάτου τῶν χρόνων δι᾽ ὑμᾶς." Rev. 13:8: "καὶ προσκυνήσουσιν αὐτὸν πάντες οἱ κατοικοῦντες ἐπὶ τῆς γῆς, οὗ οὐ γέγραπται τὸ ὄνομα αὐτοῦ ἐν τῷ βιβλίῳ τῆς ζωῆς τοῦ ἀρνίου τοῦ ἐσφαγμένου ἀπὸ καταβολῆς κόσμου."

96 Bouyer, *The Paschal Mystery*, 244.

97 Ibid.

98 Ibid., 244–45.

99 Weill, *L'Humanisme Eschatologique de Louis Bouyer*, 69–72; 70, n. 5.

acknowledged in *The Paschal Mystery*.[100] It is interesting to note that the association of the Lamb of God with an immolation as old as creation itself (and even older) is found in both the Douay-Rheims and King James versions of Scripture in their translations of 1 Peter 1:20 and Revelation 13:8. However, many contemporary Scripture scholars prefer to translate the passages differently, some to mean not that the sacrifice of the spotless Lamb was foreordained from all eternity, but that there is a book of life as old as creation itself on whose pages the names of the elect are written.

Although it is not universally acknowledged by Patristics scholars, the idea that the Lamb is predestined to immolation as the very principle of all creation can be understood to fit with attestations from several Church Fathers, including most especially Saint Irenaeus and Saint Maximus the Confessor.[101] Bouyer purported to show that this was the view of Saint Athanasius.[102] We might be inspired by these Patristic writers to raise the question of the "motive of the Incarnation" that was so prominent in the Middle Ages. Would Christ have become incarnate if Adam and Eve had not sinned? This was not in fact a question urgently raised by the Church Fathers themselves, although Bouyer seems to suggest that a proper response to it would nevertheless require a *ressourcement* of the tradition for which the Confessor is the capstone figure.[103] Maximus himself likely did not hold to the Thomist view to which Bouyer subscribes that Christ would not have become incarnate without the sin

100 See *The Paschal Mystery*, 245, n. 5.

101 See especially Paul Blowers, *Maximus the Confessor: Jesus Christ and the Transfiguration of the World* (Oxford University Press, 2016), 102–9. Blowers quotes at length Maximus the Confessor, *Quaestiones ad Thalassium*, 60 (*CCSG* 7:75–77). See Blowers, 105–6. Bouyer draws on this very passage to give his interpretation of the Confessor. Cf. *Cosmos*, 264, n. 23. The opposing view is put forth by Juan-Miguel Garrigues, "Le Dessein d'Adoption du Créator dans son Rapport au Fils d'après S. Maxime le Confesseur," in F. Heinzer and C. Schönborn, eds., *Maximos Confessor. Actes du Symposium sur Maxime le Confesseur, Fribourg, 2–5 sept. 1980* (Freiburg: Éditions du Universitaires, coll. "Paradosis. Études de Littérature et de Théologie Anciennes," n. 27, 1982), 173–92. "Le Groupe Louis Bouyer" addressed this issue throughout their 2017–2018 tri-annual meeting.

102 See *IÉ*, 132–34.

103 *Cosmos*, 189, 264, n. 23.

of the First Adam.[104] In explaining his own adoption of this position, Bouyer insists that we should affirm in the concrete order of creaturely existence God's eternal predestination of the *Logos* at once to Incarnation and to the Cross. God's plan for creation includes his "anticipation" and "prevision" for the revolt of finite freedoms against His will. In line with Balthasar, Bouyer recognizes that in the Passion of Christ, God's revelation of what is most glorious in the intimacy of the divine life is manifested. The Cross does not merely provide an accidental addition to the Son's eternal glorification of the Father but shows forth the deepest reality of divine glory as loving self-gift.[105] Moreover, he wishes to stress that God does not become incarnate only to love Himself through the mediation of an other: He comes entirely "for us," that is, "for our salvation."[106] Yet creation and Incarnation are indeed forever linked in the intimacy of a common purpose: God's eternal will for creation includes his "commitment to take upon himself the consequences of the failure and make it reparable."[107] The Incarnation is inseparable from the Cross. Balthasar showed that the orientation of the Incarnation to the Cross is a consistent theme for the Church Fathers in both the East and the West. Twentieth-century "incarnationalists" who argued that the Christian East considered the redemption to have been achieved irrespective of the Cross were deceived. In his own book on the Paschal Mystery, Balthasar provides a bevy of citations from Eastern authors which, he says, "give the lie to the modern myth . . . that Christianity is above all an 'incarnationalism': a taking root in the (profane) world, and not a dying to the world."[108]

104 See Blowers, *Maximus the Confessor: Jesus Christ and the Transfiguration of the World*, 106–7. Blowers reports that according to "a significant segment of later scholarship" the Confessor holds that Christ would have become incarnate even if humanity had not lapsed.

105 Cf. Hans Urs von Balthasar, *The Paschal Mystery*, trans. Aidan Nichols (San Francisco: Ignatius Press, 2005), 11.

106 *ES*, 365.

107 Ibid., 411.

108 Balthasar, *The Paschal Mystery*, 22. This book was originally published in German in 1970 with the title *Theologie der Drei Tage*. This was long after Bouyer's *The Paschal Mystery*. Balthasar famously dwells at length in this book on the mystery of Holy Saturday and the harrowing of hell, a theme Bouyer does not emphasize.

Creation is always "for us." In the concrete order of God's eternal "prevision" for creation, for which there is in fact neither before nor after but only an eternal present, the Son is eternally "predestined" to redeem fallen freedoms through his Cross and Resurrection by the power of the Holy Spirit.[109] Given that Christ the Final Adam suffered and died for us on the Cross, and given that the First Adam in fact did sin in the historical actuality of his existence, it is the case that the Cross and Incarnation have to be understood in the unity of the divine Wisdom as the principle of creation. In the sense of Aquinas's notion of the eternal presence of all things to God in their actuality, Bouyer can be said to agree with the Confessor (however differently he conceptualizes the point) that the divine sacrifice of Christ underlies the structure of the cosmos:

> This is the great hidden mystery, at once the blessed end for which all things are ordained. It is the divine purpose conceived before the beginning of created beings. In defining it we would say that this mystery is the preconceived goal for which everything exists.[110]

Some of Bouyer's Neothomist critics have wrongly taken him to mean that God foresees evil and orders the structure of the cosmos in accordance with evil, using it to give structure or form to His good purpose. This would—these critics maintain—undermine God's perfect innocence, which cannot foresee evil. They have sought to provide an alternative reading of the Confessor, diminishing the interpretation according to which he saw the Immolated Lamb as the principle of creation, but this reading has been disputed by many other scholars of Maximus.

It does not seem to us that it has been sufficiently noticed that Bouyer follows along the lines of Aquinas's thought rather directly on two fronts, as we have more or less just implied. First, and this much has been at

109 Cf. *Dictionary of Theology*, 143. In the entry on "eternity" Bouyer explains that eternity "is a changeless present."

110 Taken from Maximus the Confessor, *Quaestiones ad Thalassium*, 60 (*CCSG* 7:75–77). Quoted by Blowers, *Maximus the Confessor: Jesus Christ and the Transfiguration of the World*, 105–6. Again, Bouyer references this quote on p. 264 of *Cosmos*, n. 23.

least acknowledged, he holds with Aquinas that time is nothing other than the measure of created life.[111] Second, and following from this first point, he seems to hold the view, even if he does not directly put it this way, that all things in creation are present to God in their "presentiality."[112] We think that it is in this sense that Bouyer means that "God sees all the unfolding of time in the perpetual present of His eternity."[113] For Aquinas, this means that God always knows future free events, although he does not know them as future. He knows all things as present to Him in the totality of their historical existence. God knows the real existence of what we could only know to be futuribles or future contingents, what are from our perspective mere possibilities. He does not foresee non-existent evil and then make His plans accordingly. From God's transcendently elevated point of view, nothing is a future contingent.[114] He is eternal, and His relation to creation is by virtue of His eternality. He does not foreknow sin and evil as future occurrences, but He does always know them as the free actions of really existing created persons whose decisions He sees from the standpoint of His eternal present. He eternally knows, or sees in simultaneity, what are for us momentary occurrences along the arrow of time, as it flies from past, to present, to future. This knowledge includes His own freely-willed response to evil and sin, which culminates in the Incarnation and Paschal Mystery. God's response to evil and sin is not by way of a before-and-after decision on His part but is one with His eternal being and includes all of the individual responses that we distinguish from our end, including the ultimate response that He gives to us on our

111 Aquinas, *ST* I.10.1, ad 4.

112 Aquinas, *ST* I.14.13, body: "Hence, all things that are in time are present to God from eternity, not only because He has the types of things present within Him, as some say; but because His glance is carried from eternity over all things as they are in their presentiality. Hence it is manifest that contingent things are infallibly known by God, inasmuch as they are subject to the divine sight in their presentiality; yet they are future contingent things in relation to their own causes."

113 *The Paschal Mystery*, 244.

114 See Katherine Rogers, "Omniscience, Eternity, and Freedom," in *International Philosophical Quarterly* 36 (1996): 399–412; Brian J. Shanley, "Divine Causation and Human Freedom in Aquinas," *American Catholic Philosophical Quarterly* 72 (1998): 99–122; William Norris Clarke, *The One and the Many: A Contemporary Thomist Metaphysics* (South Bend, IN: University of Notre Dame Press, 2001), 240–43.

behalf in the liberality of His mercy by sending His Son to suffer and die for our salvation. This bespeaks a unity of action or response on God's part to all that we have done or will do. Given the unity of God's total vision of creation in its existentiality, cosmic being is inseparable from the Incarnation, and the Incarnation is inseparable from the Redemption. For Bouyer, the Cross and the Immolated Lamb can be seen as the principle of creation, if creation is understood, as Aquinas understood it, in the totality of its presentiality to God. The Oratorian develops the view that creation and Redemption should be taken in their total existential unity-in-distinction, with our thought of predestination centering not on individuals seen in isolation from one another but on the predestination of God's Son in his eternal humanity and historical mission. This centering act of predestination is not with respect to an evil that has not yet occurred. The persistent concreteness of Bouyer's thought indicates that he would not fall into the Molinist confusion according to which a purely ideal being, not yet really existing, can carry out a free action.[115]

It follows that the cosmos can be received anew in its living, organic unity in light of the celestial humanity of Christ, the instrument of his salvific work, the means by which he gives himself to us as Immolated Lamb, "slain from the foundation of the world." The humanity of Christ, as the fullness of humanity, is not a mere projection of our own being onto God. Bouyer shifts the focus of attention in this relation: we are a created projection from the heart of divine Wisdom. There is an effective demythologization operative here as the directionality of the mythic image is reversed. The intuition rife in mythopoetic thought that humanity is a living image of the divine is not thereby done away with but is transfigured. In light of this new, surpassing understanding of the divine image in the human being, the world can be contemplated as created cosmos, as unified and living, irreducible to the status of an object and suffused with meaning through God's ultimate communication to us of His perfect *agape* in the event of the Cross. This Christology, though not directly articulated in the text of *Cosmos*, underlies the whole of this work as all of Bouyer's developed theological, liturgical, and spiritual publications, which already began to appear, in developed form, in the 1940s.

115 See *Dictionary of Theology*, 359–62. Here Bouyer discusses predestination and rejects Molina in favor of Saint Augustine and Aquinas.

CHAPTER 6

The Loss of the Theoanthropocosmic Synthesis

THE CHURCH FATHERS, SUMMING UP BIBLICAL revelation in the light of Jesus Christ's transfiguration of the apocalyptic motif of the Son of Man, recognized the profound communion between God, the cosmos, and humanity, united without confusion in the wedding of divine and human natures in the person of the Incarnate Word. The eternal icon or image of the Father assumed flesh through the womb of the purified created image of God, the human person Mary. She embodied human personhood brought to its historical maturation and elevated to its highest plateau through the ever-pervasive incubation of the Holy Spirit in the path of salvation history.[1] This integrated, theoanthropocosmic Christian vision, gathering together the data of revelation around the theology of God's eternal icon and his Mother, did not withstand the advance of scientific modes of thought, which began to emerge already in the Middle Ages. The apocalyptic Wisdom of the tradition, with its ever-present cosmological dimension, uniting God, the human being, and the world, increasingly dissipated. How did this come to pass? This question is the central concern for Bouyer in chapters twelve through seventeen of *Cosmos*, and this is what we shall focus on in this chapter and the following, which will bring to an end the genealogical portion of our study.

The question of what it means to be a human person is inseparable from that of the meaning of the cosmos. Bouyer grasped the inseparability of these levels of meaning, which have recently entered together as a common concern in the theological reflection of the papal magisterium of the Catholic Church, as demonstrated by the developing motif in the Church's social doctrine of the link between environmental and human

1 This is the central them of *MSW*. It will be discussed in depth in our chapter 11.

ecology.[2] For Bouyer, the question of the loss of this unified perspective should cause us to question our very modes of life and manners of theologizing. Bouyer provides a genealogy of the progression of cosmological fragmentation, which we shall make the focus of the present chapter, in two preliminary sections comparing his assessment of Patristic theology, where the theoanthropocosmic synthesis is prevalent, with Scholastic theology, where it began to shatter when Nominalism took hold. Then we shall explore his assessment of modern theology, which revolves on the themes of competing humanisms and idolatry, the latter which we take to be the culprit he most identifies as the source of the modern erasure of the cosmos. Bouyer holds that a false human image or idolatrous glorification of the self, sensuous and self-seeking, was projected onto the world and onto God. Lastly, we shall suggest that our author recommends a turn to an iconic mode of perception, a return to iconography, ascesis, and the theology of the true image of God, in order to heal the breach between God, humanity, and the world. The chapters of *Cosmos* that are our focus here, as ever placed in the context of his larger work, are twelve, thirteen, and sixteen, which contain the dystopian portion of his assessment of the modern age. There is, however, another, brighter side of the story as he tells it, which will be our concern in the next chapter.

THE MYSTICAL THEOLOGY OF THE CHURCH FATHERS

The theological approach of the Church Fathers, particularly from the East, is taken as paradigmatic by our author, because he finds their writings to be exemplarily suffused with a mystical, sacramental, or Eucharistic inspiration that he thinks was threatened when the modern West made the scientific ideal normative for theology. Bouyer counsels a return to the integrated, Eucharistic vision of the Patristic writers, not in order simply to repeat their formulations blindly but to capture anew the spiritual heart and pastoral end of theological reflection.[3]

What characterizes a properly mystical or Eucharistic approach to theology? Certainly, it requires immersion in a life of prayer, "doing theology

2 See especially Pope Francis, *Laudato Si'*, chapter 4.

3 Cf. *ES*, 296–99. One should mention how deeply impacted Bouyer was by reading the seminal work of Dom Odo Casel on mystery theology and liturgy early in his career.

on one's knees," as Balthasar recommended in an essay on theology and holiness that was much admired by Bouyer.[4] This has become a famous expression, frequently used by Pope Francis. Following the etymology of the term "Eucharist" (*εὐχαριστία*), which translates the Hebrew *berakah*, Bouyer explains that a theology that is Eucharistic would be explicitly one of "thanksgiving" in response to the wonderful deeds and blessings of God on our behalf.[5] This is how theology was understood and practiced in the Patristic Age. Bouyer always insisted that revelation is given to us for our salvation and not to satisfy our curiosity, a point that he thought theologians are too often prone to forget.[6] He offers the theology of the Church Fathers as a remedy to *curiositas*, because he sees their brilliant, speculative work as uniquely inscribed within their personal responses to the salvific gifts of the Incarnate Word, who reveals himself to humanity in plenitude in the *agape* of his own eternal, Eucharistic response to the love of the Father in the Spirit.[7] Theology for the Church Fathers had the character of chant, of hymn, of lyrical praise to the Creator. This had its own proper rigor, tending always toward doxology. Their theology culminated in, as Bouyer says, "the express glorification of God in all His works as in all that He is."[8]

We concurrently employ the words "mystical" and "Eucharistic" for etymological reasons. Bouyer explains in several writings that, as used in modern theology and spirituality, the word "mystical" derives ultimately from the Pauline epistles, such as First Corinthians, chapter two, which speaks of the "Mystery" (*μυστήριον*) of Christ and his Cross.[9] This Mys-

4 See *MT*, 147. The essay from which this expression is taken was entitled "Theology and Sanctity." See Hans Urs von Balthasar, *Explorations in Theology*, vol. 1, trans. A.V. Littledale with Alexander Dru (San Francisco: Ignatius Press), 181–213.

5 See especially *Eucharist*, 16.

6 Ibid.

7 Ibid. Note that on this page Bouyer singles out Aquinas for his wedding the task of theology to the mission of the Church to lead souls to salvation.

8 *Sophia*, 196. See *MT*, 123–24. See also "La Situation de la Théologie," in *Communio* 1.1 (1975): 41–48. In this article, Bouyer explores the negative consequences that ensue when theology is too closely assimilated to the modern ideal of scientificity.

9 Cf. *MT*, 121–50; *Mystery*, 1–19. In this section, we shall follow especially these pages from *MT*.

tery is, as we have seen, the inner secret of the eternal Wisdom of God, where the love of God is self-revealed in fullness. Mystery is the nucleus of God's loving design over the whole of history and creation. It is the secret hidden "for ages and generations" (Col. 1:26) that God discloses when and to whom He chooses. In the Cross of Christ, God has willed in His perfect Wisdom to dispossess the rebellious powers of the world of their dominion over it.[10] God brings salvation and liberation in and through His Mystery, which is not reducible to a secret that we cannot and should not dare to understand. It is the content of God's revelation of His eternal Wisdom to humankind. The Mystery of Christ is the surpassing consummation of Old Testament Wisdom and apocalypse, transforming us in the Spirit of charity, "replacing," as Bouyer once put it, "the stony heart of the old Adamic humanity with a heart of flesh, a heart beating with the very breath of God's life."[11]

Against a common argument maintained in the early twentieth century by both liberal Protestant and Catholic modernist scholars, Bouyer demonstrated in his writings the specifically Christian character of "Mystery" and of the "mystical theology" that follows from it.[12] In the so-called "mystery religions" of the first centuries of the Church, mystery was understood to be a rite to which select initiates or "mystics" had access.[13] In the Christian understanding alone, Bouyer maintains, could mystery, the Mystery of Christ, be communicated and assimilated in experience. Mystical experience is thus proper to the beloved of Christ joined to him in his Eucharistic Body, and mystical theology, like the Eucharist itself, is ordered toward the ultimate end of human being, toward divine adoption in Christ or divinization.

The Alexandrian Church Fathers, according to this genealogy, held that Mystery is the key which enables one to interpret Sacred Scripture in light of its deepest meaning.[14] The language of mystery was later extended to catecheses of initiation that were composed in the fourth century, after which point the sacraments of Christian initiation, baptism, confirmation,

10 *MT*, 123.
11 *IF*, 156. See Ezekiel 36:26.
12 This is the essential argument found in *Mystery*.
13 See *MT*, 122; *Mystery*, 72–74.
14 *MT*, 123.

and Eucharist, were described as "mysteries."[15] As is known, the Latin *sacramentum* translates the Greek *mysterion*. Bouyer explains that faith discovers in the mysteries or sacraments "precisely the reality which is hidden there and which is nothing other than Christ himself."[16] Especially beginning with Saint Gregory of Nyssa, the terms mystery or mystical began to be applied to the individual believer who is assimilated in faith to the realities that are given through the sacraments.[17] Denys the Areopagite, in the sixth century, was the first to use the expression "mystical theology," which meant for him the gnosis of God given through Christ in Scripture and liturgy which brings union with God.[18] Eucharistic or mystical theology is not, then, just a form of contemplation that fills us with wonder, although that is certainly part of it. It is tied to our very spiritual progress in the life of the Holy Spirit.

Sometimes the expression mystical is construed to be so resolutely apophatic in meaning that we place overly-strict limits on what we can say about God and God's relation to the cosmos. Bouyer does not think of mystical theology in this way. He insists that if the Mystery of Christ and his Cross surpasses all that humans can reason about or imagine in their finitude, it is nevertheless true that, once given, the Mystery revealed is not enshrouded in ineffable obscurity. Instead, it is, he says, "an ocean of light that one cannot exhaust."[19] Bouyer does not propose that apophaticism should reduce one to a condition of muteness in the face of Mystery. He affirms in fact the need for theology to move between affirmations and negations, between the apophatic and kataphatic, in the sort of unending "heliocoidal" spiral of mystical discourse that is always a going-beyond, as is operative in the writings of Denys.[20]

The validity of the speculative and scientific dimension of theology is affirmed in this understanding, so long as speculation does not become unmoored from the spiritual and sacramental concreteness of embodied Christian experience. The Church Fathers are exemplary theologians

15 Ibid.
16 Ibid., 124.
17 Ibid.
18 Ibid.
19 Ibid., 125–26.
20 *IF*, 218.

for Bouyer because they manage to keep their speculations rooted in their lives as pastors, spiritual directors, and teachers of the faith. He admired the unified vision of faith that shines through in their writings and that flowed from the unity of their very lives. They neither split theology into different specializations nor sundered theology from its rooting in the liturgical assembly and in its concomitant spirituality. Their mystical or Eucharistic theology received the Mystery of Christ in its abiding unity. Modern theology, by contrast, as he understands it, tends to communicate a divided sense or interpretation of faith, having split the one Mystery of Christ into pieces detached from each other, and this reflects a certain fragmentation in the modern personality. Bouyer's writings, perhaps especially his cosmology, mimic the unified, narrative, "theodramatic" approach to theology of those Church Fathers who see "the past and future of sacred history—of *all* the history of creation—collapsed into a perfect singularity of purpose"[21] in Christ.

An important dimension of the mystical theology that Bouyer recovers has to do with the abiding place in it of the symbols through which the faith is received.[22] Like his friend Henri de Lubac, Bouyer commends the symbolic character of Patristic theology for which the concept is not elevated to a higher level of importance than the symbol and, what is more, is understood to be itself a kind of symbol. The breaking-apart of symbol and concept is for Bouyer, as we have seen, at the root of Bultmann's project of demythologization.[23] He holds that the Church Fathers operate spontaneously in accordance with an intrinsic understanding of the irreducible importance of the symbol that the best modern philosophers have only recently recovered. Bouyer stresses that the symbol is basic to humanity's fundamental experience in the world and is essential to human reception of both the natural sacred and the supernatural sacred made present in word and rite.[24]

21 Blowers, *Maximus the Confessor*, 103. See Weill, *L'Humanisme Eschatologique de Louis Bouyer*, 389–99. Weill discusses here reaction to Bouyer's *The Meaning of the Monastic Life*. Bouyer was criticized for this narrative methodology. It was deemed to be atheological and insufficiently scientific. We shall return to this criticism and address it, in light of Weill's research, in chapter 9.

22 *MT*, 110–11.

23 Ibid., 112–13.

24 Ibid., 110. On this page Bouyer and Georges Daix, the interviewer, discuss

Revelation has an inherently symbolic, sacramental structure of manifestation or presentation.

The refined symbolic intelligence of these first theologians of the Church enabled them to avoid falling into naïve literalism in interpreting Scripture. For Bouyer, recognition of this is one of the keys to grasping the difference between modern and premodern thought. He holds that Aquinas systematized the symbolic, sacramental cosmovision of the Patristic writers with his articulation of the theology of the *analogia entis*.[25] He argues that the Church Fathers as well as Aquinas and his most estimable Scholastic peers took the literal meaning of the biblical text to be decisive, but that they did not understand it in the reductive manner that modern scholars tend to do, for they could see that it harmonizes with other levels of interpretation. How did this Eucharistic, Mystery-centered, richly symbolic mode of theologizing disappear? This question is obviously significant for cosmology. Did the general Scholastic approach to theology lose sight of the unity of the divine Mystery and contribute to the disappearance of what we have been calling the theoanthropocosmic synthesis of Scripture and the Fathers? Despite Bouyer's endorsement of Aquinas, chapters twelve and thirteen of *Cosmos* lead us to wonder about the overall merits of the scientificity of Scholastic theology as a tool to communicate the Mystery of faith in its unity.

SCIENTIFIC THEOLOGY IN THE SCHOLASTIC AGE

In order to understand Bouyer's assessment of Scholasticism or the theology of the universities in the Middle Ages and early modern period, it is important to reiterate and highlight from the start the importance that he accords to the work of Thomas Aquinas. Grintchenko goes so far as to refer to Bouyer as a "faithful disciple of Saint Thomas."[26] Bouyer does make clear by the tenor of his entries in his *Dictionary of Theology* (1963) that the Angelic Doctor holds central authority in the Catholic

Hugo Rahner's seminal work on Greek myths and the Christian Mystery: *Greek Myths and Christian Mystery* (already cited). Rahner, the brother of the more famous Jesuit theologian, Karl Rahner, demonstrates in this book the use made by the Church Fathers of the mythic symbols of antiquity to express the mystery of the Church. Bouyer thinks this book is of great importance.

25 *MT*, 111, 127–29.

26 Grintchenko, *Une Approche Théologique du Monde*, 265.

theological tradition.[27] His history of encounter with the thought of Aquinas goes back to his days as a Protestant ministry student in Paris, where he was given a positive introduction to his work by the Calvinist professor Auguste Lecerf, and, as we noted in the first chapter, where he attended free lectures given on Aquinas by Étienne Gilson, from whom he was especially grateful to learn Aquinas's unique teachings on the form/matter composite unity of the human person.[28] It should not be forgotten that he was deeply shaped, after his conversion to Catholicism, by the teachings of Guy de Broglie, SJ (1889–1983) in his time as a student at the Institut Catholique of Paris.[29] De Broglie is noted for having argued in a manner that was remarkable for his time that the views of Augustine and Aquinas were in many ways compatible.[30] Bouyer's metaphysics remained so fundamentally in line with Aquinas throughout his life that it has led to annoyance in some writers, especially in that he did not follow Balthasar all the way in the latter's kenotic theology of the inner life of God.[31]

Nevertheless, Bouyer exhibits ambivalence toward the Scholastic, scientific method of expression embraced by the Angelic Doctor. He renews at points in his writings, including *Cosmos*, the critique that twelfth-century monastic theologians Saint Bernard of Clairvaux (1090–1153) and William of Saint Thierry (1085–1148) leveled at the growing Scholastic rationalism in the Church that found its apotheosis in the

27 *DT*, ix. Bouyer says of his approach in the *Dictionary*: "As brief as the articles are, we have always endeavored to give the biblical texts essential to them, together with a minimum of necessary comment, and the chief texts of the Church magisterium. Otherwise, we have limited our references to St. Thomas Aquinas, the 'Doctor Angelicus' *par excellence*. . . . Other scholars' work has been included only when a particular doctrine has been developed apart from the work of St. Thomas."

28 See *MT*, 27; *Memoirs*, 62–63, 65.

29 *Memoirs*, 113, 139–40, 142, 147. On p. 139 Bouyer says of de Broglie: "No man, I can say, will have been so effective a master for me." Broglie wrote the introduction to Bouyer's *The Spirit and Forms of Protestantism*, trans. A.V. Littledale (Princeton, NJ: Scepter Publishers, 2001).

30 *MT*, 36. Here Bouyer says of de Broglie: "He taught a systematic theology very Thomist, but of a Thomism clarified by all the Augustinian tradition, by that of the Greek Patristics, and above all, by the biblical sources."

31 See especially Antoine Birot, "Bouyer, Entre Thomas et Balthasar," in *Laval Théologique et Philosophique* 67.3 (2011): 501–29.

writings of Peter Abelard (1079–1142).[32] In chapter twelve of *Cosmos*, entitled "From the Ambiguities of Scientific Theology to Modern Science," his preference for this monastic critique is strongly implied. He gives in this chapter his most detailed assessment of Scholasticism in an exposition that bears directly on his take on subsequent developments in cosmology.

Professor Grintchenko describes this chapter, our main concern in this section, as "very thickly-wooded" (*très touffou*).[33] In her own discussion of it, she organizes its material around three motifs that we shall follow in attempting to get at Bouyer's take on this approach to theology that came to dominate in the early modern universities: 1) the originality of the three greatest Scholastic syntheses of the thirteenth century; 2) the passage from theological science to modern science; 3) the special balance that Bouyer finds in the theological work of Aquinas.

First, then, Bouyer singles out three figures whom he takes to be the greatest theologians of their day and the most eminent representatives of Scholasticism in general: Saint Albert the Great (1200–1280), Saint Bonaventure (1221–1274), and Aquinas (1225–1274). He begins by asserting that their work cannot be understood if one separates out their philosophy from its theological context or if one reads them simply as Aristotelians without further philosophical residue. The writings of these theologians, he insists, like that of the Church Fathers, do not represent strict allegiance to a single philosopher or philosophical school of thought. Each of these Scholastic masters was, he holds, fundamentally a biblical thinker who transformed the philosophical resources of Platonism, Aristotelianism, and Stoicism.[34]

Bouyer suggests that these theologians transcended their Scholastic predecessors and successors so greatly that it might be wondered if they themselves can be rightly construed as Scholastic theologians.

32 *Cistercian Heritage*, xi-xvi. Bouyer took the Cistercian movement of the twelfth century to be a model for renewal of the Church: "In France, perhaps more than elsewhere, the Church at the opening of the second half of the twentieth century is making a similar attempt to recover the sources and achieve a creative renewal. The sources just enumerated in connections with the Cistercians are the very ones which so strongly hold the attention of contemporary Catholics: the Liturgy, Holy Scripture, and the writings of the Fathers" (xii).

33 Grintchenko, *Une Approche Théologique du Monde*, 142–48.

34 *Cosmos*, 108–10.

Contemplation and the sapiential dimension of Christian experience provided the framework of their thought, as was true of the Church Fathers. Yet, he holds, the capture of the sapiential and mystical framework of theology by the emergent tide of scientificity was so powerful in its advance and appealing to the mindset of their contemporaries that their work could not defeat the ever-expanding rationalism of the age.[35] Surprisingly, Bouyer sees the work of Saint Anselm of Canterbury (1033–1109) as a precursor of this rationalist trend. The Oratorian finds his *Monologion* to be not so much an exercise in faith seeking understanding as the exemplification of a use of "reason seeking to be fiercely independent."[36] This statement should be understood in light of the Oratorian's larger complaint against overly-dolorous theories of sacrifice that emerged in the modern Church in the line of the Satisfaction Theory of Atonement. Bouyer accused this trend of thought of thinking of God as a strict accountant who spends His eternity keeping a tight balance between the sins of each of His creatures and the measure of divine reprisal.[37]

Saint Anselm should be read more sympathetically and capaciously than Bouyer does, but the Oratorian's interpretation of the Archbishop was not out of the ordinary in his day or in our own. As an example, Pope Emeritus Benedict XVI published a startling interview in 2016, granted to Jacques Servais, SJ, in which he recounted in an approving way the shift in twentieth-century theology that turned away from Saint Anselm's soteriology.[38] Many theologians, reflecting a common impulse in the Church, no longer asked, as Saint Anselm did, how humanity can be justified before God but instead wondered how God can be justified before humanity given how much evil and violence there is in

35 It is thus that Bouyer joins the Cistercian critics of Scholasticism. He dedicated a whole book to the spirituality of the Cistercians: *The Cistercian Heritage* (1955, 1958). For a more balanced take, see Pope Benedict XVI, "Monastic and Scholastic Theology," General Audience, Saint Peter's Square, October 28, 2009.

36 *Cosmos*, 110.

37 *ES*, 346–48; *Cosmos*, 110.

38 Daniel Libanori, *Through Faith: Doctrine of Justification and Experience of God in the Preaching of the Church and the Spiritual Exercises* (Cinisello Balsamo: Edizioni San Paolo, 2016). The interview was in this book. The full interview was published online: https://aleteia.org/2016/03/17/benedicts-interview-speaks-to-our-times-full-text/.

the world, especially as unleashed by the twentieth-century gulags and holocausts. How could an omnipotent and good God allow the existence of evil on such a massive scale? Bouyer does not go quite this far, but he nevertheless can be numbered among those theologians who stressed anew the primacy of divine mercy in consideration of God's redemptive generosity, and this is reflected in his critique of the Satisfaction Theory and the rationalistic accounting of sin and justice that it supposedly encourages. Bouyer sides with those theologians who emphasize the paradox and measurelessness of divine grace. As we noted in our first chapter, Bouyer's emphasis on the saturating love and mercy of God helps to account for the glowing reception that some of his first works received in the 1940s.[39]

The underlying issue in Bouyer's critique of rationalism in theology—whether as found in Abelard and later Nominalists, or, much more controversially, in the writings of Saint Anselm—concerns the unmooring of theology from the concrete reality of God given to the believer in contemplation and prayer. Jean-Luc Marion has noted that as Bouyer's work developed he became increasingly dissatisfied with Neoscholasticism, particularly the conceptualism of Francisco Suárez (1548–1617).[40] He increasingly came to hold that the attempt to turn theology into a science, even in the Aristotelian sense of the word, was fraught with ambiguities. One will thus not be prudently advised to turn to his writings in order to find encouragement to return to sixteenth- and seventeenth-century Neoscholasticism as the basis for the renewal of Catholic theology, as some are doing nowadays in the United States. His assessment of theology in this time period can be rather severe. It is succinctly stated in chapter fourteen of *Cosmos*, where in fact Suárez emerges as the best of a bunch of rationalist thinkers who failed to harmonize adequately philosophy with theology:

39 See again Zordan, *Connaissance et Mystère*, 177–82.

40 Jean-Luc Marion, "L'Unité Organique d'une Oeuvre," in *Le Théologie de Louis Bouyer*, 9–19, 15. Marion says: "There also weighed a profound mistrust toward the conceptual and particularly philosophical approach that seizes medieval thought as early the fourteenth century, which succeeds from Suárez to metaphysics and the Enlightenment, which he [Bouyer] treats, with time, with a growing scorn."

> There was no lack of Christian philosophers during the late sixteenth and early seventeenth Christian centuries. The first neo-Thomists, such as Cajetan, the Salamanca Carmelites, or John of St. Thomas were interested in philosophy at least as much as theology itself. Unfortunately, none of them did much more than extract their master's philosophy from its theological context, apparently without wondering whether the two elements were separable, or whether the philosophy they sought to define was not distorted in the process. . . . The only thinker of that period who had more ambitious aims was undoubtedly Suárez. Though exceptionally intelligent in the strictest meaning of the term, Suárez sought to reconcile the views of contemporary as well as earlier thinkers through quasi-diplomatic compromise, instead of developing a true synthesis. The result was an extraordinarily flexible philosophy, but Janus-like, affirming with one mouth what he is simultaneously denying with the other.[41]

The Neoscholastic theologians have a reputation for being the guardians of the scientific construal of the nature of theology. In dismissing their attempted syntheses as Bouyer does, one might conclude that he dismisses thereby the scientific character of theology *in toto*, as if theology must in no way bear a scientific stamp, or one might even go so far as to infer that he has it in for science in general. This is not at all the case, as the second main point of chapter twelve of *Cosmos* makes clear. Bouyer in fact defends the triad of thirteenth-century Scholastic masters mentioned above and their cosmologies as being the source of modern science, which, he affirms, has indeed borne many good fruits and should be encouraged in its development. The key distinction here is between Scholasticism and Neoscholasticism. The latter represents for Bouyer a kind of corruption, filtered through Nominalism.[42] He points to the works of Alfred North Whitehead, R.G. Collingwood, Pierre Duhem (1861–1916), and especially Stanley Jaki (1924–2009), who was a friend of his, as having shown that modern science is rooted in the

41 *Cosmos*, 131–32.
42 Cf. *IF*, 277–91.

best Scholastic theology.[43] He follows the demonstrations of the latter two which showed that the clarification of the doctrine of creation and of the God-world distinction in the Scholastic period opened the path for the development of the intellectual presuppositions needed in order to carry out investigation of the cosmos in the way that modern science has done. This is not only an historical point but an apologetical one, which indicates that he hardly thinks that science should be condemned. These two philosophers of science argued that only when it was collectively discerned by the European mind that God, as transcendent Creator, instills His creation with both rationality and contingency were the intellectual conditions of possibility for the development of modern science established. Only then, as Bouyer explains in *Cosmos*, could a science of logical analysis rooted in experimentation, focused on the object of investigation, emerge.[44] By contrast, these philosophers and historians showed, Aristotelian science had been mired in metaphysical presuppositions that did not allow the contingency of creation to be grasped. Bouyer especially champions Albert, Bonaventure, and Aquinas for having cleared the conceptual field with their science of divine things based on biblical revelation: "The three major scholastic thinkers were thus the first to show convincingly that the cosmic reality depends for its necessary principle on a personal God who remains absolutely transcendent and who therefore, in his activity *ad extra*, His creative activity, can combine contingency and rationality."[45]

Scholastic science thus provides the only coherent intellectual framework for modern physical, empirical science. Neither philosophical materialism nor philosophical idealism can credibly ground the sciences in philosophical reason. In order to undertake rational, cosmological

43 See Alfred North Whitehead, *Science and the Modern World* (Cambridge, 1920); Pierre Duhem, *Le Système du Monde Histoire des Doctrines de Platon à Copernic*, 10 vols., 1914–1959; Stanley Jaki, *The Road to Science and the Ways to God: The Gifford Lectures of 1975–1976* (Chicago: University of Chicago Press, 1978); Jaki, *The Origin of Science and the Science of its Origins* (South Bend, IN: Gateway Editions, 1978); R.G. Collingwood, *The Idea of Nature*. Bouyer and Jaki were nevertheless not in agreement on the relative merits of scientific versus mystical theology.

44 *Cosmos*, 111.

45 Ibid.

investigation, the outcome of which is uncertain, as modern scientists do, it must be presumed that the human mind is adequate to the realities under scrutiny. It must be, as Bouyer says, "capable of conceiving a structure both dynamic and precise, and then, as with Planck or Einstein, the possibility of linking this structure to a formula experimentally discovered, the formula for something fundamentally contingent."[46] This implies, contra philosophical materialism, the assumption "that the world is in effect the thought, fulfilled without impediment, of a mind that is both supremely free and totally rational."[47] Scientific reason cannot be based on a wholly subjective mode of thought, as philosophical idealism claims, because, quite simply, it can "predict, and thus control in accordance with our expectations, the operation of the universe."[48] Bouyer argues that prediction and control does not stem from a pure imposition of human subjectivity onto to the processes of nature but is possible only by first receiving the world in its givenness through our sensory knowledge.

If modern humanity can, as Bouyer says, "control in accordance with our expectations" the universe in and through science and technology, this new power is both a blessing and a curse. One cannot doubt that scientific technology has endowed us with many marvels, and that the underlying sense of the cosmos presupposed in its endeavor can inspire the sort of contemplative wonder that Bouyer himself experienced as a child in his first encounters with scientific theories.[49] Yet there is another side to the story, from which our author does not shy away, either in *Cosmos* or throughout his wider writings.[50] Bouyer chides the story told

46 Ibid., 112.

47 Ibid.

48 Ibid., 113.

49 *Memoirs*, 43–44.

50 See especially his apocalyptic "grand roman," *Prélude à l'Apocalypse*, 401–3. Scientific technology is linked to the machinations of the Antichrist. The anti-Christic figure, Orlando Brightman, is a "majestic little old man, but dripping with benignity." He is concerned with the good of everyone and promotes universal peace and prosperity through the blessings brought about by "a judicious conjunction of the most modern science and the most traditional wisdom." He imposes a factitious universal religion and ecumenism that fools the Church, which has been weakened by accommodation to the world. The link between technology and magic is a motif in this work. See Weill's discussion, *L'Humanisme Eschatologique de Louis Bouyer*, 279–87.

by modern epistemologists of the operations of the human mind that brings science into a condition of technological captivity. To what extent does Scholasticism bear blame for originating this other story? If it can be credited with having opened the pathway to modern science, with all its blessings, is it also responsible for the technocratic ideology that has unleashed many curses?

The answer, as far as Bouyer is concerned, is little in doubt: it was precisely *Nominalist* Scholasticism, in the Franciscan School, that developed especially at the end of the thirteenth century and early in the fourteenth century, which is to blame.[51] As with many Thomists, so with Bouyer, William of Occam (1285–1347) is called to the floor as the main intellectual culprit pushing the new and disastrous course of events.[52] Bouyer even questions the Franciscan School of Impetus Mechanics, inasmuch as Nicholas Oresme (1320–1382), the founder of this school, and his disciples, began to move scientific research in the direction of applied mathematics under the presumption that mathematical relations in the physical world are all-determining. Bouyer repudiates Occam's rejection of the objectivity of divine ideas in purported defense of the sovereign freedom of God, for this rejection has led to an insuperable intellectual quandary. In the wake of Occam, the formal or ideal element of reality could no longer be found anywhere else but within the human mind. The consequences have been catastrophic:

> The ultimate logic of this position, as will become evident, would lead to the conclusion that reality is absolutely devoid

51 *Cosmos*, 113–15.

52 See also *The Spirit and Forms of Protestantism*, 184–85. In this book, Bouyer argues that the negative element in the Reformation is the nominalism implicit in the thinking of the Reformers. Bouyer describes it as a "radical empiricism": "What, in fact, is the essential character of Occam's thought, and of nominalism in general, but a radical empiricism, reducing all being to what is perceived, which empties out, with the idea of substance, all possibility of real relations between beings, as well as the stable subsistence of any of them, and ends by denying to the real any intelligibility, conceiving God himself only as a Protean figure impossible to apprehend." See Silvianne Aspray, "A Complex Legacy: Louis Bouyer and the Metaphysics of the Reformation," in *Modern Theology* 34.1 (January 2018): 3–22. Aspray offers an alternative reading of the intellectual sources at the root of the Reformation.

> of meaning, and therefore unknowable, simply because there is nothing to be known in it. This would be to deny that any science is possible or to affirm that the only possible science—as indeed the Copenhagen school held—is one which merely tabulates empirical data and provides a classification of these data in accordance with the mind's spontaneous organization.[53]

First divine will, then human will, became sovereign over the ideas. This newly-sovereign human volition would grant ultimate importance to the quantitative dimension of being. The reign of quantity was thus installed in Western civilization and, eventually, expanded globally through ideological colonization.

This brings us to the third main point from chapter twelve, regarding the unique equilibrium that Bouyer professes to find in the thought of Thomas Aquinas.[54] Bouyer commends Aquinas for his "moderate realism," for the harmonization in his thought of Platonist and Aristotelian influences, and for his emphasis on the analogy of proportionality in theological predication. On this last point, he explains that Aquinas held that our predications of God cannot be made on the basis of a direct likeness to ourselves or the world but only indirectly, by way of analogy in a second degree, in which we compare God and His activities *ad extra* with the human person and his or her activities in the world.[55] In this twelfth chapter, he explains that Aquinas's doctrine of analogy teaches us that we do not reach God in our concepts but only through them.[56] If Aquinas had been truly followed, he suggests, modern science could have found its proper place and justification in the midst of a larger humanistic understanding, for his theology of analogy enables us to put science into a wider context of cosmic knowledge, in which we realize that science is not the one and only way of approaching the world in its mystery. It is one mode of thought. It is valid and important, but it is not even the most comprehensive standpoint on the mystery of the world. In order to be true to the mystery of divine, human, and cosmic being, it has to

53 *Cosmos*, 114.
54 Ibid., 115.
55 Ibid.
56 Ibid., 127–28.

allow room for other types of knowledge, particularly the religious way of knowing. Theology must itself remain leashed to its proper religious pre-comprehension, even when it assumes a more systematic and scientific form. Bouyer ends the chapter with a wise bit of counsel that he recognizes to be deeply accordant not only with Aquinas but also Cardinal Newman and the linguistic turn in twentieth-century philosophy:

> Today's science must . . . leave room for a number of other types of knowledge, all just as valid in principle, each according to its own frame of reference and its own field, such as artistic knowledge, moral knowledge, and finally religious knowledge. The latter, though it cannot replace or absorb the others, is the only one, if it is authentic, capable of uniting them. Rather, it joins them together, not within itself, but in respect to the single ultimate object of all reality, which it neither comprehends nor attains, but in whose direction it must always point.[57]

INCARNATIONAL AND ESCHATOLOGICAL HUMANISM

In chapters thirteen and sixteen of *Cosmos*, Bouyer issues a stinging rebuke to the modern mentality that has torn asunder the theoanthropocosmic synthesis. These chapters too, whose essential thrust we wish to expound in the next section, are "thickly-wooded." There are themes, genealogies, and critiques running together concurrently, articulated in the manner of a type of dystopic prophetic poetry. These chapters evoke such Christian works as C.S. Lewis's *The Abolition of Man* (1943), and Jacques Ellul's *The Technological Society* (1954), or secular works such as Adorno and Horkheimer's apocalyptic *The Dialectic of Enlightenment* (1947). It was not at all uncommon among philosophers, theologians, and

57 Ibid., 116. See page 254, n. 29. This is the footnote that Bouyer provides at the end of the passage. It reads: "It is remarkable that, as early as the middle of the nineteenth century, in his *Idea of a University*, Newman accurately outlined this conception of the relationship between theology and science. The first intimation of this idea may be found in the opening pages of the last of the *Oxford University Sermons* (whose wealth of knowledge surpasses that in the *Essay on the Development of Christian Doctrine*, which was to deepen and extend only a few of the sermon's many insights)."

critics of Bouyer's generation to write in the incendiary tone exhibited in these chapters. Much like Henri de Lubac, Bouyer thought that the modern age unleashed a special kind of drama in the world.[58] We shall comment in the next section on what we take to be the most essential points in these quite polemical chapters. However, we think they have to be read in the context of earlier debates on the nature of the Church and of Christian mission in which Bouyer was a central figure or target. In order to get at the heart of the meaning of these chapters, we want to expound in this present section some crucial points in these debates and Bouyer's contributions to them. The debates to which we refer were those carried out between Catholic theologians on separate ends of a pole regarding opinion on how the Church should relate to the world, who might be referred to, on one side, as "incarnational humanists," or, on the other side, as "eschatological humanists."[59]

Marie-David Weill has masterfully and comprehensively set the whole of Bouyer's work in the context of these debates. She persuasively argues that this context provides the best interpretive locus for his overall thought. In the portion of her book where she spells out the history of these debates, she follows Gustave Thils (1909–2000) in delineating three general Christian approaches to engaging the modern world that were present as Bouyer was beginning to make his bones as a theologian, indicative of what Thils took to be three types of Christians.[60] The first approach was that of the liberal Christians who flourished in the nineteenth century until the early part of the twentieth. These Christians resolutely separated the sacred from the profane realms of human existence and experience. They fulfilled their Christian obligations devotedly but privately, and they sought to organize the secular

58 We refer of course to Henri de Lubac's *The Drama of Atheist Humanism*, trans. Marc Sebanc (San Francisco: Ignatius Press, 1995).

59 Cf. James M. Connolly, *Human History and the Word of God: The Christian Meaning of History in Contemporary Thought* (New York: The Macmillan Company, 1965), 155–202. Connolly classifies Bouyer's work among that of the "eschatological humanists" whom he criticizes. Nevertheless, Bouyer wrote the preface for the book (pp. xiii-xiv).

60 Weill, *L'Humanisme Eschatologique de Louis Bouyer*, 307–20. See Gustave Thils, *Transcendance ou Incarnation? Essai sur la Conception du Christianisme* (Louvain: Université of Louvain, 1950), 5–17.

world according to the ideal of a profane autonomy that was distanced from direct encounter with the Gospel. The second approach was that of the so-called incarnationalists, who endeavored to bring about the restoration of all things in Christ and to show forth in daily life the visible characteristics of the Kingdom of God. Their understanding of Christian mission was given ritual validation by Pope Pius XI in his institution of the Feast Day for Christ the King in 1925.[61] In contrast to the earlier liberals, they sought to re-Christianize the world, either by conquest or by general witness, planting the flag of Christ in the center of the secular city in order "to conquer the modern world and submit it to Christ."[62] Weill explains more fully:

> We shall nuance the point a little, in speaking less of conquest than of witness, but the fundamental idea remains the same: to assure a marked presence [in the world], massive and efficacious, lifting the milieus that it penetrates in order to win it to Christ. In order to do this, a quantity of means of action and of Christian propaganda are put into work: journals, tracts, chants, marches and cortèges, mass assemblies and congresses, press, radio, and soon television. This spirituality of Catholic Action, as its name indicates, is resolutely centered on action more than contemplation. It is by the accomplishment of the needs of the state, in the world, in work, in family, that the Christian is sanctified. The principles valorized are fraternal love, service to neighbor, the solidarity which unites Christians, the justice between men and between human communities.[63]

As indisputable as some of the goals of the incarnationalists may have been, their understanding of and program for Christian mission in the modern world ultimately came under challenge by a third group, with which Bouyer was himself generally associated. This third group of thinkers embraced a more directly eschatological understanding of Christian mission. These were the proponents of the aforementioned

61 Pius XI, *Quas Primas*, Encyclical Letter, December 11, 1925.

62 Weill, *L'Humanisme Eschatologique de Louis Bouyer*, 308.

63 Ibid., 308–9.

"eschatological humanism," an expression that Bouyer himself in fact popularized, taking it from Dom Clement Lialine (1901–1958), a Benedictine monk at Chevetogne Abbey in Belgium.[64] There are four headings under which Thils summarizes the critiques leveled by the eschatologists against the incarnationalists: "Immanence without Transcendence"; "Conquest or Propaganda"; "Action or Agitation"; "Naturalist Optimism." Weill, for her part, summarizes the dispute in a series of questions: Did the incarnationalists insist too exclusively on the human royalty of Christ and collapse his Kingdom into a merely earthly one? Did they set the purpose of Christian mission as one of conquering as many people as possible through propaganda, thereby turning mission into an endeavor that is all too human? Did they exaggerate in this way the importance of the human instruments of apostolic action and fall into secularization, concordism, and complicity with modernity? In sum, did they trust too much in human resources rather than the power of the Spirit?[65]

All of these questions, and the criticisms they imply, are set forth in Bouyer's little book *Christian Humanism*, first published in France in 1958.[66] We want to expound important parts of this book here, because it is of a programmatic character in the whole of his thought and contains a pre-synopsis of certain important themes that one will find in his trilogies on dogmatics, especially regarding how to understand the modern age and the proper character of Christian mission. Moreover, it summarizes essential points that one will find in other, more historical studies by Bouyer.[67] There is one particular issue that he addresses

64 See *The Meaning of the Monastic Life*, on the catalogue information pages. The reader finds these words: "To Dom Clement Lialine, to whom he owes not only the notion and term 'eschatological' humanism, but the idea which has prompted this book, the author dedicates it as a respectful token of fraternal friendship." O'Connell refers to Bouyer as "preeminent among the eschatologists, and indeed the most vocal of them." See O'Connell, *Human History and the Word of God*, 179.

65 Weill, *L'Humanisme Eschatologique de Louis Bouyer*, 309.

66 We shall reference here the Newman Press English edition from 1959.

67 Cf. *The Cistercian Heritage* (1955, 1958); *Erasmus and His Times*, trans. Francis X. Murphy (Westminster, MD: The Newman Press, 1959); *Sir Thomas More: Humaniste et Martyr* (Chambray-lès-Tours: CLD coll. "Veilleurs de la Foi," no. 2, 1984). We confine ourselves in this list to the most pertinent books having to do with historical studies bearing on the advance of the modern age and the

near the end that bears greatly on his overall cosmological vision and understanding of the modern collapse of the religious way of seeing the world. It has to do with the gradual growth in human consciousness of the idea or imperative of "planetization," a word that might be taken as precursor to what we nowadays describe as "globalization." What is planetization? It has to do with the coming together of the entirety of the human race, united around scientific and technological endeavor, ordered by moral aspirations shaped by individualism, consumerism, and equalitarianism. "For the first time in human history," Bouyer explains, "man seems to have succeeded in fully exploring the earth and its potentialities; he seems on the verge of unifying the whole race in a common organization for which, in spite of the divergences which still remain, we are all working in unison."[68]

A new type of humanism has thus emerged, contrary to that of the ancient Greek and Latin Christian humanism for which contemplation of truth and docility to reality was of paramount importance. Now, contemplation has given way to action in building up and consolidating the new planetary society. This is perceived to be an imperative of evolutionary advance, and Christians are required to get on board. After all, it is thought, the new humanism is more biblically-rooted than the old, which was absorbed into the contemplative idealism of Plato and Aristotle. We are now in a position to bring about by political, economic, and technological transformations of the world the reconciliation of humanity with itself and with the cosmos. Salvation is now linked to creation in a new way. Bouyer explains:

> All estrangement . . . between men themselves and between man and the world, seems about to be finally overcome by this achievement of creation. What else is the salvation of man and the world but the entry of man and the world, of the world

---

need to find exemplars of Christian humanism whose lives bore the eschatological signature of authentic Christian life. See Bouyer's early article "Christianisme et Eschatologie," in *La Vie Intellectuale* 16.10 (1968): 6–38. Bouyer had a great interest in Renaissance humanism and its transformation of its Christian roots. See "L'Exemple de Pico de la Mirandola," *Communio* 1.5 (1976): 94.

68 *Christian Humanism*, 83.

> through man and of man by his knowledge and dominance of the world, into the final, supreme phase of creative evolution?[69]

Christian soteriology is reduced in this new, incarnationalist humanism to a justification for the advancement of technological society. By the time Bouyer wrote *Christian Humanism*, the incarnationalist position had morphed, at least in some of its representatives, into a type of understanding of Christian mission according to which Christianization of the world is one with the humanization or "hominization" of it as attained through technology, through a planetization or cosmicization that will enable us to achieve by our endeavors the Omega Point of history. Clearly, Teilhard de Chardin's vision is targeted in Bouyer's harsh assessment of the technocratic bureaucratization championed by proponents of this morphed incarnationalism.[70] In line with this earlier text, the Oratorian describes, in chapter sixteen of *Cosmos*, Teilhard's humanism as one which conflates technological progress with God's apocalyptic intervention in history. In Teilhard's thought, he says, "the development of technology proceeds *pari passu*, with an increasing socialization, the end result of which should be what Teilhard calls the 'planetization' of human life; this development, while not leading directly to the coming of God's Kingdom, should nevertheless promote it."[71] In both this earlier text and *Cosmos*, Bouyer directs his aim as well at both Marxian and capitalist globalists, albeit from a theological as opposed to economic or political standpoint.[72]

69 Ibid., 84.

70 See Bouyer, "Le Malheur d'Avoir des Disciples (à propos de deux ouvrages sur le Père Teilhard de Chardin)," in *France Catholique* 626, 627, 628 (1958). Available at: https://www.france-catholique.fr/-Articles-du-R-P-Louis-Bouyer-parus-dans-France-Catholique-de-1957-a-1987-.html?debut_articles=75#pagination_articles. These three articles in the French magazine contain Bouyer's first critical assessments of Teilhardism.

71 *Cosmos*, 145.

72 *Christian Humanism*, 88–89. Bouyer says: "America and Russia may indeed be at daggers drawn; but, if we wait a few more decades, till the Russians have had time to provide refrigerators and television-sets for all, and the Americans to perfect, together with their own inter-planetary 'investigators', the repression of anti-constitutional activities, and to bring to heel the intellectuals, both peoples will have come to be so much alike that a conflict between them would be meaningless."

The French Oratorian suggests that the effort made by Christians to baptize modern, immanentist programs of planetary humanization is futile, both with respect to God's Wisdom and will as revealed in Scripture and with respect to the evident limitations of our humanity, in the city of man, beset as we are by inescapable sin, suffering, and death. Immanentist humanism is in fact working surreptitiously, and perhaps unwittingly, to construct a planetary condition of servitude and dehumanization, and this is as true, if not truer, of the projects of Marxian humanism as of those of capitalist humanism. Both cannot but lead to the abolition of man, because they are each tantamount to programs for the mass reintegration of humanity that stifle our individual personalities and potential for "free harmonization" with one another. The Mystical Body of Christ cannot be built up through an uncritical acceptance and adoption of these modern paths of globalization, which are fundamentally materialistic or consumerist at root. Bouyer explains that these paths lead to "mass-regimentation, in the general setting of standardized comfort and of propaganda by radio and television moulding everywhere alike a humanity engaged solely in the same cerebral and sexual activities—is this really leading up to the fullness of the mystical body, or even preparing for it?"[73] For our Oratorian, the eschatological humanist perspective in Scripture, the Church Fathers, and the great Scholastics provides the needed corrective to certain forms of incarnationalist endeavor that seek to provide a Christian justification for the ultimately dehumanizing path of modern immanentism and its blending of capitalist and Marxian materialisms. The properly biblical view is sapiential and apocalyptic, uniting creation and redemption around the event and figure of the Cross of Christ and the not-yet-attained Parousial completion of the Kingdom of Heaven.[74] It is indeed true, as incarnationalists would want to affirm, that Christian discipleship does not mean rejection of creation but its affirmation. However, Bouyer reminds us, creation can be affirmed only by being healed and elevated around the figure of Christ in his death and Resurrection. A consequence of this is that discipleship requires an embrace of the Cross, allowing oneself to be embraced by Christ on the

73 *Christian Humanism*, 89.

74 Ibid., 97–110. These pages contain the final chapter of *Christian Humanism*. They are a plea to recover the practice of Christian asceticism.

Cross, who offers us thereby a real, adoptive filiation in his own Sonship and *reditus* to the Father. The first step into the reintegration of humanity and the world in the Mystical Body of Christ must be, then, "to break up all those pretended integrations which crush the individual and never form a real unity; and, in the whole history of the building up of the body of Christ, the various processes of assimilation are genuine only when accompanied by a continual work of detachment."[75] As we learn in *Cosmos*, Christians have a long history of refusal to allow themselves to be moved by the inspiration of the Holy Spirit to take this first step toward supernatural hominization and planetization. Indeed, it may well be said that it is early modern Christians themselves, precursors to the incarnationalists, who invented the pretended integrations, the false and fallen modern cosmopolitanisms, which Bouyer decries.

AT THE ORIGIN OF IDOLATROUS HUMANISM

Chapters thirteen and sixteen of *Cosmos* add to Bouyer's early writings on Christian humanism a genealogical portraiture of the advent of modern humanism that does indeed target Christians in earlier centuries as culpable for the emergence of the false, secularized globalism that he laments. Bouyer places the process of disintegration described in these chapters in the context of a wider dialectic of religious progress and regress that one can find throughout history. He takes the modern disenchantment of the world to instantiate with particular intensity a stage of decline that is not without certain historical analogies.[76] In the midst of the thickly-wooded pages of these chapters, we detect a primary argument: that modern humanism exhibits its own form of idolatry, a specifically Christian type, with its particular manifestation of the perennial human tendency to project its image, in its fallen condition, onto the divine. Bouyer argues that all religions, including the one and only supernatural religion, may grow for a period, reach an apogee, and then decline for a period under the weight of the inner corruption of magic and idolatry.[77] It is inner corruption in their own religious practice and their own way of seeing more than outer assault from science, philosophy, or humanist

75 Ibid., 90.

76 *Cosmos*, 117–20.

77 Ibid.

values that practitioners of religion—even Christian religion—have to fear the most.

Bouyer explains in chapter thirteen of *Cosmos*, entitled "Decline of the Religious Vision of the World," that the dialectic of growth and decline was manifest in Late Antiquity in the Greco-Roman world as life in the cities advanced to such a degree of civilizing sophistication that the illusion set in that humanity had succeeded in humanizing the world. The gods of religion were increasingly unmasked as projections of human imagination in its lowest register. In the Hellenist Age, religious cults were transformed first into agricultural techniques and then reduced to theater, having lost all genuinely religious, cultic importance.[78] Greek tragedy evolved from Aeschylus (c. 525–456 BC), through Sophocles (c. 497–406 BC), to Euripides (c. 480–406 BC), progressively expressing a desire for release from the capricious reign of the gods over human life. In the epic poems of Homer, *The Iliad* and *The Odyssey*, and the Alexandrian epics, the gods were increasingly portrayed as having only a marginal significance in the cosmic drama in relation to human existence, which was increasingly magnified in its own intrinsic glory. Virgil's *Aeneid*, Bouyer suggests, was able to reinstate the gods into human affairs only in an artificial manner. Ovid showed that the ancient myths were mere fables or mythology. Xenophanes pitted *logos* against *mythos*, though not without, as we saw in chapter four, projecting mythical characteristics onto the processes and structures of the world.[79] Bouyer argues that the religious sense or need did not disappear in this process that he succinctly summarizes but could from now on try to find satisfaction only through covert channels in a disintegrated culture.[80] Religion was split between the soteriologically unsatisfying philosophy of intellectual elites and superstitious popular practices to which the philosophers themselves, as in the case of the Neoplatonist Iamblichus (AD 245–325), turned in the end to satisfy their religious needs.[81]

The fall from the religious vision of the cosmos that characterizes the modern age from the time of the fourteenth century onward exhibits the

78 Ibid., 119–20; *Mystery*, 19–36.

79 *Cosmos*, 119–20.

80 Ibid., 120–21.

81 Ibid.

same sort of structure of decline that characterized this earlier period, although a new element was added. Bouyer explains:

> This time the crisis was not due simply to a success of civilization so marked and apparently so secure that the humanization of the universe acted as a screen against the radiance of the divine glory, so that God became as though absent if not exhausted. It was quite specifically the result of the first emergence in history of what one may call a bourgeois or middle-class civilization, one based directly and perhaps exclusively on money. Wealth became the supreme means of gaining, in this world, not only a much coveted (though rarely if ever attained) security, but also what was to be called comfort, perhaps the most novel feature of this unprecedented civilization.[82]

This new element, unknown in the Hellenistic Age, was the development of the money-lending class, of bankers, and of the reduced horizon of human aspiration in which they inscribed human life.

The development of this new type of culture, Bouyer reminds us, did not take place outside of Christian civilization but from within it. Christians themselves began to embrace the world in a new way, as it is in itself and in all that it can give to human life without respect to God, the angels, and the immortal soul. Christians themselves were the very ones who lost a sense of the imperative of the Cross and of the need for detachment from the world in order to prepare for the world to come, which is the ultimate "re-creation" (not *ex nihilo*) of this world. The monastic humanism of the Church gave way to a new type of humanism. Twentieth-century incarnationalists are the thankful heirs of this development. This new type of humanity, with the new humanism that it brought forth from within Christian civilization, quite naturally began to look at the world in a new way. It bore within itself, we might say, a new "human image," on the basis of which it projected a new "world image."[83] In an unprecedented manner, the Oratorian argues, the new

82 Ibid., 121.

83 Cf. Philip Sherrard, *Human Image, World Image*, 1–10. We draw this idea from Sherrard which, as we mentioned in the introduction, we think to be deeply

human image was tied to the quest for comfort and security. The new world image was, as a result, unprecedentedly materialistic. A quantitative approach to the cosmos became the dominant mode of scientific endeavor. As Bouyer tells the story, Nominalism arose as a defense of this new way of seeing. Occam is not without guilt in this process, but he did not cause the situation. The external form of Christian existence remained, even for centuries, but it was delusory: "Infiltrating and consolidating itself everywhere under the outward appearances of Christianity... this new civilization was progressively to inactivate all specifically Christian characteristics, simply by substituting the worship of Mammon for that of Christ, while maintaining Christian forms."[84] This new way of seeing the world, this new mode of quantitative thought founded on a proto-consumerist ideal of human action, spawned the Industrial Revolution, modern globalism, and the ecological crisis. "Man," Bouyer says, "was no longer able to see the world as a meaningful cosmos ordered to transcendence, a sense of which had previously been imparted to our lives by contemplation of the universe."[85] The cosmos was reduced in this new way of seeing to a collection of things or "soulless objects" that exist only to provide sensory satisfactions to humanity.[86]

There is much that can be said for and against this genealogy. The weight that Bouyer accords to the possibilities for human action as determinative for cosmology and metaphysics is nevertheless highly significant and, it seems to us, unique to his thought, at least among theologians of his generation. He follows to some extent the Thomist genealogy of modernity according to which Nominalism, in doing away with the reality of universal objective forms, brought about a disastrous metaphysical rupture between subject and object, reason and will, wisdom and goodness, and the modern and premodern as such. Yet his analysis has a greater level of concreteness than standard accounts of metaphysical rupture, a more convincing sense of the intertwining of metaphysics and human action, and a more realistic grasp of the determinative impact on human intellectual culture of "pre-ontology,"

---

congruent with Bouyer's thought.

84 *Cosmos*, 122.

85 Ibid.

86 Ibid.

in that he roots the advent of Nominalism in a cultural need to justify an already-existing practical materialism rather than as the first cause of modern philosophical materialism. It should be noted that Bouyer treats Marxism no better (and even worse) than capitalism in this regard. "Marxist man," he asserts, "would be but an attempt to extend to all mankind the condition of the bourgeoisie as a privileged class."[87] One is certainly justified in wondering at the blanket dismissal that Bouyer levels at modern art and of museums as "cultural refrigerators," or his Romantic take on the inherently artistic quality of peasant agricultural work.[88] Nevertheless, even here, the concreteness of his analysis seems unique among Catholic dogmatic theologians of his day. This befits the overall liturgical orientation of his thought.

As we have indicated, the unifying analytical thread that runs throughout his two most dystopian chapters in *Cosmos*—chapters thirteen and sixteen—is that corruption of human civilization and religion comes about by way of idolatrous self-projection onto the world and onto God. Bouyer describes the process through which idolatry takes hold of religion as a kind of idealism. In chapter sixteen, entitled "From Technological Magic to Cosmic Mysticisms," he gives a nod of approval to Ludwig Feuerbach's theory of the anthropological origins of religion inasmuch as Feuerbach's theory can be rightly applied to false or corrupt religion.[89] Religion may and has indeed, at different points in history, embraced a literalism of mythic representations of the divine that leads either to polytheism or to outright idolatry.[90] In the former case, the focus is on precisely the multiplicity of images that are taken to represent the divine. In the latter case, there is a confusion of divinity with humanity. Divinity is then collapsed into human representations. Idolatry is observable from the start in humanity's relation to the myths, and the best developments of critical reason have always targeted this deformation. Critical reason enables us to realize that in presuming to worship the divine we often worship only ourselves—and only what is lowest in ourselves. The reaction of critical reason, unfortunately,

87 Ibid., 124.
88 Ibid., 124–26.
89 Ibid., 156–57.
90 Ibid., 126–29. For the rest of this paragraph, we follow these pages.

can turn us away from the naïve realism of popular religious thought and practice in such a way as to submerge us in idealism, whereby we explicitly theorize that the cosmos is a projection of our own being. Idealism was present in the ancient world in forms of philosophical Hinduism and Buddhism, in late Antiquity in the Neoplatonism of the Greco-Roman world, and in the modern world in German Idealism.[91] Critical thought in these cases led to an open solipsism and ultimately to agnosticism. Based on humanity's innate instinct that we are personal beings meant to live in a reciprocity of relations, the realization eventually dawns on philosophers and the society that both influences them and is influenced by them that a religion of the self and the self alone is ultimately untenable.[92]

Idolatry, Bouyer insists, is present in the Christian world in the modern age not only in idealism but in the collapse of Christian thought into a preference for so-called "biblical literalism," which has its source in the very materialism embraced by bourgeois humanity. It is promoted by "those who in fact no longer seek comfort and strength in the God they confess, but rather in the material things of this world."[93] Humanity no longer contemplates the cosmos, and it loses touch with the depths of symbolic intelligence. This leads to the temptation to give greater importance to the formal expression of faith than to its real object. Bouyer suggests in chapter sixteen that Aquinas's rendering of the *analogia entis* would have contributed to the rescue of humanity from the turn to "conceptual idolatry" that confuses our concepts of the divine with the divine itself, in that his analogy of proportionality, properly grasped, would have tempered our obsession with the concept through which we know transcendent reality. But Nominalism won out. The fourfold meaning of Scripture—literal, mystical, tropological, and anagogical—was lost from Christian practice in the context of the new materialism. The pathway was opened for a fragmentation of faith, the loss of a sense of the organic unity of the Mystery of Christ, the devaluation of the symbol, and ultimately to the anthropological reduction of modern theology that collapses the content of the Gospel

91 Ibid., 127.
92 Ibid.
93 Ibid.

into the transcendental conditions of possibility inherent to human subjectivity.[94]

It is in the light of this analysis of conceptual idolatry that Bouyer places in chapter sixteen of *Cosmos* the mainstream modern philosophers René Descartes (1596–1650) and Francis Bacon (1561–1626). The common appeal made by these two thinkers to the importance of religion should not, Bouyer warns, deceive us. Both of them are, he argues, proponents of the technocratic ideal, of the reduction of science and the exploration of the cosmos to its use-value for human self-gratification. These are not the true founders of modern science, as they are sometimes taken to be. The French Oratorian does not place them in the same class of intellectual dignity as Galileo and Isaac Newton, who are, he says, "legitimate heirs of the fundamentally biblical and Christian cosmology propounded by the great scholastic masters of the thirteenth century."[95]

Bacon and Descartes are taken to be decisive spokespersons for Nominalist voluntarism and idealism, in that they are held to secure and consolidate the trajectory of modern thought and life in their respective philosophies, which have been influential to all subsequent mainstream modern philosophers, including the Marxists. The technological civilization that they promote is the product of "a self-centered and alienated mankind."[96] These philosophers tend to justify any means that can bring about the end that this alienated humanity seeks above all other ends, even beyond comfort and security: the creation and stockpiling of wealth.[97]

How absurd, then, are Christian "incarnationalist" attempts to fit the Gospel into the ideals of modern civilization, with its technocratic idealism and lust for wealth creation? Interestingly, Teilhard, the preeminent theorist of this sort of incarnationalist Christianity, is often criticized nowadays for lacking a proper sense of alarm at the social, political, and economic forces that led modern civilization to a situation of ecological catastrophe. There is clearly no such lack in Bouyer's eschatological humanist rendering of the development of the human race. The Oratorian

94 Ibid., 128.
95 Ibid., 152.
96 Ibid., 153.
97 Ibid.

even evinced in the 1970s and 1980s, when *Cosmos* was written, a certain sympathy for the contemporary ecological reaction, then still in nascent form, although this sympathy went only so far, as is made abundantly clear in chapter sixteen of *Cosmos*. He detected in much of the ecological movement of our day not so much the smashing of idols as a new way of giving homage to the old idols. These "illogical ecologists," as he calls them in *Cosmos*, promoted the rejection of all technology to go along with a withdrawal from all civilization. This is clearly not the path forward that the humanist Bouyer promotes. Seeking to overcome modernity's "spurious humanization of the cosmos," he says, the ecologists have gone on a Rousseauian search for a delusory "pristine nature untainted by man," while engaging in advocacy of a decidedly polluted form of sexual activism.[98] If they pretended to reject the money-seeking ways of their technocratic parents, they nevertheless worshipped at the same altar of the self as their parents did. A truly effective ecological reaction would, he insists, embrace the cosmos as divine creation. The *de facto* ecological reaction was, instead, "but a projection onto all things of rootless man, stripped of mystery and denied his true being as a creature called to divine filiation."[99]

In recent years, as we noted, the papal magisterium of the Catholic Church has developed social doctrine in explicit recognition of the inseparable link between human ecology and environmental ecology. Proper care for the world, we are taught, requires proper direction in our own moral pursuits. The link between the two levels of ecology was deeply understood and well-articulated by the French Oratorian. He explicated much that is entailed in this linkage. His assessment of the modern fall is hardly more apocalyptic, in the popular meaning of this term, than what we see in Pope Francis's *Laudato Si'*, and, as we shall find in the next chapter, Bouyer's assessment of the modern age is not tantamount to a univocal condemnation of it. He shows that there is an alternative modernity that exists in the line of the Church Fathers and Scholastic masters as well as the medieval monks and mystics. It embraces the link between the human and the cosmic, recovering a sense of the world as "meaningful cosmos ordered to transcendence." It provides a needed

98 Ibid., 159–60.
99 Ibid.

resource to replace, with an iconic mode of thought and perception, the idolatrous modes of thought and perception that have characterized the mainstream currents of culture described in this section.

TOWARD AN ICONIC WAY OF SEEING

If Bouyer's stinging critique of the practical and philosophical materialism of bourgeois civilization bears analogy to a much wider group of thinkers who have labeled our age the "age of nihilism," this does not mean that his thought is incapable of being targeted by thinkers of a decidedly more incarnationalist cast of mind. He would, even today, in the era of radical ecologism, and of radical Catholic critiques of modern liberalism that are becoming more and more prevalent, be subjected by some—perhaps increasingly few in number—to attack for an apparent failure to embrace "courageously" and in full the arrow of evolutionary advance and technological progress, especially in regard to the sexual revolution. Moreover, transhumanism, the melding of human and machine, finds its Christian advocates, who would doubtlessly see Bouyer as a reactionary or even as a Gnostic who seeks to escape from the world as it has been unveiled by modern science and opened to our control by exponential increases of technological power.

Yet for those who reasonably question whether or not the sexual free-for-all of our brave new world or the advent of the cyborg heralds the dawning of a glorious new era of cosmic being in which *anthropos* and cosmos will attain a greater wholeness of being—for instance, in overcoming what is perceived by many to be the violent duality of the "social construct" of sexual differentiation—Bouyer's genealogy of the modern has much to teach us. It is not simply irrational reactionaries who wonder if the technological manipulation of sexuality or the ideal of the cyborg does not in fact portend a condition of depersonalization or of the evacuation of living personality from the cosmos, which would leave as its residue "a totally fossilized universe resembling an immense concentration camp."[100]

It seems that the prospect of a posthuman future, or the reality of a posthuman present, should strike one as rather more threatening to the

100 Ibid., 159.

integrity and harmony of cosmic being than as its salvation. At the very least, the advent of transhumanism should cause us to question anew our understanding of what constitutes the fullness of humanity, and Bouyer's work can help us with this question, because he gets to the root of our assumptions about what it means to be human or humanist. The Christian should recognize with the eyes of a charitable and hopeful faith that in order to embrace creation and humanity as a whole, in the genuine plenitude of being, we require transformation in the New Human, the Eschatological Adam. It is not in reactionary fear but in the confidence of such faith that Bouyer says: "Christianity is finally the only true humanism, because it is an integral humanism, which is to say a humanism which opens to the other and the gift of self, which is the most perfect reflection of the divine life, and not moreover only a reflection but a true participation in it."[101]

"Outside Christianity," Bouyer tells the reader at the end of chapter sixteen in *Cosmos*, "an outstanding example of the revival of the human spirit in a rediscovered cosmos is evidenced by the flowering of Chinese painting during the Sung period."[102] A new form of humanism arose, in a very different time and place, yet marked by religious crisis, with a greater sense of the religious and personal meaning of the cosmos. If Bouyer appears unstinting in his assessment of the bourgeois mentality at the heart of modern civilization, he does not in fact see the modern age as nothing more than a barren wasteland of irreligiousness, and if there are countercurrents of thought at the heart of the modern age, there are also signs of reawakening within the Catholic Church itself. One of these signs of reawakening was explored in a beautiful picture-book that Bouyer published at around the same time as *Cosmos*, and we want to turn briefly to consideration of this book in conclusion of the chapter.

In *Vérité des Icônes* (first published in 1984), the Oratorian provides an appreciative though critical assessment of the recovery in recent decades of the tradition of Christian iconography. He opens the reader to the Church's tradition, East and West, of contemplation of the Holy Face of Christ, of his Mother, and of the followers who immediately surrounded them. He seeks in the book to give a theologically correct account of the

101 *MT*, 200.
102 *Cosmos*, 160.

meaning of icons, which he thinks is often lacking among enthusiasts who support this recovery. His major point of emphasis is that the icon transforms the directionality of human intentional consciousness in that it renders one truly open to the other by first being rendered open in oneself by the Spirit of God to the ultimate Other. The idolatrous gaze projects the untransformed image of the self onto God and onto the world, while God, in the iconic gaze, imprints His living image on the one who contemplates Him in and through His eternal image given in the flesh of the Incarnation. The Christian icon effects an inversion of perspective. It converts our gaze, enabling us to see all reality, including our own selves, in light of the image of the Eternal Son.[103] In the icon, we do not look at or encounter Christ so much as Christ gazes at or encounters us, training us to perceive ourselves and the world in the light of his transfiguring Mystery.

The rediscovery of the truth of icons is necessary for articulating anew a Eucharistic or sacramental ontology. Bouyer shows that the iconodules in the early Church recognized the close connection that obtains between veneration of icons and celebration of the Eucharist, and they surrounded the Eucharist with holy images. These images

103 *Vérité des Icônes*, 34. Quoted by Jean-Luc Marion in *The Crossing of the Visible*, trans. James K.A. Smith (Stanford University Press, 2004), 98, n. 12. Bouyer gives the example of the Daphni monastery in Athens: "One can rightly observe, for example, the effectiveness of inverted perspective in the alcove above Daphni, where the Transfiguration is reproduced; it appears to push us toward our own encounter with the Christ of Glory, between the adoration of Moses and Elijah." See also *RC*, 101. Here Bouyer stresses the primary importance of shifting our sensibility through a transformation of imagination in order to see God and the world in a Christian light. We need, he says, a "drastic purification, not only of our representation of the world, not only of our activity in the world, but in the sources of our action as of our thought, on the most profound, most intimate, most abyssal plane of what we can call our sensibility. Without this correction, properly radical, of all our instincts, beginning with our imagination, in the sense that Coleridge, among others, has given to this word, and which concerns together the prospective of our acting as of all possible prospection of our universe, it is perfectly vain to hope to act, or rather to be acted on, according to *agape*, and for a stronger reason to hope to encounter the divine signification in it." This passage points to the need for ascesis and transformation of imagination. It seems to us that it is clear that Bouyer holds that this transformation of imagination requires, in part, iconic transformation of the images that we behold.

helped to remind the faithful that, in the Eucharistic celebration, the high deeds of God are made present again to be represented to Him in prayer and supplication, so that the fulfillment of His promises may be accomplished. The Eucharist gives us God's very presence and communicates His work to us and in us *ex opere operato*, while, in the iconic gaze, Christ's own gaze penetrates us to the depth of our souls. It is not we who peer into him but he who enters into us, purifying the image of him that we bear within. We become practiced in receptiveness to the grace of iconic counter-intentionality. It is clear that for Bouyer this way of seeing is needed if we are to discover the cosmos anew in its religious meaning. The iconic way of seeing enables us to know the world as creation, as gift of God, irreducible to a self-sufficient nature. If we return to our first section of this chapter, we cannot help but think that it is necessary to develop anew the practice and theology of the icon in order to see the world in its living unity as the self-expression of the triune Creator. *Vérité des Icônes* was appropriately published around the time of *Cosmos*, indicating the movement and maturation of the thought of a theologian for whom the theology of the image and imagination were increasingly taking on paramount importance. The Oratorian's précis on the theology of the icon is central to his overall quest to inspire us to seek transformation from self-involution and the projection of the idolatrous gaze to which we are all too easily accustomed by allowing God to open us to the purifying counter-intentionality of this truly religious way of seeing.

CHAPTER 7

The Recovery of Poetic Wisdom

MODERN MECHANIST MATERIALISM, AND THE distortions in Christian life and thought that ultimately enabled it, come under fitting attack in *Cosmos*, but, as we have just suggested, not every modern form of thought is thereby condemned. As much as Bouyer puts practical and philosophical expressions of Nominalism in their rightful place, he recognizes that the Nominalist hegemony in modern culture is not absolute. There are directions of thought that resist the urge to reduce physical creation to the status of a collection of "soulless objects" controllable by human will, and our author seeks as much as possible to baptize the alternative cosmologies that these traditions set forth. Chapters fourteen, fifteen, and seventeen of *Cosmos* contain explorations of the often hidden channels of modern thought that approach more or less closely to the cosmological implications of the Mystery of Christ. Theologians, philosophers, scientists, and especially poets are called upon as guides to discover a potentially healing wisdom. The thinkers to whom Bouyer gives preference in these chapters are not recognized as of yet in the official corridors of ecclesial policy-making in the Catholic Church, but this should not exclude them from fellowship in the capacious embrace of the Christian humanist. There is one particular current of thought, directly or indirectly responsible for some of the most audacious theses of modern theology, which Bouyer takes to be particularly important, the "sophiological" current to which we have made reference in this study.

In this chapter, we shall explore Bouyer's genealogy of modern sophiology, his rendering of its pertinence to theological cosmology, and its implicit as well as explicit presence in the final chapters of the first part of *Cosmos*. Bouyer had a great love for the English cosmic poets of the nineteenth century, particularly William Wordsworth, and he focuses on their work in chapter seventeen, the capstone chapter in this first part of the book. However, we think that even these eminent figures find their place in Bouyer's thought only in the context of the work of the theologians and philosophers for whom the biblical figure of Wisdom was paramount. We shall explore, in a first section, the connection of the theology of Wisdom to the poetic vocation. In a bit of an inversion unique to our present chapter, the seventeenth chapter of *Cosmos* will be our starting point but will be placed in the framework of Bouyer's exposition of the unifying, poetic theology of the Russian sophiologists Bulgakov and Florensky found in writings outside of *Cosmos* and ultimately in connection with a brief discussion of Jacob Böhme (1575–1624) as found in one of the Oratorian's early books on the history of spirituality. In a second section, we shall turn to Bouyer's brief exposition in chapter fourteen of *Cosmos* of certain nineteenth-century Catholic sophiological philosophers of nature whose work exists in the line of Böhme's thought and preceded that of the Russians. We shall suggest in a third section, following chapter fifteen of *Cosmos*, that Bouyer sees the most recent advancements of modern science as potential paths opening human thought to a unifying, "sophianic" view of creation, particularly with the advent of depth psychology. Exploration of these themes will enable us finally to begin, in a concluding section of the chapter, to address directly Bouyer's answer to the determining questions that he poses at the beginning of *Cosmos* concerning the relation of spirit to nature as well as of human history to the cosmos at large. He promises the reader at the beginning of the book that he will address these questions in the chapters of *Cosmos* that are our concern here. He does not do so directly and explicitly, and in this final section we shall outline the contours of his implied answer, which requires turning to the second part of his book in order to be filled in.

THE POETRY OF WISDOM

There is no way to receive empathetically Bouyer's vision of the cosmos without recognizing the fundamentally poetic character of his thought. This point has certainly been emphasized in our exposition of his phenomenology of mythopoetic thinking and his argument for its foundational presence in all of humanity's genuine wisdom traditions, but his final chapters in the genealogical portion of *Cosmos* move us to dwell even more directly on his attempt to plumb the poetic mode of human perception at the basis of all intellectual endeavors. The literary inspiration which moves him to recover the poetic unity of *anthropos* and cosmos in a religious light comes not only from the doxology of the Church Fathers but from the modern cosmic poets. Indeed, the words of these poets nurtured his soul from the time of his youth until his final illness. Jean Duchesne says that Bouyer, at the end of his life, let himself be "lulled by the rhythms and sounds of the verse of Wordsworth" that his friends read to him while he stayed with the Little Sisters of the Poor in Paris during his bout with Alzheimer's disease.[1] Bouyer left behind an unpublished manuscript entitled *Religio Poetae* in which he explores the writings of several nineteenth- and twentieth-century cosmic poets.[2] He summarized his intention in writing *Religio Poetae* in another unpublished manuscript held at Saint Wandrille Monastery:

> [Poetry for me] is something of very great importance, and I would say that in culture it is what is closest to religion. Poetry is only the discovery, or rediscovery of the world as fundamental revelation of the one who made it. It is itself, as word of man, fundamentally a word of exaltation, of praise. The poetic intuition, in the case of the poet himself, is like a rediscovery of lost paradise under the banal surface of things and tends to become a confident expectation, even an anticipation, of paradise regained. It is because of this that all true poetry, if it does not name God, evokes Him, because it speaks to us of things and of beings in a manner which restitutes them in transparence to a light which comes from above and must lead us, or in any

1 Duchesne, *Louis Bouyer*, 99. Duchesne reported to us in an email exchange that Fr. Bouyer suffered anxiety in his final years while suffering from this illness. The words of the poets were a kind of medicine that helped to ease the anxiety.

2 Ibid. In *Cosmos*, Bouyer briefly explores the works of Henry Vaughan (1621–1695), William Wordsworth, Friederich Hölderlin (1770–1843), Percy Shelley (1792–1822), John Keats (1795–1821), and others. Interestingly, *Religio Poetae* contains a two-chapter treatment of the works of Gerard Manley Hopkins (1844–1889), but he is not mentioned in *Cosmos*. See Weill, *L'Humanisme Eschatologique de Louis Bouyer*, 266, n. 1. *Religio Poetae* is forthcoming for publication by Ad Solem. The manuscript is 172 pages, the bulk of which was written in the 1950s and 1960s.

> case draw us, further. I have rewritten several times the chapter of *Religio Poetae* where I try to develop and explicate this view of things. But it is very delicate, subtle even, and I am not yet entirely satisfied with my work.[3]

In a more truncated manner than in *Religio Poetae*, Bouyer returns again to these poets in chapter seventeen of *Cosmos*, entitled "Renewal of Poetic Experience." He tells us there that the essential vocation of the poet is to provide "a symbolic vision of the world in God."[4] The specific task of poetic creation is, he holds, to preserve by the exercise of creative imagination in and through living symbols the unity of human experience, including of the divine presence/absence. If, Bouyer suggests, the classical Hellenist poets intuited the harmony between the cosmos and the human soul and were inspired to sing the mystery of existence in its inherent unity, the modern cosmic poets had the task to create prophetic songs of lament in the face of the destruction of humanity's unified perception of the cosmos by the Industrial Revolution and its after-effects. Wordsworth, the seminal cosmic poet from Northern England, is given a special place of significance in *Cosmos*. His "cosmic mystique" and "panentheism" is interpreted as a movement toward a genuinely Christian contemplation. Clearly, his poetry helped to shape the Oratorian's way of seeing the cosmos.[5]

There is no doubt that Bouyer wished for theology to discover anew the poetic, contemplative heart of human existence in order to rediscover the cosmos, and his fondness for the modern cosmic poets seems generally to have exceeded his esteem for many modern theologians. This was not always the case, however, because he recognized that there was a certain type of theology in the modern age, with a semi-continuous, traceable genealogy, that in fact was in some ways an unfolding or development in the light of divine revelation of authentic poetic experience or imagination, not unlike the theology of some of the Church Fathers. We refer, of course, to the loose tradition of theological sophiology that

3 Quoted by Marie-David Weill, *L'Humanisme Eschatologique de Louis Bouyer*, 266.

4 *Cosmos*, 161.

5 Ibid., 173.

we have discussed a little, particularly although not exclusively as found in the writings of Bulgakov.

We have seen that the Oratorian was first made aware of this tradition of reflection on biblical and cosmic wisdom as a consequence of encounters with Bulgakov that left an indelible impress on his soul. Bouyer became well acquainted with the Russian theologian's life and thought. He does not comprehensively discuss Bulgakov in any of his theological volumes and hardly at all in *Cosmos*.[6] However, he did take the time to summarize the totality of the Russian's work in his article that we highlighted in the first chapter.[7] He emphasizes, as we saw, the inherence of the cosmological motif in all of Bulgakov's thought and argued that the uniqueness of his theology lies in its joint embrace of cosmology and anthropology in light of Christ's death and Resurrection.[8]

A quotation from Bulgakov describing a mystical experience he had while visiting the Church of Saint Sophia in Constantinople was dear to Bouyer and is quoted at length in this article. It calls to mind his own experiences in the Loire Valley of the unity of God, humanity, and the cosmos. It doubtlessly had an impact on his interpretation of these experiences. Bulgakov said:

> An ocean of light explosively spreads from on high and dominates all this space, enclosed and yet free. The grace of the columns and the beauty of their marble embroideries, the royal dignity—not luxury but royalty—of the walls of gold and of their marvelous ornamentation: all this captivates, melts, subjects, and convinces the heart. There results from it a feeling of interior transparence. The heaviness and the limitations of the mediocre and self-sufficient self disappear: this self is no longer—the soul is healed by melting and losing itself among these arches. It becomes the world: I am in the world and the world is in me. And this feeling of weight which weighed on

6 Bulgakov is referenced on only four pages of the text: 140, 185, 216, and 217.

7 "La personnalité et l'œuvre de Serge Boulgakoff," 135–44. Bouyer gives in these pages a thorough overview of the life and theological/philosophical trajectory of Bulgakov's work.

8 Ibid., 136.

> the heart dissolves, the liberation from gravity, of being like a bird in the blue sky, gives not only happiness, nor even joy, but beatitude. It is the beatitude of the final knowledge of all in all and of all in one, of the infinite plenitude in multiplicity, of the world in unity. It is there in truth "Sophia," the real unity of the world in the *Logos*, the co-inherence of all with all, the world of the divine ideas....[9]

This holy dome, embodying created Wisdom, is itself a microcosm or a true macrocosm, "a great world in the small world." Here liturgical space draws together the whole of creation which it interpenetrates. The viewer, the liturgical participant, would enter at Hagia Sophia into the All-unity of God and creation. Heaven descended to earth and enraptured the viewer. Cosmic liturgy found its axis on the terrestrial plane. These descriptions are true to Bouyer's own understanding of the beauty of liturgical form when rightly ordered, and his own experience of the Hagia Sophia later in life mirrored Bulgakov's, except that Bouyer explicitly noted the eschatological nostalgia that the dome evoked in him, given that it is no longer a site for Christian liturgy.

The theme of wisdom is naturally unifying, poetic, and imaginative, so much so that we might say that the poet even more than the philosopher is the true lover of wisdom, inasmuch as, in Bouyer's words, "the poetic view of reality has led man to see in the world the result and the permanent manifestation of God's loving wisdom, which is His glory."[10] If theology, then, is to be a wisdom discipline, it must remain diligent not to lose sight of the unity of God and creation centered on the Mystery of Christ. It is not only Bulgakov who shows a way forward for theology in this regard. In reviewing Pavel Florensky's *Pillar and Ground of Truth* for the journal *Communio* in 1975, Bouyer marveled at the unity of vision contained in this book, which he describes in very similar terms to his description of the many good aspects of Bulgakov's work:

9 Ibid., 144. Bouyer does not tell the reader where this passage comes from. It parallels a passage in the most recent English edition of Bulgakov's *Sophia: The Wisdom of God*, 1.

10 *Cosmos*, 161.

> The heart of this book, very clearly, is in its treatment of "Sophia"—: the Holy Wisdom of God, which is to say, His eternal design over creation and man in particular, included from all eternity in the thought by which God thinks Himself in His Word, coming to realize itself progressively in the Virgin Mary, in the proper human individuality of the Savior, then, after the resurrection, in the constitution of his mystical body, the Church and all the redeemed cosmos, from the Eucharist.... This admirable, unifying perspective on the divine life and its communication is grasped and proposed as the theological problem par excellence, from the mystery of the Trinitarian life and its opening to creation.[11]

Bouyer recognized that Florensky was both theologian and poet. The two dimensions of his personality and vocation were united in the task of writing this monumental book. In consequence, his speculative theology was not detached from spirituality and asceticism: "From one end of the book to the other, the speculative, or rather contemplative, aspect is never separated from the ethical and ascetical aspect, and meditation does not cease to continue in dialogue with the reader, as a passionate exhortation, which is reminiscent of the tone of the Macarian Homilies, surpassing mere thought in an experience of adoring prayer and total life in faith."[12]

As remarkable and brilliant as Bouyer understood the greatest Russian sophiologists to be, he knew that they did not invent the theology of Wisdom from whole cloth. There is a much longer tradition at work that they inherited and developed. We have seen that Bouyer thinks this tradition ultimately goes back in its Christian form to the controversy with the Arians and the writings of Saint Athanasius and Saint Augustine. In *Cosmos* and *Sophia*, he points out that this tradition developed through the Middle Ages by way of practical channels rather than in the writings of the major theologians.[13] It is present, he suggests, in the Marian liturgies

11 "Le Colonne et le Fondement de la Vérité de P. Florensky," 95. Pavel Florensky, *The Pillar and Ground of Truth*, trans. Boris Jakim (Princeton, NJ: Princeton University Press, 2004).

12 Ibid.

13 *Sophia*, 115–22; *Cosmos*, 135–36.

of both Eastern and Western Christendom.[14] It surged forth in some of the visions of Saint Hildegard of Bingen, now a Doctor of the Church, in her *Scivias*, one of which we described in our introduction. We find this tradition advanced, Bouyer claims, in the cosmology of Bernard Sylvestris (c. 11th century), and in the *Ampitheatrum Sapientiae* of Heinrich Kuhnrath (1560–1605) as well as in the writings of Jacob Böhme and his followers. There are ambiguously Christian or extra-Christian sources exemplifying this tradition as well, especially the Jewish Kabbalah and Florentine Neoplatonism. Bouyer recognizes that it was often connected with schools of empirical medicine that were magical or alchemical in nature, inspired by Paracelsus (1493–1541) and practiced by the Cambridge Platonist Henry More (1614–1687).[15]

In modern times, the humble and diminutive figure of Böhme looms like a colossus over Germanic, French, Russian, and English traditions of sophianic mysticism and philosophy, and Bouyer was himself quite aware of this fact.[16] The history of Böhme, his followers (the Behmenites), and their collective influence on modern Romantic and Idealist thought is being recognized more frequently in our day than when Bouyer could say in the 1960s that Böhme's thought was the source of Hegel's dialectic as well as his idea of God "realizing Himself in the world and ultimately in the collective consciousness of mankind."[17] In the ecumenical book on various Christian traditions from which this quotation is drawn, Bouyer highlights three elements in the thought of the cobbler from Görlitz that he thinks set his work apart: 1) the significance he gave to the body and to the materiality of the cosmos, against certain excessively Hellenist currents of Christian thought; 2) the tragic sense of human, cosmic, and even divine life; 3) the communication of divine life to humanity by way of the participation of God in the life of the creature.[18]

These three emphases are hardly absent from Bouyer's own thought, although he does not portray the mystery of the divine essence in the

14 *Sophia*, 103–13.

15 *Cosmos*, 136.

16 *Orthodox Spirituality and Protestant and Anglican Spirituality*, trans. Barbara Wall (New York: Burns and Oates, 1965), 164–68.

17 Ibid., 166.

18 Ibid., 167.

theodramatic manner of Böhme. These Böhmian emphases speak to a profound, indeed, poetic sense of the unity of divinity, humanity, and the cosmos in the mysterious figure of Wisdom through which the divine love is mediated to the cosmos. There is surely a deeply and specifically Christian inspiration that makes it possible for Böhme to understand the involvement of God in creation as one of interpenetrating intimacy with the materiality of the world and its dramatic becoming. God is not, for him, a remote prime mover who neither knows nor loves His creation, and he draws out implications of the divine intimacy with creation that are unique. Böhme's development of the wisdom tradition colors it with the agonistic hues of kenotic love. He gives impetus to a tradition that Russians such as Soloviev and Bulgakov assimilated in their own thought. Yet there resides a fundamental ambiguity in the humble mystic's visionary expostulations, at least as these are communicated by many of his descendants, which not only heresy-hunters but theologians of good faith and real depth have marked for criticism. Bouyer did not fail to notice this ambiguity. It has to do with the thought that God must empty Himself for the sake of His creature in order to fulfill His own being.

It has been compellingly argued that it is a category mistake to read Böhme, the visionary mystic, as if he were doing theology or philosophy in his writings.[19] It may well be the case that the systematic theologian is the person least capable on earth of entering into his thought and understanding it. We might admit that Bouyer, however much he extols the dignity of the vocation of the poet and the mystic, is himself too much burdened by his training as a dogmatic theologian to be able to assess the genuine character of Böhme's work. Nevertheless, Böhme's mystical experiences became fodder for the technical works of many theologians and philosophers whose views are not free of this aforementioned ambiguity. Perhaps it is the attempt of these theologians and philosophers, however speculatively brilliant, to fit the expansive visions of Böhme into the straitjacket of reason that is to blame for these distortions.

19 Cf. Michael Martin, *The Submerged Reality*, 39–61. See also Caldecott, *The Radiance of Being*, 251–52. Yet Caldecott warns: "Nor should we assume that Boehme, just because he was a visionary, was always correct either in what he saw or in the way he interpreted and expressed it. His work is something of a stylistic mess, and full of real or apparent self-contradictions."

Whether Bouyer himself makes the category mistake of treating the writings and legacy of Böhme through the heresy-hunting lens of dogmatic theology, we can surely commend the cautious stance that he does indeed take in his dialogue with Behmenite thinkers in *Cosmos* and elsewhere.[20] His entrance into the tradition of modern sophiology is prudent and mindful of the pitfalls of any dialectic that would understand the movement of divine life as an unfolding and actualization of Absolute Idea. The Oratorian's thinking is in line with the Behmenites inasmuch as he seeks to express the loving unity of God with the world recapitulated in the salvific humanity of Christ through the suffering of the Incarnate Lord on Golgotha. He sees the Immolated Lamb as the very foundation of the cosmos and seeks to develop the Patristic tradition beyond its largely implicit sense of the presence of creation in God's Wisdom. He recognizes that the Behmenites, particularly in Russia, have to a large extent uniquely uncovered this fundamentally biblical vision. Nevertheless, he is surely correct to warn us, as he does, that if this development projects the agonistic character of the fallen cosmos into the Godhead then it might cause us to relapse into a pre-biblical, mythic view of the world.[21] The mythopoetic instinct requires sifting by *logos* on this point.

THE WISDOM OF NATURAL PHILOSOPHY

One can see in Bouyer's writings that he does not find the sophiologists to be entirely unproblematic. He nevertheless accords them a place of special importance in modern thought. In *Cosmos*, he shows a particular sympathy for nineteenth-century Catholic sophiologists, whose writings predate that of Soloviev and his followers. Bouyer gives a largely uncritical overview of key dimensions of their mystical theology in chapter fourteen of *Cosmos*, entitled "From Positive Theology to Philosophies of Nature," our focus in this section. The shared efforts of these visionaries to erect a bridge between the nineteenth-century scientific worldview and

20 See especially *Sophia*, 118–22. We note that in this text Bouyer argues that Böhme should not be given quite as much credit as is the case for being the originator of the trend of thought in question. This, he asserts, should go instead to Heinrich Kuhnrath (1560–1605).

21 Caldecott, *The Radiance of Being*, 255.

Christian faith do not, as far as he is concerned, belong in the dustbin of history or relegated to a museum of intellectual curiosities. They may provide the philosophy lacking in the projects of earlier proponents of "positive theology," whom he discusses in an admiring light in this chapter but not without criticism.[22] Three thinkers stand out for him as particularly noteworthy in the recovery of Christian cosmology in this time period, and he focuses on them as the main work of this chapter: Franz von Baader (1765–1841), Louis Bautain (1796–1867), and Anton Günther (1796–1837).[23] It behooves us to pay close attention to his exposition of these thinkers, because he assimilated some of their ontological positions or descriptions in his own thought, which will be apparent in what we expound in our forthcoming chapters.

Franz von Baader was a successful and innovative mining engineer as well as theologian and philosopher. He was a highly influential figure, perhaps even more so in his day than Schelling, whose reputation subsequently surpassed his own, whether justifiably or not.[24] Bouyer once described Baader as a seminal Catholic thinker, a man whose

22 *Cosmos*, 130–31. with regard to the positive theology of 17th-century theologians Denys Petau and Louis Thomassin, Bouyer says: "The only weakness in Petau's admirable efforts is that in wishing to avoid scholastic theology's entrapment in a philosophy which became increasingly abstract and less Christian, he did not sufficiently recognize the need to provide any theology with a philosophical instrument adapted to its own requirements and suitable for communicating with one's contemporaries."

23 Ibid., 135–42. It should be pointed out that Bouyer intends to provide an overview of nineteenth-century philosophy of nature and begins by briefly describing the work of Johann Wolfgang von Goethe (1749–1832) as the starting point. He recognizes Goethe's forays into a science of qualities to be "the most worthy of truly scientific consideration among all the endeavors of German *Naturphilosophen*." He also points to Hermann Lötze, but not in as much detail as these others. The positions he summarizes in regard to the Catholic sophiological cosmologists gets at some important aspects of his own cosmology. Goethe is not as singularly decisive for Bouyer as he is for Balthasar.

24 Cf. Eugene Susini, *Franz von Baader et le Romantisme Mystique: La Philosophie de Franz von Baader,* 2 vols. (Paris: Vrin, 1942). Bouyer references this work in *Cosmos*, 258, n. 21. See also Emmanuel Tourpe, *L'Audace Théosophique de Baader: Premiers Pas dans la Philosophie Religieuse de Franz von Baader (1765–1841)* (Paris: l'Harmattan, 2009). Tourpe's work has the advantage of connecting Baader's thought to Balthasar, who is given considerable treatment in the text, but also to Bouyer.

opinion holds some intellectual heft.[25] In prefacing the exposition of our author's generally flattering albeit succinct summary of Baader's cosmology in chapter fourteen of *Cosmos*, we might take note that the quality of Baader's work as an engineer should give us pause if we want to dismiss his known connection to occult schools of empirical psychology or medicine as being the product of an unstable mind without capacity to see the world as it is in its effective functioning. The idea of Baader's that Bouyer highlights for special consideration has to do with the "relative supernaturalism" of human identity. He quotes in this chapter of *Cosmos* at length from a passage of Baader's that is impossible to paraphrase justly:

> The world is therefore composed of scattered fragments animated by the divine breath and spirit. We cannot grasp the whole. There is in this regard a lack, a gap, whether the fault lies within or outside us, and this is reminiscent of original sin, which is both internal and external to ourselves. Be that as it may, we find in nature only verses to be completed or set straight, *disjecta membra poetae*. And this is consequently how the shares of the sciences, of philosophy and of poetry are apportioned: the role of the scientist is to assemble these fragments; that of the philosopher is to interpret them; and that of the poet is to imitate them, along the lines of ancient poetics, or still more boldly to put them in order... and reestablish their new poetic unity.[26]

Here we have a direct statement of the aesthetic and poetic interpretation of the cosmos and of the human vocation that sophiologists share in common. All creatures, Bouyer says in explanation, are constituted by the region or the circle in which they exist. They are essentially relational: they draw nourishment from below in the matrix of creation and life from above by the power of an active, cosmic source. With respect to

25 *Sophia*, 93. The precise quotation is: "un penseur catholique aussi seminal que Baader." Baader is taken as an estimable authority who showed the importance of developing an orthodox sophiology.

26 *Cosmos*, 136, 259, n. 24. The quotation is taken from Baader's *Aesthetica in Nuce* II, 265.

this bi-directionality of cosmic being we can speak of a supernaturalism of the human with respect to animals while maintaining God's supernatural plenitude with respect to humans.[27] All creatures are capable of existing, organically and dynamically, in a relation of free obedience to the will of God and of being consummated in a supernatural, second birth.[28] They can be enlivened or quickened by a germ of life given from above and thereby empowered to stretch out to that life as the seed of a plant stretches out to the sun. The gift of second birth, according to Bouyer's explanation, can be rejected, at which point organic creation below the level of man becomes mechanical and the path to life above him closed off.[29] Bouyer insists that Baader does not propose a philosophical monism for which physical nature would possess its own pure activity or creativity. Rather, he recognizes that there is a passive, subjective principle in nature that depends upon a transcendent "object" in order to be actualized: the subject depends upon the object and the creature upon the Creator.[30]

Although Bouyer's exposition does not make the point clear, it seems that it is only man and angels to which this discussion of free acceptance of divine gift is pertinent, even if Baader recovers the Pauline view that the entirety of creation is deeply affected by the choice of created freedoms to accept or reject the gift of new life in Christ. Bouyer does explain that should this rejection occur in the case of man, as in fact happened, salvation would require a gift of suffusing love from above to fill the world through humanity. Baader sees authentic knowledge as tied to this grace-infused, higher life in organic unity with God, through a nuptial communion of subject with object. Bouyer summarizes:

> This implies the death of our hardened or armored mode of existence, so that life may recover its ductility. Instead of being consumed, we will then be fulfilled in our higher individuality, while divine transcendence will yet be maintained.[31]

27 *Cosmos*, 136.
28 Ibid., 137.
29 Ibid.
30 Ibid.
31 Ibid.

Salvation brings mystical, infallible knowledge to the human society (the Church) that is organized in obedient relation to God, and the individual member of this society is granted a contingent, communicated infallibility.

The philosopher Louis Bautain, who founded the Institute of Saint Louis in Juilly, is the next sophiological philosopher of nature to whom Bouyer turns his attention in chapter fourteen of *Cosmos*. Bautain follows the organicist and vitalist direction of Baader's thought, and Bouyer exonerates the French philosopher from the charge of heretical fideism, seeing him as a forerunner of the philosopher Maurice Blondel (1861–1949), whose influence on twentieth-century Catholic theology was immense.[32] It is the dimension of Bautain's philosophical thinking directly influenced by his medical studies that most interests Bouyer. Our author suggests that Bautain—himself sometimes referred to as the "French Newman"—may possess by virtue of his medical training the most genuinely scientific cast of mind of all the thinkers in the nineteenth-century Romantic stream of natural philosophy outside of Johann Wolfgang von Goethe (1749–1832). W.M. Horton, Bautain's foremost commentator, says that the mother idea in his thought is the unity of life: "Life, one in itself, one in all the universe."[33] According to Bouyer—assessing the whole of this eminent commentator's book on Bautain—this is to say that "all life is a reciprocating movement between subject and object."[34] The subject, he explains, after encountering the object, is from then on in continual relation of intussusception and polarization with it. Life develops on the basis of being a subject, passive with respect to an active object. Bouyer further explains that in clarifying the need for the object in the development of life in the way that he does, Bautain sets himself apart from the Idealism of Johann Gottlieb Fichte (1762–1814), which he had at one time followed.[35]

This philosophy of subject and object becomes the basis for a nuptial ontology of the whole creation. There is, according to Bautain, a universal,

32 See Oliva Blanchette, *Maurice Blondel: A Philosophical Life* (Grand Rapids, MI: Eerdmans Press, 2010).

33 W.M. Horton, *The Philosophy of Abbé Bautain* (New York: New York University Press, 1926), 122. Bouyer draws heavily on this book in his exposition of Bautain. See *Cosmos*, 259, n. 25.

34 *Cosmos*, 138.

35 Ibid.

male, objective life-principle, and a universal female, subjective, receptive principle. These dual principles, intrinsically ordered to one another, are features of biology, thought, moral life, and spiritual life. "Every living being," Bouyer summarizes, "has its center of indifference, on an axis with one pole pointing toward stimulative life [male] and the other toward the matrix [female]." "In man," Bouyer continues, "the result is head, heart, and abdomen, just as in plants we have the shoot, the seed, and the rootlet."[36] Bautain argued, we learn, that even inorganic being bears analogy to organic life, and he compared the development of inorganic being to embryogenesis through which the living being grows in accordance with cosmic mutual polarity.

Drawing on Horton's analysis, Bouyer briefly expounds four "daughter ideas" (*idées-filles*) in Bautain's thought.[37] The first is that nature and spirit are conjoined. This means that the nature of any created reality is in fact the divine idea of all its possibilities that it contains within itself. Created spirit is drawn by life from nature as subject and exists in two kinds, earthly (animals, plants, minerals, Mother Earth) and heavenly (the celestial spirits). Earthly spirits are fecundated by the rays of the sun, and the celestial spirits or angels are filled with life by God Himself in the Spirit. With Gustav Fechner (1801–1887), Bautain held that there is a matrix or Earth-Spirit from which derives, as Horton says, "the physical life of all earth's creatures."[38] This is distinguished from the World-Spirit or Macrocosm, which, along with man the microcosm, is a "mixed spirit." It is "half of earth and half of heaven, half intelligent and half physical."[39] The third and fourth main consequences are quickly summarized by Bouyer. The third has to do with Bautain's postulation of a trifold structure of the human spirit: intelligence, animal spirit, and the mixed spirit through which intelligence orders sensory data. The fourth has to do with the seminal importance of relationship in his system, which is in fact a key feature of all of the systems of natural philosophy in which Bouyer takes interest.

36 Ibid.

37 Ibid., 138–39. Horton, *The Philosophy of Abbé Bautain*, 124–41. Bouyer does not reference these pages but summarizes them in his usual condensed manner.

38 Horton, *The Philosophy of Abbé Bautain*, 129. We find it necessary to turn directly to Horton's text to understand better Bouyer's exposition.

39 Ibid. Bouyer puts this in his own words in *Cosmos*, 138.

Our author spells out Bautain's second main "daughter idea" in greater detail. It has to do with the being of God the Trinity. Bautain, we learn, links creation very closely with the Trinity, and he associates the divine ideas with the angels. These are both points of crucial importance in Bouyer's own systematic cosmology. His triadology can be understood to be a development of Bautain's theology of triune "unity, duality, duality returning to unity."[40] His description of Bautain on the link between angels, divine ideas, and divine Wisdom is worth quoting in full and might well summarize his own position, with due qualifications regarding the origin of space and time:

> God is totally self-sufficient, Bautain emphasized, and did not need the universe, which was nevertheless created out of pure love, and which is therefore entirely dependent on divine love. However, Bautain held, God cannot have created this world bearing the blemishes of time, space, death, and suffering. The heavenly world is perfect and changeless, eternally conceived and begotten in the bosom of divine Wisdom, and forever fertilized by the influx of divine life, by the Spirit. Each point of the sphere of Wisdom is a potential center of life. Fertilized by divine life, it reaches eternal existence and becomes an immortal spirit. In other words, the ideas harbored in the divine mind are actually angels. The spatial and temporal universe derives from the contemplation of an angel falling back upon itself: the spirit of the world, in which our organisms are all involved. For evil is but the isolation of the particular in the universal. The first effect of evil is time, produced by the introversion of the creature, and the product of isolated reflection is space.[41]

The Bautainian spatial description of the divine Wisdom, a heavenly world of divine ideas, is imaginative. One can speak only analogically of "points" of potency in the sphere of Wisdom. The description captures the inner connection between divine and created being, the latter first of all existing as pure spirit or actualized idea through the breath of the

40 *Cosmos*, 139.
41 Ibid.

Holy Spirit. Do space and time as we know them exist only because of the fall of the angels? This seems to be a consideration that Bouyer takes very seriously, even if he does not embrace the idea in full, and we shall explore it further in chapter ten.

Our author concludes chapter fourteen of *Cosmos* with a succinct description of key ideas in the philosophy of nature of the German Catholic philosopher Anton Günther which he will incorporate into his own systematic retrospections in the final chapters of the book. In fact, among all the Romantic cosmologists, Bouyer suggests that Günther may be the one to have provided "the deepest insights."[42] He insists that this Austrian Catholic priest and theologian endeavored to overturn Hegelian pantheism by employing its own conceptual weapons against it. It is of interest to note that although nowadays largely forgotten, Günther inspired in his own day a large and influential following. Several of his disciples earned chairs in Catholic philosophy at the university level in Europe. Eventually, though, his system was called into question by the ecclesial authorities in Rome, and in 1857 his books were placed on the *Index Librorum Prohibitorum* (Index of Forbidden Books). One of the fundamental charges against him was that he promoted rationalism, failing to distinguish properly the knowledge of God made available through natural reason from the knowledge of God made available through the light of faith.

Although a fair assessment of his work indicates that the charges against him were hardly baseless, Bouyer remarkably seems to suggest in *Cosmos* that he was entirely misunderstood on this score, although his assessment of Günther is a little more mixed elsewhere.[43] In *Cosmos*, Bouyer claims that Günther was not a rationalist seeking to attain faith on the basis of reason but a man whose operative principle was in line with Saint Anselm: *fides quaerens intellectum*. In associating Günther with Saint Anselm, both thinkers are given a boost, and the latter is exonerated for a time from the taint of rationalism. There is, Bouyer maintains, a distinction that one has to uphold for proper interpretation of the Germanic theologian and philosopher:

42 Ibid., 141.
43 See *IF*, 294–95; *Sophia*, 165.

> It is ... true that Günther may have inadvertently distorted the revealed truths whose supernatural nature he sought to emphasize. On this question, however, of the interpretation of a cosmos created by a transcendent God, whose transcendence is above all that of a boundlessly generous love, it must be recognized that Günther has many important insights. His entire thinking must be seen as an effort to assimilate the element of truth in nineteenth-century German idealism, and mainly in Hegelianism, while decisively correcting its extreme immanentism. Much more so than Barth, his doctrine may be described as a dialectical theology.[44]

The elements of Günther's thought that Bouyer picks out for consideration may indeed help us to understand a little better the Oratorian's own way of establishing the connection between God, humanity, and cosmos. Günther, Bouyer explains, sought to explore the nature of the cosmos, including our own bodies and sense appetites, from the standpoint of the entrance of the world into human consciousness and to see it from within the experience of our reflective selves. This led him to postulate an analogy between human and cosmic life. "If therefore," Bouyer explains, "I see myself as the microcosm through which the nature of the cosmos can be revealed to me—as an idea coinciding with reality, and not merely as a concept replacing reality—I must necessarily view the world as composed (as I am myself) of spirit, of nature, and of their synthesis in consciousness."[45]

In this understanding, matter is inseparable from life and nature universally exhibits a vitality and propensity to develop in the manner of the human organism. Both humanity and the world, we learn, are comprised of trifold substantiality that is one in form, as body and soul are united in consciousness. Neither humanity nor the world is self-subsistent. All finite and dependent being depends on the metaphysically prior existence of an independent *Urgrund*, or perfect being, in order to be. There is, Bouyer explains, analogical correlativity between God and creation, but this relationship of analogy is not, like that between humanity and

44 *Cosmos*, 141.
45 Ibid.

cosmos, a direct one. Instead, it is a "reverse analogy." The world is "the counter-position of God" (*der contraponierte Gott*).[46] If the world is, we are told, trifold in substance yet one in form, God is one in substance, yet trifold in form. God is Subject, transcendent Object, and their relational Unity—Father, Son, and Holy Spirit. The *analogia entis* as articulated within the dialectical scheme of Günther's thought is explicitly Trinitarian. We can readily gather from Bouyer's exposition of Günther the Oratorian's recognition that it is desirable to pay heed to nineteenth- and twentieth-century efforts in Catholic theology to provide a more comprehensive Trinitarian basis and understanding of the history of creation than had been common in mainstream Neoscholasticism.

SCIENCE ON THE WAY TO WISDOM

However stimulating some of the ideas proposed by nineteenth-century philosophers of nature may be, it cannot be denied that the works of these figures did not attain genuine and lasting scientific success. Bouyer uses their ideas in his retrospective overview of theological cosmology that we shall expound and analyze beginning in the next chapter, but it is in a theological or philosophical mode of presentation that he does so. In the genealogical portion of the book, he finally turns to the more seemingly assured accomplishments of twentieth-century science in order to open our eyes to deeper contemplation of the cosmos. Modern science, he realized, has become itself a potential source for discourse on wisdom.

In chapter fifteen of the book, entitled "From Positivist Science to the Rediscovery of the Spirit," our concern in the present section of this chapter, Bouyer details shifting understandings in physics, biology, and psychology that marked the twentieth century, opening science to the reality of spirit, and we want to treat each of these in turn in the present section. Grintchenko cautions that one should keep in mind that the Oratorian was, in interpreting these matters, a man of his times. We must be careful, she advises, to practice discernment regarding "the scientific propositions then held to reconcile science and faith . . . on which he bases himself."[47] Furthermore, she suggests, the progress of science

46 Ibid., 142

47 Grintchenko, "*Cosmos*: Une Vision Liturgique du Monde," *La Théologie de Louis Bouyer*, 116.

since his time has to be taken into account in order to deepen and renew dialogue between science and faith.

All of this is, of course, quite true, and Bouyer himself would surely agree. At the same time, it should be noted that the Oratorian was in dialogue with some great minds on the question of science and faith whose insights were not simply reducible to passing fads. In addition to his already-mentioned connection with Henri Ellenberger, we would point especially to the kinship between Bouyer's thought and that of the physicist and philosopher of science Olivier Costa de Beauregard (1911–2007), who had been a student of the eminent French physicist Louis de Broglie (1892–1987).[48] Bouyer initiated a correspondence with Beauregard and read several of the physicist's writings. He met with him to discuss what the two men apparently held to be a shared endeavor to give a new, properly spiritual interpretation of the world in light of the implications of twentieth-century physics.[49] Bouyer in fact dedicated *Cosmos* to Beauregard, although the English translation of the book does not show it.[50] We take this to be a highly significant point and want to explore the connection between these two thinkers in some depth in chapter ten. For now, we think it suffices simply to mention this connection in order to suggest that Bouyer's succinct exposition of revolutions

48 See especially Olivier Costa de Beauregard, *Le Second Principe de la Science du Temps* (Paris: Éditions du Seuil, 1963); *La Physique Moderne et les Pouvoirs de l'Espirit* (Paris: Greco, 1981). Davide Zordan says that Beauregard's work is singular for having included certain traditionally philosophical themes in the domain of physics which had been excluded by mechanist materialism, including the role of consciousness in the world, the inversion of time, and the reciprocity of cause and effect. See Zordan, *Connaissance et Mystère*, 729, n. 2.

49 Beauregard's daughter, Evelyne Tritsch, kindly emailed us copies of two letters that Bouyer sent to Beauregard found in the latter's archives. These were previously unknown to scholars of Bouyer. One is dated November 27, 1964. The other is dated November 23, 1971. In the first, Bouyer expresses his gratitude to Beauregard's work for posing the question of how freedom, creation, and "spirit" can be posed in the context of contemporary science, and he queries Beauregard on whether and to what extent modern physics has demolished the crisis of modern determinism. In the second, he expresses regret at not having responded sooner to one of Beauregard's letters to him, and he commends Beauregard again for the importance of his work.

50 This dedication is in the French-language edition but is missing in the English-language edition.

in twentieth-century physics in chapter fifteen of *Cosmos* is not simply the product of his own, isolated, non-expert thought. We shall treat, in three subsections, first of the new physics, second of the new biology, and third of the new psychology, as interpreted by Bouyer.

The new physics

The Oratorian is in the same current of judgment as Beauregard and many others in suggesting that twentieth-century revolutions in physics have exploded the Newtonian, mechanist world-picture that the nineteenth-century philosophers of nature tried less reliably to overcome by reviving in sophianic context ancient forms of organicism and vitalism.[51] The mechanist world-picture proposed, as is known, the atom to be the fundamental constituent of being. As Bouyer explains in chapter fifteen, this basic reality was thought to be a stable solar system in itself, with heavy, positively-charged protons in its nucleus orbited by light, negatively-charged electrons on its inner periphery. The transfer of electrons from one stable atom to the next was thought to be able to account for electrical charge and all possible chemical reactions.[52] In principle, these transferences could be predicted *in toto* by precise measurements. Radiation in x-rays or light was construed as a projection of waves emitted by the vibration of electrons shifting from their regular orbits propagated through a universal ether or elastic medium.[53] Bouyer relates the often-told story that physicists in the late nineteenth-century confidently asserted that they had discovered the precise physical constituents of the world and were now ready to forecast the future development of the world on the basis of the laws that govern the structure and action of these constituents. "It did not seem unrealistic to hope," he summarizes, "that Newtonian physics of the macrocosm and the new physical chemistry of the microcosm could be unified in an all-embracing formula of universal determinism."[54] This unification would make it possible at last to carry out in full the much-hoped-for reduction of the qualitative realm of direct human experience to the quantitative domain studied by physics.

51 *Cosmos*, 143–48.
52 Ibid., 143.
53 Ibid., 143–44.
54 Ibid., 144.

The dream of the reductionists was not in fact a source of hope and a project for the future of the human race shared by all distinctively modern thinkers. The nineteenth-century philosophers of nature that Bouyer favors certainly understood this project to be dubious. They were not alone or the first thinkers to have cast the modern dream in an alternative light. At the very beginning of modern philosophy, George Berkeley, the Anglican Bishop of Cloyne (1685–1753), saw it as a deep aberration. Bouyer briefly summarizes Berkeley in chapter fourteen of *Cosmos* (and elsewhere in the book) to help understand the epistemological muddle prevalent in interpretations of modern physics. Much as Alfred North Whitehead did, our author thinks that Berkeley got to the heart of this muddle that contributes mightily to the vitiation of modern thought and culture, the so-called "bifurcation of nature."[55] Against the growing mechanist materialism already emerging in his day, Berkeley held that the concrete, objective existence of the world is irreducible to so-called "primary qualities" such as mass, extension, and velocity. As Bouyer explains, Berkeley realized that one cannot maintain consistently that these inaptly described *qualia*, which have no phenomenality, are more real than "secondary qualities," such as color, sound, and taste, which are, as phenomenologists have noted, in fact the only *qualia* rightly so-called.[56]

The Anglican Bishop of Cloyne did not hold that extramental realities are inexistent, but that they exist only in the context of genuine

55 See Alfred North Whitehead, *The Concept of Nature* (Cambridge: Cambridge University Press, 1920), 26–48. The bifurcation of nature has to do with the Cartesian splitting apart of the world into extended bodies (*res extensa*) and the human soul (*res cogitans*). Inevitably, the latter gets reduced to the former, and inappropriately described "primary qualities" are fallaciously accorded the status of ultimate reality. There are remarkable similarities between Bouyer and Whitehead in their analyses of the advent and error of modern scientific materialism.

56 *Cosmos*, 132–33. Bouyer argues that the true meaning of Berkeley's philosophy of knowledge has been recovered by the editors of a critical edition of his collected works, Arthur Aston Luce (1882–1977) and Thomas Edmond Jessop (1896–1980). See *The Collected Works of George Berkeley, Bishop of Cloyne* (London: Nelson, 1964). Bouyer commends Luce's independent studies of Berkeley. He cites in particular *Sense Without Matter, or Direct Perception* (London: Nelson, 1954). See *Cosmos*, 258, n. 15, and 269, n. 1. Bouyer maintained a personal correspondence with Luce. See *Memoirs*, 55.

qualia, in the rich manifold of perceptible being. The implication of his argument, rightly understood, is that it is the philosophical materialists, not Berkeley himself, who succumb to subjectivist-idealist deformation, for they were the ones to perpetrate the fallacy that mistakes mental abstractions for concrete realities.[57] Berkeley's axiom "to be is to be perceived" (*esse est percipi*), on this reading, does not lock nature into the inner cabinet of the human mind but expresses instead the truth that the matter of the world is inseparable from the plane of qualitative existence and ultimately from the realm of spirit. Bouyer explains that the world for Berkeley is a language through which spiritual beings communicate with one another.[58]

This interpretation of Berkeley places him as a forerunner of the most sophisticated philosophical interpreters of modern physics. Bouyer suggests that there is confirmation of Berkeley's so-called "immaterialism" in these scientific advances rightly understood, which have compelled some philosophers of science to take account of the levels of abstraction that are necessarily at play in the scientific enterprise, moving us to confront the symbolic character of our theories and formulas.[59] The particular developments that Bouyer focuses on in chapter fifteen are well-known in popular literature. One such development is theories of "quanta" which seem to show that there is a fundamental indeterminacy in our physical descriptions. We cannot, as we learn from the Heisenberg Uncertainty Principle, know at one and the same time both

57 Alfred North Whitehead, *Science and the Modern World* (Cambridge University Press, 1925), 64, 72.

58 *Cosmos*, 133. Bouyer points especially to Berkeley's *Siris* as his most developed reflection on this point. We shall discuss Bouyer's interpretation of Berkeley further in chapter 10.

59 See Aimé Michel, "La Gnose de Princeton," *France Catholique-Ecclésia*, 207, no. 1487 (June 13, 1975). This article is a review of Raymond Ruyer's book by this title. In addition to the works of Beauregard already cited, Bouyer sees this book by Ruyer and a book by Michel as important French contributions to understanding the relationship between theology and science in light of the new physics. See Ruyer, *La Gnose de Princeton* (Paris: Fayard, 1974); Michel, *Métanoia: Phénomènes Physique du Mysticisme* (Paris: Albin Michel, 1973), 133–35. It should be noted that the thought of both Whitehead and Ruyer is subject to recent recovery by thinkers influenced by the French philosopher Gilles Deleuze (1925–1995).

the position and the velocity of particles.[60] Any determination that we make exists together with an even greater indetermination. At the same time, we are compelled by this line of physical explanation to reckon with the fact that our observations of the world modify physical nature, at least on the microcosmic domain.[61] Science, then, does not confront a pure object that remains unaffected by the scientific observer. The objectivity of science conforms to the conditions in which the scientist places him or herself in relation to the world. Bouyer briefly points in chapter fifteen to another development that shattered classical physics—Einstein's theory of relativity, which proposes that space and time are wedded together in the configuration of space-time curvature.[62] This should, Bouyer suggests, disabuse cosmologists of the illusion of being able to develop a coherent world-picture corresponding to clear-cut subject-object duality.

The upshot, as Bouyer describes it in this chapter, is that science has come face-to-face with the inescapable presence of symbolic discourse in its operations.[63] The waves and particles of modern physics are not, he explains, realities that can account for the perceptible world but symbolic figurations that enable us to abstract a dimension of reality from the whole. Our author insists that this is not to say that the objects of physics are wholly "un-objective," but that their objectivity is partial, fragmentary, conforming to a subjective mode of thought, a particular way through which humanity seeks to bring greater order to the world. The idea of pure matter, he says, "has vanished as far as its materiality is concerned."[64] Really existing material being present to human experi-

60 *Cosmos*, 145.

61 Ibid.

62 Ibid.

63 Whitehead, *Symbolism*, 30–59. Recall that Bouyer commends the great importance of this book. See also E.L. Mascall, *Christian Theology and Natural Science* (Archon Books, 1965), 47–90; I.T. Ramsey, *Religion and Science: Conflict and Synthesis* (London, SPCK, 1964). Both of these latter books discuss the role of language and symbol in the development of scientific theories. Bouyer was a friend of Mascall and Ramsey, both Anglican theologians, and he thought their contributions in this area, especially the latter's, were important. Ramsey was especially cognizant of advances in linguistic philosophy from the time of Wittgenstein. Mascall was an Anglo-Thomist.

64 *Cosmos*, 146.

ence subsists only as "in-formed." Matter does not produce thought but is reducible to it inasmuch as it has no pure, unmixed existence except as an abstraction. As Bouyer had already explained in *Rite and Man*, matter is "a complex of intelligible qualities, of mathematical relations between forces that are themselves immaterial, considered in abstraction from the mind that thinks of them."[65] The cosmos in its true, total, "objective" reality is qualitative and fit for contemplation by created spirits. Ultimately, it is indeed true to say that to be is to be perceived: creation comes from the thought and will of God "who 'thinks' the universe into existence by being the object of his own thinking [perceiving]."[66]

The new biology

The historical, developmental, and temporal constitution of human and cosmic life as scientifically discerned was likewise at the forefront of Bouyer's thinking on science. He often invoked the language of evolution in his writings and explores Darwinism briefly in chapter fifteen of *Cosmos*. He did not, unlike various disciples of Teilhard de Chardin, take evolution to be the needed master and controlling category of human thought, as if this theory of human origins should lead to fundamental changes in Catholic doctrine. He even once went so far as to suggest that the evolutionary view of humanity may amount to little more than an *a priori* form of sensibility in the Kantian sense, "without," he wryly suggested, "even the compensation, in this case, of there being a 'transcendental aesthetic' awaiting discovery at the root of human consciousness."[67] The Oratorian evinces caution or prudence with respect to theological adoption of modern scientific theories. His discussion of modern scientific theories in chapter fifteen of *Cosmos*, as richly suggestive as it may be, is in fact truncated, which is indicative of his cautious approach. Even with respect to modern physics he prudently warns that we may await a future development that would invalidate and supersede our present scientific cosmology.[68]

65 *Rite and Man*, 22.

66 Ibid.

67 *IF*, 5–6.

68 *Cosmos*, 147. Yet he holds that these future possible returns will not be by way of the recrudescence of mechanist materialism.

That being said, Bouyer does attempt to integrate evolution into the overarching story of creation that he ultimately tells. In his final book, he asserted that it is odd that theologians in the eighteenth and nineteenth centuries should have found the idea inimical to Christian faith.[69] If Bouyer does not think that we have to reorder theological doctrine or human culture in light of theories of evolution, he does recognize the need to take evolution into account in theological cosmology. His theology of creation thus provides an alternative to usual efforts by some contemporary evolutionary scientists to give a "Big History" of the cosmos.[70] For these scientists as for Bouyer, history does indeed extend to the farthest reaches of time and space in the past. Thinkers who engage in Big History have the merit of possessing an implicit understanding that narrative is essential to ontology to the extent that they implicitly think ontology in narrative terms. Bouyer's story of development extends just as far as those of the raconteurs of Big History, but it does not presume that created spirit is, at best, an epiphenomenon of material processes. Created spirit has been copresent with the cosmos from the moment of the Big Bang (or before). It has been present throughout the whole course of cosmic development, from its "state of maximum concentration through an irreversible process."[71] The cosmos is ever in development, but it is not infinite in temporal expanse. This development, Bouyer urges, is mysterious from an ontological standpoint because the Second Law of Thermodynamics suggests that the universe is "running down," subject to entropy and decay. Is new life and movement possible only because there is spontaneous generation of matter from nothing, as Fred Hoyle (1915–2001) proposed in Bouyer's time and Stephen Hawking

69 *Sophia*, 158: "It is strange in this regard—and this signals an undeniable weakness of the theological thought of the eighteenth and nineteenth centuries—, that many Christians are frightened at this point before the discovery of evolution in general (does it not correspond to the very vision of Genesis?) and more precisely of theories proposed from this evolution, as those of Darwin and his successors."

70 Cf. John Haught, *The New Cosmic Story: Inside Our Awakening Universe* (New Haven, CT: Yale University Press, 2017), 1–6. In this book, Haught, a disciple of Teilhard de Chardin, attempts to provide a Teilhardian alternative to the usually atheistic accounts of "Big History" in contemporary literature.

71 *Cosmos*, 147. Big Bang theory originated with Fr. George Lemaître (1894–1966), a professor of physics at the Catholic University of Leuven.

in our own? In this chapter of *Cosmos*, Bouyer is content to suggest that the development of life on earth stubbornly challenges the principle of entropy or of the degradation of energy in the cosmos.[72] He argues here that the influence and power of a transcendent final cause has to be affirmed and understood in order to elucidate the causal basis of the movement of creation to the development of new life, which is not a mechanism of collected parts but instantiated through the organism, the whole of which is greater than the sum of its parts.[73] As we shall see, he has much more to suggest in this regard in later chapters.

Our author does not think it makes sense to speak of the transformation or evolution of species without the allowance of formal and final causality in one's explanatory apparatus. The postulation of blind chance or pure randomness as causal power is meaningless to him. Darwin's principle of natural selection by chance variation, reduced to the struggle of the fittest for survival, does not, he thinks, even rise to the level of a *petitio principii*. It is simply an unverifiable definition.[74] With the best of the French Structuralists when at their best, Bouyer pointed out the extent to which Darwin's understanding of evolution was a form of idolatrous self-projection onto the screen of the cosmos. "For the 'survival of the fittest' implies as a basis for its assumed rationality, an underlying divinity who is not only in the image of men, but in the image of typically Victorian man."[75]

None of this should be taken to imply that Bouyer denies the reality of evolution. There is indeed, he affirms in chapter fifteen of *Cosmos*, an evolution of "living beings toward forms better and better integrated in their increasing complexity."[76] But this is an "emergent evolution," which is to say that it cannot be explained on the basis of an inexplicable condition of initial disorder. We can understand the development of life only by reference to a final order immanent and transcendent to the world. Evolution is unintelligible in purely materialistic terms. Bouyer explains: "Consistently and more and more clearly as one ascends the

72 Ibid.
73 Ibid., 147–49.
74 Ibid., 148.
75 Ibid., 149. See also *IF*, 50.
76 *Cosmos*, 149.

scale of integrated complexity, it is therefore the spirit rather than either chance . . . or matter considered to be alien to any spirituality, which can make the development of the world intelligible."[77]

In *Rite and Man*, Bouyer explained that "the philosophy of emergence" should impel us to take account of the development of the world in terms of its discrete, definite, irreducible stages, culminating in the emergence of human mind.[78] The world is not, on this view, formed by an absolutely continuous process. There is discontinuity as well as continuity. There is continuous development from "inanimate matter to living matter, from living matter to man, and from man to his elevation to the supernatural order."[79] At the same time, the different stages in this process each have unique, ontological integrity, irreducible to that of previous stages. The higher, more advanced stages of intelligible structure are explicatory of what came before rather than the reverse: "each essentially new link in the evolutionary process is necessary for a reflex knowledge of the preceding one."[80] Only the appearance of the highest reality can explicate the meaning of preparatory factors: "instead of any higher stage being reducible to the lower, like the parts of a telescope, it is only the higher stages that reveal all the potentialities of the lower, for it is only then that these potentialities emerge fully activated."[81]

From what has just been said, it is clear that in some ways Bouyer's thinking on evolution aligns with Teilhard de Chardin's. Like the Jesuit paleontologist, he sees meaning in the development of physical life in light of the development in man of reflective consciousness, which enables the cosmos to be known from within its own materiality. Both Teilhard and Bouyer agree that consciousness reaches a new plateau of perfection in the Incarnation of Christ and the communication of God's agapeic, Trinitarian love in the communion of the Church. Yet, to return to our previous chapter, the Oratorian reorients Teilhard's incarnationalist humanism along eschatological lines. Unlike Teilhard, he does not

77 Ibid.

78 *Rite and Man*, 22. The philosophy of emergence is associated with Whitehead and John Dewey (1859–1952).

79 Ibid., 23.

80 Ibid.

81 Ibid.

propose that humanity waits for a new dawn of consciousness in the emergence of "religionless Christianity," a secular Catholicism that would surpass the supposedly Neolithic religiosity presumably all-too-present in the ritualism of the Church. The cosmos, Bouyer holds, is indeed evolutionary, and Christ is the Omega Point, but as we await his Parousial work of ultimate reconciliation, we are not entering a new axial age of transformed consciousness in which the Church can be refounded according to the dictates of the secularized consciousness prevalent in late capitalism, whether in individuals of overtly capitalist or overtly Marxian persuasion. Christ's redemptive work has already been definitively accomplished. He already founded his Church.

The new psychology

In various points in his writings, including the penultimate section of chapter fifteen of *Cosmos*, Bouyer suggests that insights from twentieth-century depth psychology are significant with respect to the future "evolution" of human civilization. From what we have just said, we do not want to be taken to imply that Bouyer does not think that societies can develop or progress or that science cannot at all assist in this process. Science and technology become problematic for him when they replace religion in magical fashion. When this happens, we fall into an anti-Christic distortion of the role of science and technology, as his pseudonymous novel *Prélude à l'Apocalypse* especially teaches us. He suggests in his writings that depth psychology might in fact play a role in helping to loosen the hold that the distorted use of science and technology has over our lives by reacquainting us with the irrational, with the surd of existence that we have to integrate in our wisdom in order to have a unified view of reality. Bouyer advocates a critical though appreciative reception of this psychology even by theologians. He had already done so in the 1950s, thereby outpacing most other theologians. By the 1970s, theology was awash in the concepts and lingo of depth psychology, uncritically received, and Bouyer's opening to the discipline might have appeared antiquated by that time, because he did not, unlike many in the new breed of theologian-psychologists, suggest that the new psychology could govern theology and adjudicate the soundness of its propositions. Bouyer was sternly critical of what would become a ubiquitous conflation of theology or spirituality with depth psychology. Nevertheless, he thought that advances in psychology especially with regard to the discovery of the unconscious mind might be immensely important to the future of humanity. Indeed, he went so far as to suggest that these may be even more important in the long run than the discovery of nuclear fission or cybernetics. He wrote about this in an article that appeared in *France Catholique* in 1971. We shall follow this article as a point of departure to end the section, which gives a fuller portrait of his view on these matters than his brief overview toward the end of chapter fifteen of *Cosmos*, which we shall nevertheless also reference.[82]

The French theologian suggests in this article that not only has the new psychology discovered a depth to the mind inaccessible to behaviorist modes of analysis, it has opened the "science of the soul" to a more artistic and ultimately sapiential perspective. He claims that it "represents the invasion into the scientific form of medicine of pre- or para-scientific medicines."[83] Even if the practitioners of depth psychology have not always been cognizant of the point, they have recovered the human need for an art of living, for a body of wisdom teachings that is more than a science and can never be exhaustively reduced into clear and distinct ideas without losing its substance. Bouyer recognizes that the immediate roots of twentieth-century depth psychology lie in the nineteenth-century philosophies of nature that brought to light realities that science cannot fully grasp in its nets.[84] These philosophies are, he suggests, as much poetry as science or rational knowledge. They are more attuned to *mythos* than to critical *logos*. The most important of these are surely to his mind the Catholic sophiologies that he summarizes in *Cosmos*, although he does not mention these in his article from 1971. Bouyer contends in this article that the philosophers of nature and their twentieth-century heirs

82 "La découverte de l'inconscient," in *France Catholique* (December 31, 2011): running text, https://www.france-catholique.fr/LA-DECOUVERTE-DE-L-INCONSCIENT.html. The article first appeared in this magazine on March 5, 1971. It is a review of Ellenberger's *The Discovery of the Unconscious.*

83 Ibid.

84 See also *Cosmos*, 133–34. Bouyer suggests here that the thought of Leibniz (1646–1716) is at the root of modern theories of the unconscious. We saw above that Bouyer recognized the decisive influence of Böhme in Hegel's theory of the collective consciousness.

in the field of depth psychology recognize correctly that the scientific comprehension of truth does not exhaust the totality of being. There is a return to the "premodern" irrational in these thinkers, and this is not wholly to be deplored. It is better, he concludes, to acknowledge with the depth psychologists in all frankness the limitations of human rationality and to draw as rationally as possible the consequences of this than to persist in concealing the irrational with the mask of science.

In chapter fifteen of *Cosmos*, our author emphasizes that the work of the best depth psychologists gives credible testimony to the reality of psychic phenomena that reside beneath the rational mind, affecting its operations. He suggests that these explorers of the human mind have shown that psychological phenomena are not epiphenomenal to man's very physiology and biology but are primary, capable of governing the latter. Tracing the depths of human consciousness to the level of a collective unconsciousness, these psychologists have inadvertently pointed to the existence of a divine mind and will that can alone make sense of the structuring of phenomenality by which the world is given to us:

> Man's psyche now appears irreducible to any other reality and capable of influencing and even governing his physiology, and through it his biology, and therefore of having a decisive impact on the actual appearance of the living body, and even of inanimate bodies. These conclusions increasingly compel recognition as we work our way backward from immediate individual consciousness to its wellspring in the subconscious of the individual past, then to experiences completely submerged in the unconscious memory of early childhood, and finally to some form of collective unconsciousness which is actually the atavistic consciousness of the entire race, finally emerging from the preconscious state of a physical universe which must now be recognized as having no meaning or existence unless they derive from a mind or will without which the phenomenality of matter would be inconceivable.[85]

85 *Cosmos*, 150. See also *Rite and Man*, 23.

Here we get at Bouyer's own view of the unity in mutual implication of the history of the individual, the history of the human race, and the history of the cosmos. Depth psychology takes up the direction of nineteenth-century philosophy of nature according to which "nature" exists as a potentiality for the development of consciousness. Nature is the objective Fichtean "not-I" required in the polar structure of being for the emergence of self-reflective consciousness. The "collective unconsciousness" to which Bouyer refers is a concept that has its origin in the modern age ultimately in the Böhmian view of the unconscious, the divine abyss out of which God Himself emerges in the drive of His will to attain self-knowledge.[86] Introducing the "Retrospections" section of *Cosmos*, Bouyer adopts the language of *Ungrund* to describe his own understanding of the divine life. This Böhmian expression originally referenced an unconscious will as the origin of all things, an impersonal, "dark ground of spirit," even in the divine life. Bouyer himself speaks of "God's unfathomable depths . . . from which everything not only in ourselves but even in God derives."[87] This is an audacious correlation for such an orthodox theologian to make. Does it not depersonalize God? On the contrary. For Bouyer, these unfathomable depths of divine being are associated with the Invisible Father, with the first person of the Trinity, the fontal source of divinity, on whom everything in creation as in the divine life depends, who Himself has no further origin.[88] They are not associated with an impersonal, primordial ground, swathed in darkness. Bouyer affirms the eternity of God in perfect, self-conscious, personal existence, in the dynamic generation of the eternal present in triune personality, and does not think that God has need to create in order to become self-conscious or personalized. Yet this depth dimension of divine being in the Invisible Father is, he holds, reflected in creation precisely in the matter from which human life emerges and in the collective unconsciousness of humanity through which individual human spirits, in their reciprocal interplay in the maternal matrix of society, are deeply intertwined with physical creation. This dark and "irrational" matrix of

86 See S.J. McGrath, *The Dark Ground of Spirit: Schelling and the Unconscious* (New York: Routledge, 2012), 44–81.

87 *Cosmos*, 181.

88 *ES*, 406.

cosmic life is not the polar object through which divine subjectivity is brought to the perfection of personalization but is the deep ground of fleshly human spirit within creation through which ultimately the created soul of Christ is given being in the womb of Mary. Explaining the findings of the Hungarian Jewish psychiatrist Leopold Szondi (1893–1986) and his "progressive discovery" that human creativity is characterized by "the conscious assimilation of all the [unconscious] determinants from which it [the individual] emerges," Bouyer says toward the end of chapter fifteen of *Cosmos*:

> This can have no other conceivable meaning than to reflect the primordial inclusion of these psychophysiological determinants in a creative liberty that is prior and superior to all these organisms and to the cosmic system from which the body and soul of each individual human being finally derive, within what we call the social body and the collective animation of all mankind. Prompted by an "oversoul" remaining enigmatic as long as it does not become manifest as the divine Spirit, calling from the beginning of time and throughout the history of the cosmos for the appearance of the soul of Christ, through which all human souls are to be reconciled with each other and with the invisible Father, this collective and universal animation will finally show itself in the creature becoming one, when its evolution is completed, with the eternal Word in which God reveals and offers himself forever.[89]

SPIRIT AND NATURE

This just-quoted passage articulates Bouyer's own view much more than that of Szondi or of any other eminent scientific authority on the unconscious depths of the human mind. Within this passage we slip from the Szondian view of the creative impulse of the unconscious mind to the Bouyerian theological view of universal cosmic history. This passage succinctly brings together in a single view a Trinitarian theology of salvation history, an emergentist construal of cosmic development, and an

89 *Cosmos*, 150.

invocation of the theme prevalent not only in depth psychology but in the Behmenite tradition of a collective consciousness and unconsciousness, placed in the context of a nuptial ontology as we described it in our second chapter. This passage indicates a vision of creation and human history seen in their unity, as God's transcendent Spirit cooperates with angelic and human spirit(s) in the development of physical creation throughout the vast scope of space-time. The special cosmic mission of the Holy Spirit comes to the fore. The second part of *Cosmos* will connect the "phenomenology of the Spirit" invoked here with divine and created Wisdom. Balthasar summarizes the Bouyerian view that this passage begins to call forth: "Spirit, right from the first creation, is the One who orders the world of nature; he shows himself to be the natural-supernatural Wisdom of God poured out over the entire world, yearning for the New Creation (Rom. 8) that is initiated through the Spirit's overshadowing of Mary and in the divine-human fruit of her womb; in him, its Firstborn, the New Creation is straining toward its eschatological fulfillment."[90] Wisdom or Sophia, on Balthasar's interpretation of Bouyer, is representative of the person of the Holy Spirit, who, as consubstantial with the Father and the Son, is supernatural but is also united with the mysterious grace which truly becomes an accidental form of created spirits by divinization in and through the Church.

We can begin to see now in retrospect and more precisely the transformative manner in which Bouyer takes up the crucial philosophical problematic that inspired development of the philosophy of nature in the nineteenth century, having to do with the intrinsic relation between spirit, nature, and history. Hegel's system, as Bouyer explains in *The Invisible Father* though not in *Cosmos*, especially left the task to Christianity "to show itself capable of integrating the truth within it: its sense of the essential kinship between the divine and the human, as well as the need of lifting everything human, and even everything cosmic, into the divinizing work of grace."[91] In general, Idealist modes of thought understood the history of nature to be the total progress of human civilization, which the early Schelling and the whole Hegel took to be the decisive fact of

90 Balthasar, *Theo-logic III*, 60. Balthasar does not reference this particular passage.

91 *IF*, 294.

existence that needed to be explained.[92] Although the Catholic philosophers of nature that Bouyer prefers did not do so, the early Schelling and Hegel thought that Absolute Spirit comes into the fullness of being through the plenary development of human existence. Hegel promoted a historical immanentist worldview—a "pantheism of progress." History was thought to be a dialectical process through which God attains self-consciousness by the unfolding of an absolute logic that can be uncovered through the study of the philosophy of history. Nature was held to be an essential dimension of this process, because it was needed as the other of self-consciousness that makes it possible for reflective subjectivity to emerge in fullness. Creation is conceived in necessitarian terms. It is in the context of this understanding of objective nature that Schelling developed his theory of the unconscious, the pre-reflective, absolute presence of subjectivity, whose inner drive leads both to objectification of the world and overcoming of it through the achievement of subjective idealization. The world was understood to be a single tissue, thread, or chain of being, with nature and history woven together in a unified process. The perfect realization of humanity was understood to be the final purpose of this process. The human was the God of the world, the God-world, and progress was the history of this God.[93]

This story could be clothed in a Christian veneer. Christ could be presented as a symbol of the perfection of humanity through which Absolute Spirit is born. And so it was with Hegel's system, as Bouyer explains in *The Invisible Father*:

> [T]he whole of reality is nothing but the unfolding of a thought which has to objectify, project, situate itself outside itself, and hence alienate itself in order to find itself once again in the human mind. This development is accomplished by mind entering into full self-knowledge as it progressively assimilates intellectually first, the nature in which it is born by a process

92 See Charles Secrétan, *La Philosophie de la Liberté*, vol. I (Paris: Chez L. Hachette, 1849), 250. We reference this book because Bouyer read Secrétan in his youth to much profit and originally drew his method of "positive theology" from him. See *Memoirs*, 54, 63.

93 Secrétan, *La Philosophie de la Liberté*, 250.

> of opposition, and, then, eventually, human society in which it perfects itself by taking it to its true, freely realized perfection, thereby realizing its own freedom. This became objectively visible in the Christ-event (the summit of human and cosmic history) when Christ consummated his self-projection in his death to rise so far as the Spirit draws out the full effect of that death in us. By understanding the ultimate meaning of this event, our mind will manifest the *synthesis*, the supreme *Aufhebung* (a provisional denial which safeguards the original datum by ultimately going beyond it) in which the one Idea from which everything has come is finally reconciled with itself.[94]

The Hegelian system ultimately interiorizes the Kingdom of God, bringing it down to our level through the instrument of philosophical rationality, reducing theology to anthropology. If it has the advantage of recognizing a need, in order to recover a sense of the unity of the cosmos, to relate metaphysics and cosmology to the polis and kingship, it was, Bouyer judges, nevertheless "ultimately a catastrophe."[95] The later Schelling, we would add, put this philosophy of nature and history onto a somewhat more orthodox theological footing. He recovered the Patristic theology of humanity as reconciler of the world, of spirit with nature, without holding that the becoming of the world is equivalent to the necessary self-perfecting of God. The Russian sophiologists developed and corrected Schelling's thought at this point, and their corrective work is the launching ground for Bouyer's developed theological cosmology. This poetic, sophiological alternative to the modern Idealism that sought to overcome the dualism and fragmenting philosophy of mechanist materialism is not brought forth in full ontological and phenomenological scope until the final chapters of *Cosmos*. Bouyer faithfully follows in this book his method of "positive theology." He conveys the history of the questions that the Church has confronted in order at last to show where we are today and suggest how we should proceed. In these final chapters of *Cosmos*, the unifying, sophiological perspective is put forth as the way forward theologically, although it is

94 *IF*, 292.
95 Ibid., 294.

not received uncritically, and Bouyer avoids in this text the language of sophiology and Sophia.

The nineteenth-century problematic has the advantage of linking together God, humanity, and cosmos, but its immanentism has to be overcome. Sophiology, particularly in the developed work of Bulgakov, recognizes the need both to see the unity of creation anew and to show the rooting of our deeply personal world in the triune Wisdom of the transcendent Creator. In their view of unity-without-confusion of divine and created Wisdom, the best exemplars of the sophiological trend of thought may enable us to espy the unconscious depths of humanity, its fundamental intertwining with cosmic being, and its drive, under the inspiration of the Spirit of Wisdom, for eschatological perfection, without having to reduce the process that leads from unconsciousness to the perfection of being as a logical and necessary unfolding of the Absolute seeking its own expansion or as the completion of an inevitable process within nature. We saw in our second chapter that Bouyer is careful to avoid reducing individual consciousnesses to "atoms of consciousness" that melt together at the Omega Point of cosmic completion. In all that we have said, we should keep in mind his commitment to the idea of unity by way of mutual implication and reciprocity. Our perfection, he insists, is the perfection of the exchange of consciousnesses in a fundamental, real relationship with one another, whose irreducible alterity in unity is made possible by the continuance of individual existence in the perfection of supernatural society. Bouyer shows that it is only with reference to the eschatological Christ and the incubating presence of the Holy Spirit in history that this process has sense and meaning. He understands the history of creation more so than Bulgakov (we think) in terms of apocalyptic interventions, of ruptures, of the inspiration of the Holy Spirit infusing with a transcendent light the prophetic witnesses to God's holy Word and Wisdom throughout the ages.[96] Decline and

96 See Davide Zordan, "De la Sagesse en Théologie: Essai de Confrontation entre Serge Boulgakov et Louis Bouyer," *Irénikon* 79, no. 3 (2006): 263–82. One of the ways in which Zordan distinguishes Bouyer from Bulgakov is to stress that Bouyer's thinking is more eschatological, oriented toward the future, more cognizant of cosmic evil, whereas Bulgakov's thinking is more protological, oriented to human origins.

fall are mixed in with progress. Our own age hardly attests to its being the pinnacle of history. Yet God, history, and nature are seen holistically from the Bouyerian vantage point just as surely as in the Church Fathers or in the great proponents of Christian Wisdom in the modern age. The human being is indeed the reconciler of nature and spirit, but only in the person of the God-Man and his Body, the Eschatological Adam, the eternal image of God, who supernaturally elevates our human image and with it our image of the world through the Eucharist. Physical nature and spirit are seen in this Christological and Trinitarian way as a unity without division, yet not mixed together or confused with one another. This view of things is expounded more fully in the second part of *Cosmos*, to which we now turn.

PART II

EXITUS-REDITUS

CHAPTER 8

Trinitarian Wisdom

IN THE SECOND AND FINAL PART OF *COSMOS*, Bouyer builds upon his exposition from the first part to give a prospective synthesis of the theology of creation, the stated purpose of which is to help us to rediscover "the world in God, whose Love produced, redeemed and adopted it."[1] He describes this second part as "Retrospections," and these serve as a summary of the whole of his work on the economy of salvation. The ordering of themes parallels that of Aquinas in the *Summa Theologiae*. Like Aquinas, Bouyer begins by showing that all cosmic being is brought into being in and through the eternal, triune God, and ends, as Aquinas intended to do, at the end of history, with the Second Coming of Christ and the general resurrection.

The sophiological inspiration in Bouyer's thought is especially evident in this second part, which clarifies the whole book as "a phenomenology of that divine Wisdom which is the spirit of the redeemed world."[2] The genealogical overview of cosmology in the first seventeen chapters was necessary to lay the groundwork for these retrospections, because it brought us to the state of cosmological questioning as we face it today. The first part of the text is not merely a historical overview but "positive theology" in Bouyer's meaning of the expression, that is to say a kind of phenomenology of cosmology in all of its developmental lines and pertinence for theology, clearing a path for future theological research.

In this second overall part of our own study, we shall focus on Bouyer's retrospective chapters in *Cosmos*. In the present chapter, we narrow in on Bouyer's Trinitarian theology of Wisdom as found in chapter eighteen of the text, entitled "Wisdom in the Trinity," putting it, as always, in the

1 *Cosmos*, 181.

2 *IF*, 310. Bouyer describes in this earlier work the approach that he thinks theology needs to follow. What he says here in *The Invisible Father* describes well what he would in fact accomplish in *Cosmos*.

whole context of his trilogies on dogmatics.[3] This will give us the opportunity to develop in a first step an analysis of Bouyer's way of uniting and distinguishing uncreated and created Wisdom. This is an especially important chapter in the book, because it is here that we find our most direct access to the Trinitarian dimension of his theology of creation. It requires its own focus, albeit with contextualization provided by other books of his. The chapter is broken up into five sections. We shall begin by exploring Bouyer's deeply personalist rendering of the triune essence of God as love by summarizing the main themes in the first three sections of the chapter. Given the succinctness of Bouyer's presentation, we shall be required in a second step to turn to some of his wider writings in order to show the line of Trinitarian reflection in the history of theology with which his triadology is associated. Third, drawing from and developing clues from the fourth and fifth sections of chapter eighteen, we shall give an overview of his vision of the world in the love of God, precisely in the divine, uncreated Wisdom of the Word, which is the ground of the Wisdom of creation. We shall in this context next explore the kenotic aspect of Bouyer's theology of creation, which he does not treat until chapter twenty but that is appropriate to consider at this juncture given the dialectic of transcendence and immanence that he sets forth in chapter eighteen. In a final stage of exposition and analysis, we shall focus on what Bouyer means when he speaks of "created Wisdom," particularly in light of his suggestive deployment in the final two sections of the chapter of Aquinas's concept of "notional distinction" and of the essence/energies distinction in the Eastern Christian tradition. In this final step, we shall make an initial assessment of the importance of his sophiology with respect to the question of the relation of grace to nature. The exposition and analysis in this chapter will help us as we move forward to understand Bouyer's vision of Wisdom as applied to the central personal figures through whom the history of salvation is actualized.[4]

3 Chapter 18 of *Cosmos* is entitled "Wisdom in the Trinity." This parallels the first chapter of Bulgakov's *Sophia: The Wisdom of God*, entitled "The Divine Sophia in the Holy Trinity." See Lesoing, *Vers la Plénitude du Christ*, 292. Lesoing draws the connection between these two chapters on this page.

4 Cf. Grintchenko, *Une Approche Théologique du Monde*, 201. On this page, Grintchenko gives a helpful pre-summary of what Bouyer means by Wisdom,

TRINITARIAN LOVE

If Greek philosophy demythologized ancient mythic cosmologies by freeing them from degraded concepts of divinity, it did not succeed in escaping from the mythic propensity to confuse God with the world. This is the first point that Bouyer makes in the second part of *Cosmos*, at the beginning of chapter eighteen, as a way of linking together the two parts of the work. Bouyer thus maintains a phenomenological character to the book in continuity with the first part, in that he begins by placing the Christian understanding of God against the horizon of pagan philosophy and myth, as well as against some misguided, post-Christian attempts to recuperate a spiritual or religious understanding of the cosmos.

In order to set forth the intelligibility of the Christian doctrine of creation, it is helpful to show the way in which Christian revelation reverses the oppositional dialectic between God and the world prevalent in ancient pagan thought and in modern materialism and Idealism. Bouyer develops his theology of Wisdom in *Cosmos* with precisely this concern in mind. He asserts at the beginning of the chapter that both Greek pagan philosophers and modern philosophers in the line of Baruch Spinoza (1632–1677) failed to grasp God's omnipresence in creation because neither could fully acknowledge the distinction between God and the world that alone makes possible their mutual co-presence.[5] Bouyer says in the first section that Plato's ideas and idea of the Good were "nothing but the substance of the sensible world, the other side of it, as it were," while Aristotle's Unmoved Mover was an entelechy within the world, an unobtainable perfection that exists only for itself and not for anything else.[6] The Stoic god, as *pneuma* and *logos*, even in its pure state in the fiery sky, was "only the outer envelope of the world,

---

as she begins her own journey through the retrospective meditations in *Cosmos*. She describes Wisdom "as the projection or prolongation of the love that God is in Himself, as a feminine figure, participating from the very beginning in the mediation between God and a creation endowed with the freedom to respond to this love."

5 *Cosmos*, 182. The three first sections of chapter 18—our concern in the present section—are entitled "Immanence and Transcendence" (182–83), "Person, Love, and Trinity" (183–85), and "Immutability, Impassibility, and Agape" (185–88).

6 *Cosmos*, 182.

inseparable from it."[7] Bouyer suggests that Neoplatonism gave a subtler way of thinking about the dialectical interplay of transcendence and immanence, but it too failed to escape the mythic confusion of cosmogony with theogony. The divinity in Neoplatonism, the One beyond being, was in the end only "the world considered exclusively in terms of unity."[8] The world in its multiplicity was, alternatively, only the One in a condition of dispersion and could not partake of the One. It could be distinguished from the One only depending on the point of view that was privileged by the sage who contemplated it.[9] The transcendent One and the immanent Many were, in confused manner, considered to be identical but at the same time not coextensive. Neither of them had anything of the other:

> To the extent that they appear to be distinct, they become alien to one another. Either God makes Himself known to us and His identity changes, or we return to Him and are no longer ourselves. The alternative is inescapable.[10]

This brief summary in chapter eighteen of Greco-Roman philosophical dialectic captures a central argument of the whole text of *Cosmos*, which is that only biblical revelation summed up in Christ enables the human race to discover what it is to be a person, whether human, angelic, or divine.[11] Bouyer does not discuss in detail in this book modern, post-Christian attempts to think anew the dialectic of transcendence and immanence, although he was certainly well aware of these. His compact discussion of Hegel in *The Invisible Father*, which we briefly expounded at the end of the previous chapter, demonstrates that he accords great significance to these modern efforts.[12] Hegel's "catastrophic" attempt to rescue Christian Trinitarianism by immanentizing the life of God in the

7 Ibid.

8 Ibid. We recall from our discussion in chapter 4 that Bouyer thinks that Neoplatonism got a little closer to the truth because it was influenced by biblical revelation.

9 *Cosmos*, 183.

10 Ibid.

11 Ibid.

12 *IF*, 291–94.

becoming of the world is an implicit target for Bouyer's reformed sophiological Trinitarianism in *Cosmos*. It seems that he saw validity in projects to surpass Hegel from within, and this is why he looked kindly upon the work of the nineteenth-century Catholic sophiologists, who, in his view, set forth an exemplary path in this regard. Our author shows that the specifically Trinitarian God is indeed the source and end of all creation and gives meaning to the flow of history, as Hegel wanted to affirm in his own manner, but the biblical God is certainly never a mere Idea who self-actualizes as Spirit in the cosmogonic process. Hegel realized that Christian thought cannot do without cosmology, and that Christian faith should issue in a vision of the unity of the end of human and cosmic being through God's Trinitarian self-communication to the world. The tragedy of Hegel's thought seems to be, in our author's view, that he understood this ultimately inescapable desideratum better than many of his opponents but addressed it in such a devastatingly deceptive manner. Christian theology needs to exhibit the theoanthropocosmic unity of the vision of faith, but Bouyer insists that it has to do so, contra Hegel, by affirming the biblical revelation of God's triune kingship and sovereign transcendence, as only the God who is transcendent to the world in the manner of the biblical God can be omnipresent in His immanence to it. Only God's transcendence to the world opens the metaphysical space for creation to exist within Him in His Word and Wisdom.[13]

The Trinitarian cosmology on offer in *Cosmos* is a form of Christian "personalism" that runs counter to the inevitable "impersonalism" of pantheism or dialectical Idealism and materialism. Bouyer does not categorize his own work with the label "personalism," and, indeed, recognizes the ambiguity of the expression, lamenting in the second section of chapter eighteen of *Cosmos* that many Christian thinkers have

13 Cf. *Sophia*, 148–60. See also Jean-Luc Marion, *The Rigor of Things: Conversations with Dan Arbib*, trans. Christina M. Gschwandtner (New York: Fordham University Press, 2017), 28. Marion speaks here in a way that is deeply resonant with Bouyer's thought: "That is why it remains difficult to call oneself Hegelian from a Christian point of view: When Hegel claims that God must externalize himself in order to become real, he makes the most anti-Christian statement possible because nothing remains exterior to God, not even the nothing. For the world is not created *outside* of God but inevitably *within* the Trinity, more exactly *within* the Principle, the Word, the Son."

rendered the term ideological, having co-opted it as part of what he calls their "fawning and unseemly apologia for Marxism."[14] He lauds Maurice Nédoncelle (1905–1976) for offering a corrective personalism, a truly Christian personalism, which does not oppose the individual to society in the manner of the dialectical materialism to which many self-proclaimed personalists have wedded the term. Standing apart from his peers, this French philosopher "developed with subtlety and depth the theme of the reciprocity of consciousnesses as the leitmotif of his personalism."[15]

As we have seen, Bouyer assimilates this theme from Nédoncelle's work and, we would say, elevates it. The Oratorian recognizes that it develops the Christian understanding of the person in a fully Trinitarian light, moving beyond the somewhat insufficient, classical, Boethian definition of the person as an individual substance of a rational nature. Human persons are, Bouyer insists, intrinsically relational. Each created person, angelic or human, is, as he says in *Sophia*, "only an obediential potency for opening itself to the grace of no longer living otherwise than God Himself, in the perfect reciprocity which makes of the three divine persons three inseparable as well as complementary aspects of this unique love that God is."[16] Bouyer teaches that God's gift of grace perfected eschatologically fulfills the individual person as relational on the model of the divine persons who come to in-dwell him or her.

Bouyer's précis of Trinitarian theology in the second section of chapter eighteen of *Cosmos* highlights God's being as transcendent, interpersonal *agape*, the supereminent love that enables God to be fully present to the whole of creation in relations of communion. He follows those many theologians of stature in the twentieth century who reminded the Church that "the God" (*ὁ θεός*) referred to in the New Testament just is "the Father."[17] He argues throughout his trilogies as in *Cosmos* that the Father is "first," and that the nomination "God" does not refer first and foremost to an "essence." This does not mean, as Bouyer explains

14 *Cosmos*, 183.

15 Ibid. See Maurice Nédoncelle, *La Réciprocité des Consciences* (Paris: Éditions Montaigne, 1942).

16 *Sophia*, 141.

17 Cf. Yves Congar, "Le Père, Source Absolue de la Divinité," *Istina* 27 (1980): 237–46; Karl Rahner, "Dieu dans le Nouveau Testament. La Signification du Mot 'Theos,'" in *Écrits Théologiques*, t. I (Paris: Desclée de Brouwer, 1959), 11–111.

in *Cosmos*, that only the Father is God, for God's paternal essence can be perfectly "actualized" only by His self-communication and self-projection in a Son who is consubstantial with Him.[18] God cannot be perfectly the Father or the One from "whom all paternity in heaven and earth is named" (Eph. 3:15) by virtue of His relationship to creation alone. Bouyer appropriates Goodness to the Father, Wisdom to the Son, and Beauty to the Holy Spirit, but he does not make these appropriations in a rigid and exclusive manner.[19] He recognizes with Bulgakov that none of the three divine persons is any less good, wise, or beautiful than the others. These appropriations help us to understand a little better the distinctive mode of subsistence that each person is in its relation of origin, but Bouyer insists that nothing we can say about the divine persons whether by way of appropriations or by articulating the metaphysics of relation can exhaust the reality of their personal exchanges. These are, Bouyer says, "so rich and deep as to defy totally our attempts at analysis."[20] God is, he insists, "personality" par excellence, supereminently so. He is the one to whom personality can be attributed in its most rigorous meaning, although His personality is beyond our comprehension. We, on the other hand, are persons "only because," as Bouyer says, "we carry within ourselves, as conscious and spiritual beings, an analogical participation in the divine mystery."[21]

Bouyer argues in section two that the production of any and all possible worlds could not adequately reciprocate God's essentially paternal essence as the *Ungrund* of all uncreated and created life.[22] Only the begetting of His consubstantial Son could do so—but this only by the simultaneous procession of the Spirit of sonship in the Son. In the Holy Spirit, the third person of the Trinity, the Father gives the Son "the soul

18 *Cosmos*, 184.

19 Ibid., 185: "The Spirit itself is therefore the luminous effulgence of the divine life in the Son, as Beauty shining in the Truth of the Goodness which is its source and the source of all things." The Father here is Goodness as source. On p. 191, Bouyer says that the divine essence and existence in the divine processions appears in the Son as Wisdom and in the Spirit as Glory.

20 Ibid.

21 Ibid. This contention is a rejection of Rahner's view that we should no longer speak of the Trinitarian hypostases as "persons."

22 Ibid.

of His own life: the ability to give himself as He does, to answer the paternal love with a love that is not only its reflection, but its living image."[23] Without suggesting that this viewpoint could be established on a purely philosophical basis, Bouyer argues in the manner of Richard of Saint Victor (12th century) that the community of love between the Father and Son could not be perfect if it were inward-looking but must go outward, communicate itself in a third, in the gift of the Holy Spirit who, like the Father and Son, is both the subject and object of divine love but in personalized fruitfulness.[24]

God's perfect, interpersonal *agape* does not require creative supplement in order to be made complete, as Bouyer insists in section three of chapter eighteen, where he argues that the attributions of immutability and impassibility must be applied to God, because these derive from the revelation of God's eternal kingship:

> How could He be God unless He were absolutely perfect? Or if anything outside Himself could have any effect on Him? The biblical assertion that He is King, the King of heaven and earth, the King of ages, the only King, would then lose all meaning. Either God, the God of biblical and evangelical revelation, is unchanging forever and the one exclusive source of every being—since He is the source of all being and indeed is being itself—or He is no longer Himself.[25]

If these attributes are indispensable to a truly Christian understanding of God, Bouyer contends that they nevertheless have a very different meaning than what the Greek philosophers gave to them, whether in the case of Aristotle's Unmoved Mover, who does not know and is therefore unconcerned with whatever is drawn toward him, or in the case of the Platonic Good and Neoplatonic One, from which beings necessarily emanate without an act of sovereign will to call them forth.[26] The biblical God's immutability and impassibility is one with His interpersonal

23 Ibid., 184.
24 Ibid.
25 Ibid., 186.
26 Ibid.

love. This recognition casts a very different light on the divine essence than Greek philosophers could give it. Bouyer insists that the God of biblical revelation made Himself known in a surpassing manner in the Cross of Christ as the unchanging perfection of loving self-gift, that is, as the One who loves the friends He has created with a love than which nothing is greater.[27] He is critical of some translations of Aquinas which present the Common Doctor's thought as if he embraced an uncritical Aristotelianism. In line with biblical tradition, Bouyer insists, Aquinas in fact held that the perfect, unchangeable happiness, knowledge, and love of the triune God is precisely what enables Him to create other beings in pure gratuitousness.[28] This is an affirmation which far surpasses Aristotle's understanding of the divine. The biblical God delights in His creatures, "takes pleasure in their welfare and happiness, as no other conceivable lover would or could do for the beloved."[29] God relates to His creatures by "infinite sympathy," not by cold indifference or violent opposition.[30]

Modern Nominalists pushed a decidedly un-Christian understanding of God when they explained that God's inner being is *potentia absoluta*, as though God could be sovereign "only by holding for naught everything other than Himself."[31] They did not grasp the real infinity of the triune God, whose glory is all the more perfectly reflected in creation to the extent that the latter is "filled in" by greater numbers of finite entities in the hierarchy of being.[32] Bouyer accepts Aquinas's view that God relates

27 Ibid. See Romans 5:8: "greater love has no man than this, than to lay down his life for a friend." Bouyer references this passage in *Cosmos* to demonstrate God's love for His people, who are His friends.

28 *Cosmos*, 186.

29 Ibid., 187.

30 Cf. *LC*, 438. Without quoting him directly or saying where this can be found in his works, Bouyer points to Marius Victorinus as the source of this expression.

31 *Cosmos*, 187.

32 Ibid. Cf. Thomas Aquinas, *ST* I.47.1: "God . . . brought things into being in order that His goodness might be communicated to creatures, and be represented to them; and because His goodness could not be adequately represented by one creature alone, He produced many and diverse creatures, that what was wanting to one in representation of the divine goodness might be supplied by another. . . . The whole universe together participates the divine goodness more perfectly, and represents it better than any single creature whatever."

to creation only as cause and not as effect, by way of "mixed relation," and he shows that this allows the full affirmation of secondary causality.[33] God produces everything in the creature, and the creature produces nothing in God. The Oratorian holds that this position supports rather than undermines the belief that God relates to creatures in infinite sympathy, because it is one with Aquinas's recognition that God knows and loves His creatures in His own life. "God the Father," Bouyer summarizes, "knows and loves us, delights in us, only by knowing and loving His Son, in whom He is eternally well-pleased."[34]

TRINITARIAN *EXITUS-REDITUS*

The foregoing points are quickly and successively summarized in chapter eighteen of *Cosmos*, and we have followed just now, in our own staccato way, Bouyer's order of presentation. He highlights God's immutable transcendence and perfection in love as the precondition of His perfect, loving concern for creation, but the discussion is truncated. In order to grasp with greater nuance the accents manifest in this presentation, we must stop to consider the Oratorian's wider Trinitarian thought, which he presumes but does not explicate in full in *Cosmos*. In his wider thought, Bouyer brings to the fore a certain tradition of Trinitarian reflection that he makes his own and develops in conjunction with the theology of Wisdom in order to keep fully in view the simultaneity of God's sovereign transcendence and all-encompassing immanence to creation. He presents this line of thought as an alternative to the Neoscholasticism of Cajetan (1469–1534), Suárez, and their followers, which was in the first half of the twentieth century representative of the mainstream of Catholic theology. Like de Lubac, Balthasar, and many other theologians of his generation or a little older, Bouyer found the Neoscholastic outlook on creation to be extrinsicist and insufficiently Trinitarian.[35] It too greatly, in his view, separated God from the world,

33 *Cosmos*, 187. Cf. Thomas Aquinas, ST I.28.1, obj. 3. In this text, Aquinas distinguishes real relations from logical relations. He explains that because God creates by His intellect and will and not by the necessity of His nature He does not have a "real" relation to creation, whereas creatures have a real relation to God because they are dependent on Him as contained under the divine order.

34 *Cosmos*, 188.

35 See *IF*, 263–76; *Sophia*, 50; *Women Mystics*, 13–86, 173–97.

and it did not sufficiently recognize the tri-personal manner in which God communicates His creative and redeeming grace to the cosmos.

One of the guiding motifs in Bouyer's trilogies, as we briefly indicated above, is that of the monarchy of the Father. This is a typical emphasis in Eastern Christian theology and was an emphasis in the work of Newman as well.[36] Bouyer shows its presence in a line of Western monastic and mystical theologians whose Trinitarian thought he aims to recover. He finds these theologians to be especially attuned to the triadology of the Christian East, especially as summed up in the work of Denys the Areopagite. He calls forth the testimony of Marius Victorinus, William of Saint Thierry (1085–1148), the Beguine Hadewijch of Antwerp (13th century), Meister Eckhart (1260–1328), John of Ruusbroec (1293–1381), and a tradition of women mystics that includes Saint Teresa Benedicta of the Cross (Edith Stein).[37] These mystics possess greater normative authority for Bouyer than the standard Neoscholastic theologians, but he does not oppose them to Augustine and Thomas Aquinas.[38] It is

36 See *MT*, 210–11. In these pages, Bouyer points out the guiding motifs in his trilogies. In addition to the theology of wisdom drawn from Bulgakov and the monarchy of the Father, he points to the theme of the mystical body of Christ that he drew from the writings of Emile Mersch (1890–1940). He tells us that he draws his emphasis on the primacy of the Father in part from the teachings of his friend and mentor Dom Lambert Beauduin (1873–1960), a Belgian monk who founded Chevetogne Abbey in Belgium. See also Théodore de Régnon, *Études de Théologie Positive sur le Dogme de la Trinité* (Paris: V. Retaux et fils, 1892). Bouyer read these studies by Régnon as a young man, which gave him "a priceless initiation" into the writings of the Greek Church Fathers. See *Memoirs*, 91. Régnon's influential work is known and nowadays criticized by some for pitting a presumably "essentialist" Western theology of God that begins with the divine essence against a "personalist" Eastern theology that begins with the monarchy of the Father. On Newman, see Ker, *Newman on Vatican II*, 57–59.

37 See especially *Women Mystics*, 51–86, 173–97.

38 See *IF*, 267–76. Other theologians from the Western tradition that Bouyer calls upon as resources in need of rediscovery in this regard are John Scotus Eriugena (815–877) and Nicolas of Cusa (1401–1464). At the origin of Western Trinitarian theology, Bouyer finds the ultimate source of this trend of thought in Marius Victorinus (290–364), who deeply shaped Saint Augustine's theology, and who, having once been a Neoplatonist philosopher, grasped at its root the difference between Christian and Neoplatonist understandings of the divinity. See especially *Sophia*, 50. The intrinsic sympathy of this line of thought with Denys the Areopagite is an important factor in Bouyer's preference for it.

truer to say that he understands his preferred sources to be capable of actualizing untapped potencies in the theologies of these latter two figures, who hold such canonical weight in Roman Catholicism. What he ultimately draws from these mystics is an understanding of Trinitarian communion as an eternal *exitus-reditus* that spills out freely into creation and is reflected there.[39]

Interestingly, Bouyer especially privileges the writings of Meister Eckhart, whose work he takes, contra much popular and even some scholarly opinion, to be irreducibly Trinitarian.[40] In a telling exposition of the Rhenish master in *The Invisible Father*, he especially latches on to a scheme of Trinitarian appropriations that Eckhart proposed whereby the Father would be *intelligere* or consciousness, the Son *vivere* or the content of God's absolute self-consciousness, and the Holy Spirit *esse*, or the unity of being of the divine persons manifested eternally at the term of the divine processions.[41] Why does Bouyer exhibit affection for this unusual schema of appropriations, which reverses Saint Augustine's vastly more influential one, according to which *intelligere* is appropriated to the Son and not to the Father?[42]

Guillaume Bruté de Rémur took this question as a central puzzle in his doctoral dissertation on Bouyer. He argued that we can understand Bouyer's fondness for Eckhart if we see the way he develops the mystic's Trinitarian scheme in connection with the advances in modern depth

39 Cf. Rik Van Nieuwenhove, *Jan Van Ruusbroec, Mystical Theologian of the Trinity* (South Bend, IN: University of Notre Dame Press, 2003), 77–99. Nieuwenhove explores in this chapter Ruusbroec's understanding of Trinitarian *regiratio*, the flow of the divine persons back to their source and shows Ruusbroec's uniqueness as well as his connections to a larger tradition. It is precisely this idea of *regiratio* that Bouyer draws upon, without naming it.

40 See *IF*, 306–7. In these pages, Bouyer recommends Eckhart's Trinitarian scheme as the way forward for a theology of the God-world relation.

41 Ibid., 269–70.

42 See Meister Eckhart, *Parisian Questions*, trans. Armand A. Mauer, C. S. B. (Toronto: Pontifical Institute of Medieval Studies, 1974), 39. Mauer references the appropriations in Eckhart that Bouyer will adopt, but he interprets Eckhart very differently from Bouyer, arguing that Eckhart thinks that there is a unity of the divine nature transcending the Trinity. See *Cosmos*, 185. Here Bouyer recognizes the biblical warrant for Augustine's association of the Son's begetting through intelligence and the Spirit's proceeding through will, although he does not see this association as absolutely binding on theologians.

psychology that he thinks have definitively surpassed nineteenth-century scientific materialism. Recall from our own second chapter that Bouyer held that modern sensationist epistemology tricked humanity into thinking that human experience is a synthesis of a multitude of basic, elementary sensations, and that this epistemological confusion bolsters scientism and its concomitant philosophical naturalism. We saw that Bouyer disputes sensationism in favor of the position according to which the unity of consciousness is primary and for which sense-perceptions are inseparable from the center of human affectivity which unifies our sense of the true, the good, and the beautiful. The human spiritual soul is not, according to this anthropology, simply a part of our being but the encompassing unity and totality of it where all of our sensory experiences are received.

Rémur argues that Eckhart's scheme of Trinitarian appropriations allows Bouyer to apply this understanding of consciousness analogically to God. According to this triadology, the Father, the fontal source of divinity, is already consciousness, the source of all thought "who thinks Himself in His own life, and who thinks the universe in creation, but also places it at the root and the source of divine life."[43] In developing this scheme, Bouyer effects a reversal of German Idealism by going back to its source (before Böhme) in the writings of Eckhart, whose thought the Idealists mangled by taking his idea that the divine Spirit is born in human consciousness to mean that God is at last made real there. Bouyer insists that if, as philosophical realists rightly claim, for human beings thought has reality to the extent that it is conformed to the transcendent realities toward which conscious intentionality is necessarily directed, the reverse is true in God. For Him, "beings *are* only to the extent that they are conformed to the thought of God over them, to the idea that He has of them."[44] For God Idealism is true—but for God alone. The thought that God has over being and beings is already in the Father, who, as the sole ultimate principle and conscious source of the Trinitarian relations, is able to include the materiality of created being in His consciousness as the ultimate productive source of all divine and created life.

Rémur suggests that for Bouyer God the first person of the Trinity is consciousness, while the very essence of God is love. Yet, as we saw in

43 Rémur, *La Théologie Trinitaire de Louis Bouyer*, 321.

44 *LC*, 440. Emphasis ours.

the previous section, in *Cosmos*, Bouyer associates the fontal personality of the Father with goodness, and this is why it is true to say that He cannot be alone. *Agape* has to do not only with God's relation to creation or to His immanent nature communicated by the Father in the Son and manifested in the glory of the Spirit but with the fontal personality *in se*, which is always a going out to the Son and the Spirit. Bouyer moves in the company of those theologians for whom the Johannine divine nomination of love (1 John 4:8) must qualify the Mosaic divine nomination of being (Ex. 3:15), and this would be downplayed if the consciousness of the Father were granted processional priority to the paternal *agape*. He follows Étienne Gilson's recovery of Aquinas's metaphysics of Exodus, but he also urges that the specifically Christian dimension of this metaphysics be highlighted, infused as it is with the Gospel message of divine charity. God's essence, in this view, as interpersonal consciousness, is sharing of love, the total self-donation by which the Father speaks Himself in the Son in whom the Holy Spirit proceeds as the gift of reciprocity by which the love of the Father gives itself and returns to itself in the thanksgiving of the Son. The divine life is thus "Eucharistic."[45] Nevertheless, it is true to say that it is in the gift of the Spirit that the unity of divine consciousness is cemented. In the whole of Bouyer's writings, God's *agape* is said to be established not by directing His consciousness to creation but on the basis of His own inner élan as the intrinsic goodness of His being.[46] In turning Himself to His creation, in knowing and loving creation in Himself in the divine Word on whom rests the Spirit of filiation, the Father knows and loves all things in accordance with who He is in the depths of His own loving consciousness and life.

In Bouyer's direct exposition of Eckhart in *The Invisible Father*, we are given penetrating insight into the dynamic character of the *agape* that constitutes the divine essence as a circling or flowing communion of love. God's eternal perfection is not that of static, objective being, of the frozen immobility of an essence beyond personal relation, but of *perichoresis*, *circumincessio*, or divine movement in interpersonal relationality without change. The divine essence is the *bullitio* or bursting forth of generous life that enables the *ebullitio* of creation. "God," Bouyer explains in this

45 *IF*, 231–33.
46 Rémur, *La Théologie Trinitaire de Louis Bouyer*, 321.

text, "completes Himself in being [in the procession of the Holy Spirit to whom the attribute of *esse* is appropriated] and then projects Himself, so to speak, beyond Himself into the being of creatures."[47] Inversely to God, creatures begin only as beings, but are called to "flower into total, unified consciousness in the Father."[48] Eckhart says that our vocation involves absorption into the unity of the *deitas* or *Gottheit*. Bouyer insists that he does not mean by this that the unity of God is an essence beyond the divine persons or an ocean of undifferentiated being where the reciprocity of consciousnesses no longer obtains and where we would disappear in deification. *Deitas* is instead "the dynamism, communication, communion which is simply identical with that 'pure being,' which is the one being of God."[49]

In summarizing Bouyer's adoption and development of Eckhart's Trinitarian scheme, Rémur says that his dynamic vision of Trinitarian *exitus* and *reditus*, of descent and re-ascent, enables him to evade a subordinationist reduction of the divine monarchy, because it affirms the truly transcendent character of God's unity, which transcends both unity and multiplicity.[50] We add that in the history of Trinitarian theology we find that God's unity is recognized to be ungraspable; it can be fittingly approached only by seeing it from different perspectives, for instance, with regard to the unity of God in the Father as the sole principle of divinity, or in the unity of the one divine essence, which is a *perichoresis* of a single life, or in the unity of the Holy Spirit, who is the bond of love between the Father and the Son.[51] In the widest scope of Bouyer's thought, Trinitarian *ressourcement* is crucial for bringing to light the ontology of love and personhood. God's personality and love are not foreign to our own. The revelation of God's being as personal *agape* gives transfiguring meaning to our own love and loves. God's love is not set apart from human love in radical opposition. God's *agape* is communion of love, and

47 *IF*, 270.
48 Ibid.
49 Ibid.
50 Rémur, *La Théologie Trinitaire de Louis Bouyer*, 321.
51 Certainly, it is necessary to prioritize the unity of the three divine persons as "one substance." These other ways of conceiving divine unity are supplementary and help us to avoid taking the unity of substance as a foundation on which the relations are extrinsically built.

we may even affirm that there is a kind of divine *eros*, akin to human *philia*, in that God possesses the supereminent desire to have communion with His creatures.[52] God is not an indifferent Unmoved Mover or One beyond being who is all alone and can withstand no difference "beside" Him. His divine immutability and impassibility are one with the generation of the Son and the procession of the Spirit and, as all the divine attributes, require interpretation in light of His interpersonal love and generosity, whose most powerful evidencing in creation is the love of the Son poured out for sinners on the Cross. This Trinitarian vision sees creation as existing within the flow of the triune life, known in the Father's loving projection of His own consciousness in the life of the Son, resplendent with the glory of the Holy Spirit. Creation is in the divine Word, in the Wisdom that is communicated from the Father to the Son. It is irradiated by the light of the Spirit of Christ, who personalizes the mutual love of Father and Son in its subjective and objective aspects as loving and loved.

THE WORLD IN GOD

It is thus Christian thought alone that can reverse the oppositional dialectic of transcendence and immanence that vitiated the achievements of ancient Greek philosophy and crippled modern philosophy in the train of Nominalism, resulting either in God's diffusion into the plurality of beings in the world, in pantheist fashion, or in His projection into an inaccessible transcendence.[53] Operating a recovery and development of Trinitarian creationism in chapter eighteen of *Cosmos* as in the whole of his retrospective meditation in the second part of the book, Bouyer weaves together scriptural interpretation, metaphysics, and dogmatic theology.[54] This interweaving is apparent in the distinction he makes in section four—entitled "Preexistence in God of His Creation"—between Name and Wisdom in the Word.[55] This distinction has roots in a double

52 *LC*, 436.

53 *Cosmos*, 191–92. Grintchenko, *Une Approche Théologique du Monde*, 202.

54 Grintchenko, *Une Approche Théologique du Monde*, 211–12.

55 *Cosmos*, 190; Grintchenko, *Une Approche Théologique du Monde*, 211. Grintchenko points out that Bouyer completes the idea of this distinction in *Mystery*, where he speaks of the double revelation to Moses on Mount Sinai. For Moses alone, God's revelation was of the divine Name. For the people of God, it

aspect of divine revelation detectable in Scripture and parallels Bulgakov's distinguishing between, on the one hand, the Word's eternal revelation of the Father and, on the other, his expression of the content of the divine thought as the ideal ground of creation.[56] The divine life, for Bouyer as for Bulgakov, is interpersonal self-revelation, with the Father eternally revealing Himself and all that He will create in giving Himself to the Son.[57] From one point of view, the Son, eternally generated by the Father, is the very mirror of the Father's hypostasis, while, from another point of view, he uniquely contains the whole of creation in his personal subsistence.[58] From the first vantage point, he is the divine Name, while, from the second, he is Wisdom.

We saw in our fifth chapter that Bouyer follows Aquinas's existentialist rendering of God's predestination of all things in the Son in and through the divine ideas. In section four of chapter eighteen of *Cosmos*, he strikingly says that "by begetting His Son and Word, the Father produced him with the intention that he would become incarnate in our flesh, in our sinful humanity, so as to bring it back to its origin and its original purpose, and thus to complete our interrupted adoption, which is literal and not figurative."[59] This passage makes the point very clearly that God's eternity does not include a before and after, and that God does not beget the Son only subsequently to create and to redeem. He projects and gives Himself eternally in the consubstantiality of Son and Spirit, who share His will perfectly, and all together eternally create and redeem, each in the special manner of their constitutive relationality.[60] Bouyer develops

---

was of the Torah. God is revealed in a double aspect as "I Am who I Am" and as the One who makes to live.

56 Cf. Bulgakov, *Sophia*, 41–43. Bulgakov distinguishes the image of the Father in the hypostasis of the *Logos* from the Wisdom in the *Logos* that "stands for the wisdom and the truth of all that is worthy of participating in divine being."

57 Ibid., 39. Bulgakov distinguishes the Trinity understood as self-revelation from the Trinity understood as relations of causal origin. This distinction is not as firmly sustained in Bouyer, although Bouyer too maintains a phenomenological rendering of the divine life.

58 *Cosmos*, 190.

59 Ibid., 188.

60 Cf. *IF*, 230–32. Bouyer does not highlight this point as explicitly in *Cosmos* as he does in *The Invisible Father*. But the point is presumed in the text of *Cosmos*.

the Thomist understanding that God's being is *ipsum esse subsistens* ("being itself subsisting") in a Trinitarian way. He holds that God knows and loves us eternally both in His perfect actuality, which is "pure act" (*actus purus*), the unity of essence and existence, and in our own, in which there is metaphysical composition. Creation and redemption are, as he describes this doctrine, both eternal, Trinitarian deeds, one with the generation and procession of the Son and the Holy Spirit. God knows and loves us in His eternal present specifically in the giving and receiving of His interpersonal life.

In section four of chapter eighteen of *Cosmos*, we are taught that the divine essence is one and actual and that the unity of divine being, unrestricted by limiting essence, is at once necessary emanation of love and perfectly free.[61] Bouyer distinguishes the eternal life of the Trinity from the life of creation in this section by taking note of the paradoxal, reciprocating interplay of necessity and freedom in God. The divine life, he says, "directly expresses the necessity inherent in His essence (which is love)," while the life of creation "reflects foremost its sovereign freedom."[62] We might, from our limited point of view, juxtapose the attributes of necessity and freedom, but they are perfectly one in God: "Just as the necessity of the eternal unfolding of His love is in no way a constraint, the freedom of its communication to us is absolutely not fortuitous or accidental."[63] The necessity of God's relational being is thus not abridged by His forever freely willing our creation and salvation in His Incarnate Son. Our creation and redemption are not by chance or secondary considerations. It is the Cross of Christ alone, Bouyer emphasizes, that enables us to fathom the unity of necessity and freedom in the divine life with respect to our own being in the drama of history, because it shows forth *in concreto* the operation of the divine will in its free eternity responding to our original, self-imposed exile from participation in divine blessedness.[64]

The Wisdom of God, as we have learned in our study, centers on the Mystery of the Cross, which is eternally willed by God in His actual

61 *Cosmos*, 189.
62 Ibid.
63 Ibid.
64 Ibid.

response to our historical free actions. Bouyer develops this theme in section five of chapter eighteen of *Cosmos*, which is entitled "Wisdom in the Word."[65] Davide Zordan argued that Bouyer so identifies divine Wisdom with the eternal humanity of the Word ordered to the Cross that he is uniquely able to see the unity of God and the world without doing away with God's transcendence or rendering as aleatory events the production of creation and its salvation. There is much truth in what Zordan says, even if he goes too far beyond Bouyer's texts by strictly identifying uncreated Wisdom with the Son's eternal humanity.[66] All that Zordan says can be affirmed without making this identification as strictly as he does.

The unity-in-distinction that Bouyer emphasizes in regard to the relation of the divine Name and divine Wisdom implies the inner correlativity between God and the human that was articulated in Jewish mysticism and that we have noted. As Bouyer says in *The Eternal Son*:

> In God there is no real distinction between the Wisdom through which He disposed all things, and especially foresaw or preordained our human history, and the *Logos* in whom He speaks Himself revealed, as Father in His eternal Son. This is why in history, in which the divine Word speaks itself in human words, there is an intimate correlation and concordance between the revelation of the divine Name and God's Design for us. Is not His design only His Name imprinted at this point in our flesh and our whole being, that as adopted children of God in the Son, we become a living praise to His glory through the Holy Spirit? But properly, in God, Wisdom is distinguished from the *Logos*, in the *Logos* itself, as that most intimate aspect of His life by which God turns towards the world and His life bursts forth into creation, the creation of humanity in particular.[67]

65 Ibid., 190–91. The section is entitled "Wisdom in the Word."

66 Zordan, "De la Sagesse en Théologie," 280. We follow Dom Bertrand Lesoing on this point. See Lesoing, *Vers la Plénitude du Christ*, 295, n. 1. It should be noted that Bulgakov, for his part, does make this identification.

67 *ES*, 402–3.

Wisdom, God's eschatological plan for His creative work, though indistinguishable in the divine *esse* from God's *Logos* and Name, is taken to refer to the depths of the divine life as it is turned toward creation. Eternal Wisdom is communicated by the Father to the Son in the Spirit in order to impart the Trinitarian life to creation, although the completion of the divine design is accomplished only through a joint endeavor that involves the common work of divine and created freedoms.[68] Divine Wisdom is in the Son but as ordered to become, in time, by the incubation of the Holy Spirit, "the Mother of his [the Son's] redeeming incarnation and thus of redemption fulfilled and bearing fruit in the universal adoption of the new mankind, as the new Eve in the Second Adam."[69] Wisdom represents an eternal feminine in God.[70] She is "Daughter of the Father in the Son,"[71] both spousal and maternal in character. She is not a fourth divine hypostasis but tends to become personalized as created Wisdom in the history of finite freedoms. Her preeminent individual expression is in Mary, and her collective fulfillment is in the eschatological Church.[72]

In the present dispensation, the Mystical Body of Christ, the Temple of the Holy Spirit, manifests created Wisdom, the latter of which Bouyer speaks of as the truest depth dimension of the world through which the Spirit had always been at work to prepare the way for the Incarnation.[73] Creation was given being from the very beginning in divine Wisdom in order to be "'Christified' in the Christ of the Parousia."[74] The Immolated Lamb through whose eternal humanity Wisdom is communicated is the principle of the whole creation.[75] The Mystery of the Lamb is the key of God's eternal Wisdom, but divine Wisdom metaphysically contains a kind of active-potency (Bouyer does not use this term) for actualization *ad extra* in the being of created Wisdom. As Bouyer says in his book on *The Holy Spirit*, in a passage that helps us to understand the final sections of chapter eighteen of *Cosmos*:

68 *Cosmos*, 190.
69 Ibid.
70 See Grintchenko, *Une Approche Théologique du Monde*, 217–18.
71 *Cosmos*, 190
72 Ibid., 190–91. See also *CG*, 142.
73 *Cosmos*, 191.
74 Ibid.
75 Ibid., 231. See also *LC*, 445.

> Turning back to the principle of this sublime process, to that which alone could bring it about and account for it, Wisdom is, in God and from all eternity, first and foremost everything which will ever exist outside Him, and is destined to revert fully to God, not to be reabsorbed and lost in Him, but to live in Him with only one life, the divine life itself.[76]

In His Wisdom God has a place for creation within Himself, even in its history and becoming. God is neither changed nor contracted by creation, but He seems nevertheless "to include change even in His immutability and thus recapitulates, without confusing it, the created finite into His uncreated infinity."[77] Divine Wisdom is a mirror reflecting the communion of the divine persons. Rémur explains:

> Because the essence of God is *agape*, it can be the subject of a history including multiplicity, because it is not only in itself but in other things. Uncreated Wisdom being the divine reflection of the Trinitarian life is, at the same time, the content of the history of recapitulation, as created Wisdom.[78]

Rémur does not make the point, but this description aligns Bouyer's theology of Wisdom with the earlier modern tradition of sophiology, which emphasized the significance of Wisdom 7:26: "For she is a reflection of eternal light, a spotless mirror of the working of God, and an image of His goodness." This tradition presents an alternative dialectic of the relation between the infinite God and finite being to that of Hegel, for whom, with respect to the Absolute Idea, Spirit comes after nature. In this Behmenite tradition, especially as represented by Baader, Spirit and Nature are reciprocal, eternal suppositions of one another, in the circular movement in repose of Trinitarian self-donation of the *Logos* and Sophia.[79] Bouyer effectively connects this sophiological tradition with the earlier tradition of Western mysticism that we described above and ultimately with

76 *LC*, 445.
77 *ES*, 413.
78 Rémur, *La Théologie Trinitaire de Louis Bouyer*, 337.
79 See Tourpe, *L'Audace Théosophique de Baader*, 197.

the mystical theology of Denys the Areopagite. His distinction between Name and Wisdom has to be maintained in this in order to be true to his thought. Uncreated Wisdom can be spoken of as a spotless mirror of the Trinitarian life but only as ordered to its communication *ad extra* in a way that we would not speak with regard to the divine Name.

There are a couple disputed points in the themes that we have discussed in this section that we would like briefly to note in conclusion. First, the theology of Wisdom in the Word that Bouyer deploys emphasizes so much the enfolding of creation in the divine Word that it seems to privilege the idea of *creatio ex Deo* over that of *creatio ex nihilo*, and the publication of *Cosmos* in France generated controversy on this score. As we have seen, at the beginning of *Cosmos*, Bouyer affirms the doctrine of *creatio ex nihilo*, saying that this article of faith entails the freedom both of God and creation, "an insight," he says, "no one had ever considered [before biblical revelation], either to affirm its validity or to deny it."[80] Yet his sophiological positioning, which moves him to envision all of creation *in* God, in the divine Wisdom, led one reviewer of the book to criticize him sharply for downplaying the doctrine of *creatio ex nihilo*, which, this reviewer argued, needs to be affirmed in order to safeguard the holiness and transcendence of the Creator.[81] Davide Zordan defends Bouyer on this disputed point by arguing that the latter's emphasis on creation by the Word alone affirms these attributes of God just as much as the doctrine of *creatio ex nihilo*. He reminds us that "nothing" is an abstract concept, that there exists in reality only "relative nothingness" (*néant relatif*), "included in the state of dependence of the creature."[82] As Jean-Luc Marion has put it, "nothing exists outside of God, not even the nothing."[83] In the papal encyclical *Laudato Si'*, Pope Francis says that the "universe unfolds in God, who fills it completely," a statement fully congruent with Bouyer's thought.[84] Nevertheless, the textual evidence

80 *Cosmos*, 7.

81 J.-M. Maldamé, "Sciences de la Nature et Théologie," *Revue Thomiste* 86 (1986): 283–308. The article includes a severe criticism of *Cosmos*. In spite of what Bouyer affirms at the beginning of *Cosmos*, the Dominican Neothomist critic holds that the book is not based on a proper metaphysics of *creatio ex nihilo*.

82 Zordan, "De la Sagesse en Théologie," 281–82, n. 33.

83 Marion, *The Rigor of Things*, 28.

84 Pope Francis, *Laudato Si'*, 6.VI.233.

shows that Bouyer affirms the doctrine of *creatio ex nihilo* while also emphasizing creation in the divine Word, and that he does not see these as contradictory affirmations.

The criticism just noted is connected to a second issue that one may raise, concerning the role of the Scholastic concept of metaphysical causality in Bouyer's cosmology. Zordan rightly notes that Bouyer is circumspect in employing this concept in theology, and that Bulgakov's thought deeply influenced him in this regard.[85] Yet the Oratorian does not absolutely reject analysis by way of causality. In his *Dictionary of Theology*, he gives an accurate overview of the Scholastic delineation of causes, which implies that he recognizes the importance of clarifying these in theology.[86] In his trilogies, he is reticent to apply the category of efficient causality to God, at least inasmuch as he wants to ensure that it does not lead us away from a sense of participation of creation in the Creator or that God really relates to us in a Trinitarian way in the order of grace.[87] Nevertheless, his particular emphases do elicit for us the question of whether he had a sufficiently differentiated understanding of analogical predication of the divine or of subtler nuances in Scholastic thinking on causality. For instance, it is not unhelpful to note, as Bouyer does not do, that for Scholastic theology, it is precisely with respect to *material* causality that creation is from nothing, in that creation has no prior substratum in matter.[88] As far as Bouyer's exposition of the concept of causality in *Cosmos* is concerned, the ultimate point is a rather limited one—God's causality is uniquely His own and the source of all created or secondary causality: "God's inaccessibility to any causality other than His own derives from the fact that the latter, being the first cause of all the others, remains the basis of them all."[89]

85 Ibid. See Slesinski, *The Theology of Sergius Bulgakov*, 147–54. Slesinski gives a thorough and concise, critical but appreciative, overview of Bulgakov's understanding of causal metaphysics and its application to theology.

86 *Dictionary of Theology*, 76–77. See the entry "Cause." He references here Aquinas's *Liber de Causis*.

87 See especially *IF*, 63.

88 See Slesinski, *The Theology of Sergius Bulgakov*, 150–51. Slesinski raises this very criticism with respect to Bulgakov's controversial dismissal of Scholastic definitions of causality.

89 *Cosmos*, 186.

DIVINE KENOSIS

We want at this point to jump ahead a bit in *Cosmos* and draw on other texts to consider another piece to the puzzle that can help us to explicate the nuances of Bouyer's theology of creation in God, in the divine Word, with its tension between the assertion of God's sovereign, immutable transcendence and the sophiological vision of divine immanence. The Oratorian embraces the view that God and creation are related to one another by virtue of what Aquinas called a "mixed relation" and argues for the immutability and impassibility of God, but he nevertheless experienced the allure of nineteenth- and twentieth-century theologies of divine kenosis that challenged the application of these latter attributions to God. He drew elements of this kenotic theology into his own thought.[90] Given what we have said about Bouyer's emphasis on divine, paternal *agape*, on the constitution of the divine essence in the self-giving of the persons, and on God's infinite sympathy for His creatures, it should come as no surprise that he did so.

The language of kenosis comes from Saint Paul in Philippians 2:7, where the apostle says that Christ "emptied himself, taking the form of a slave." Some theologians have gone far beyond Paul's literal text and projected kenosis into the very eternity of God or interpreted the event of the Cross as a diminution of Christ's divinity. In taking the form of a slave, some have argued, Christ renounced his share in the Godhead, or, alternatively, that the Christian figuration of kenosis in Christ effects the ultimate symbolic action by which human religion has finally renounced its perennial claims to bear a divine status. Bulgakov has been an important influence and inspiration in the spread the Gospel of the kenotic Christ in some quarters of academic theology, and this may be the most influential aspect of his thought to this point.[91] The Oratorian

90 Ibid., 187. On this page, Bouyer defends the view that "any relationship between the creator and the creature is real only in and for the latter." See also *Cosmos*, 214–15. This is to say that God is not changed by creation. On these pages, found in chapter 20, Bouyer discusses his understanding of divine kenosis.

91 Cf. Bulgakov, *Lamb of God*, 313. On this page, Bulgakov speaks of the "diminution" of Christ's divinity on the Cross where the divinity "co-dies" with the humanity of Christ. This is a very interesting idea, but Bouyer resisted this sort of language or approach to the unity of divinity and humanity in the person of the incarnate Christ. The influence of Bulgakov's kenoticism may be especially true

left little doubt about the sway that Bulgakov's kenotic theology held over him when he described it in *The Eternal Son* as "the shadow of a powerful vision of human and cosmic divinization."[92] In this book on Christology, Bouyer took the care to delineate, in a sympathetic manner, the various sorts of divine kenosis that one finds in Bulgakov's work, of which he distinguishes three: first, a kenosis of creation by which God limits or contracts Himself; second, a kenosis within the Trinity itself which the kenosis of creation reflects; third, a kenosis in the Incarnation through which there is diminution of the divinity of the second person of the Holy Trinity.[93] Bulgakov, on Bouyer's assessment, avoids the excessive dolorism of earlier kenotic theologies by virtue of his manifest, fundamental hope in the Resurrection, which shines forth throughout the pages of his books. Bouyer describes as "splendid" Bulgakov's demonstration that the Word discloses God's agapeic kingship in and through his kenotic, suffering humanity.[94]

Marie-David Weill has summarized the extent to which Bouyer adopts Bulgakov's kenotic theology. She argues that he does not accept Bulgakov's idea that there is intra-Trinitarian kenosis by which the Father projects Himself into the Son by losing His own being.[95] Certainly, our author's triadology does not entail a moment of self-loss or of potentially competing wills in God. Yet we might want to exercise a bit of caution in interpreting Bouyer on this point. Clearly, he does not go as far as Bulgakov or Balthasar and speak directly or at length of distance and

---

in regard to Balthasar. Cf. Katy Leamy, *The Holy Trinity: Hans Urs von Balthasar and His Sources* (Eugene, OR: Pickwick Publications, 2015); Jennifer Newsome Martin, *Hans Urs von Balthasar and the Critical Appropriation of Russian Religious Thought* (Notre Dame, IN: Notre Dame University Press, 2015). While Leamy argues that Balthasar rejects Bulgakov's sophiology while drawing on his Trinitarian theology of divine kenosis, Martin argues that Bulgakov influenced Balthasar even with respect to the sophiological theme. See especially Martin's essay "True and Truer Gnosis: The Revelation of the Sophianic in Hans Urs von Balthasar," in *Heavenly Country*, 339–54.

92 *ES*, 382. With this quotation, Bouyer summarizes the description of Bulgakov's kenotic theology by Paul Henry, but this is clearly Bouyer's own assessment as well.

93 Ibid.

94 Ibid., 381–83.

95 Weill, *L'Humanisme Eschatologique de Louis Bouyer*, 65.

potential loss in the eternal divinity, but he does, in *The Invisible Father*, commend the idea that there is a kind of withdrawal or contraction in the eternal life of the divine persons, and that God reveals Himself eternally in making room for the other in Himself.[96] As Bouyer sees it, at least in this writing, the idea of divine contraction becomes problematic only when theologians follow Böhme and conflate the kenosis of the Trinity *ad intra* with the kenosis of creation.[97] He affirms nevertheless in this text that there is supereminent disappropriation of self in God and thus takes up, albeit very cautiously, the mystical tradition that speaks of Trinitarian *exitus-reditus* as a simultaneous perfection in God of poverty and wealth. He affirms that there is in God self-giving movement in rest. His emphasis on God's agapeic self-reflection in Wisdom may highlight the always-present unity of divine life in a way that is a helpful complement in these discussions to Balthasar's Trinitarian theology of ever-greater distance in loving difference.[98]

Weill clarifies the extent to which Bouyer explicitly adopts in his work, in readjusted mode, the idea that there is a kenosis of creation. The key text to which she refers is chapter twenty of *Cosmos*.[99] The first kenosis in creation is, this text indicates, with relation to the angels, on whom God makes Himself freely dependent for loving response, whose perfection in the good He eternally wills but cannot guarantee, in that the angels were given the capacity to choose for or against His will. There is a second kenosis that Bouyer embraces, with respect to the human being, which, Weill notes, goes beyond Bulgakov, in that he suggests that in creating the human being God confided the redemption of the fallen world to him.[100] With Bulgakov and Saint Paul, Bouyer speaks of a kenosis of God in the Incarnation of Christ, who identifies himself with fallen humanity to the point of becoming the Second Adam in the flesh. He voluntarily assumes

96 *IF*, 306.

97 Ibid.

98 Cf. Tourpe, *L'Audace Théosophique de Baader*, 270–76. Tourpe describes Baader's dialectic of mediated love as just such a complement to Balthasar on this point. We think that there are striking similarities to Bouyer's triadology in what he describes of Baader, although he downplays the sophiological dimension of Baader's thought.

99 *Cosmos*, 214–15.

100 Weill, *L'Humanisme Eschatologique de Louis Bouyer*, 67; *Cosmos*, 215.

the consequences of our fallen condition and freely submits himself to suffering and death on the Cross. There is a deeper kenosis still, involving the person of the Holy Spirit. Bouyer himself says:

> The kenosis of the Son would bring about, however, the possibility of an even more mysterious kenosis of the Spirit of love. The ultimate kenosis will end in the death and transfiguration of the physical universe, with redeemed mankind following the steps of its redeemer. In the ultimate Parousia of the Son of God, the Spirit of God will appear as the Glorifier of the Father in His Son and in all things, when the eternal Wisdom will appear with the Saving Word in the divine glory which . . . will gather back into God everything which has ever proceeded from Him.[101]

Weill suggests that both Bulgakov and Bouyer take Christology and sophiology to be inseparable aspects of a fully integrated theology of creation. Bouyer, she says, remains marked by the Russian thinker with respect to his "splendid vision of the vocation of man—and of all the cosmos—in divinization, in a permanent tension, almost unsustainable, between the most kenotic accents and the vigorous optimism which always sustains his theology."[102] The idea of a kenosis of the Holy Spirit underscores this tension. Bouyer understands the Spirit to be in God eternally the divine personality revealed as love in its term as glory, just as the person of the Father is the eternal, super-essential source of the divine consciousness and love. Through the Spirit, the Wisdom of creation is made distinct from the creative Word of God and is called to be joined through time and history with the Word in spousal and filial union, where it will at last rest "with the Son, by the Spirit, in the bosom of the Father."[103] The Spirit is joined with the Son in the event of the Cross. The Spirit, the personalized glory of the divine love, is united with the Son in the depth of the latter's self-emptying in creation. It is in the Spirit that the glory of the Resurrection on the day of Parousia will be made fully manifest and faith at last transformed into accomplished vision. The

101 *Cosmos*, 215.
102 Weill, *L'Humanisme Eschatologique de Louis Bouyer*, 66–67.
103 *LC*, 446.

Spirit is thus at once the heart of the divine kenosis in the event of the Cross and the power through which God's royal *agape* is fully disclosed.

UNCREATED AND CREATED WISDOM, REAL AND NOTIONAL DISTINCTIONS

The kenotic relationship of the divine persons to creation extends, on our author's view, throughout the entirety of space-time, as the Father's two hands are always at work in creation seeking to draw it at last through the Paschal Mystery made present in the Eucharist of the Church into the divine life. As we saw in chapter three, Bouyer, following Saint Irenaeus and other Church Fathers, particularly in the Christian East, does not limit the presence of the Trinity in the world to vestiges or to static and unhistorical forms of being.[104] He holds that though God is transcendent to and independent of creation, all of creation is an echo of the divine Word, traversed by a "vital flux awakened by the Holy Spirit, by the very breath of God."[105] God the Father is like a potter who uses His two hands to form and prepare the ground for the coming of humanity, which was in gestation in the whole cosmos from the very beginning, and then to form and to prepare it to receive the very Word of God in the flesh in the Incarnation.[106]

God's Wisdom for creation embraces the whole of its materiality and centers on the human as the predestined center for its recapitulation. Toward the end of *Cosmos*, Bouyer describes the essence of creation in terms of the Wisdom at its heart:

> All creation is ultimately a development of Wisdom from the initial stage when the distinction between it and the Word is purely rational, to the final stage, when it becomes real. But Wisdom reaches this separation only to strive—through the entire history of the created world, taken up finally into the flow of divine life, of which it is only a reflection—to be reunited to the Word again, this time forever.[107]

104 Ibid., 368–70.
105 Ibid., 369.
106 Ibid., 370–71.
107 *Cosmos*, 231.

Wisdom becomes really differentiated from the eternal Word in the eschatological completion of creation in order to have perfect communion with the Father in the Word for eternity. The very basis for our communion with God is contained within the unity of the divine Word and essence in Wisdom.

We shall now begin to deepen our analysis of what Bouyer means by the relationship between uncreated and created Wisdom, which is a necessary step as we move forward in these final chapters. We draw from the final two sections of chapter eighteen of *Cosmos*, the first entitled "Divine Essence, Wisdom, and Glory," and the second "Wisdom and the Divine Energies."[108] There are two preliminary issues, not yet discussed, that it is crucial to point out as we advance in this endeavor. First, Bouyer's theology of Wisdom, created and uncreated, is in part a response to a debate between Eastern Orthodox and Roman Catholic theologians on how we should understand the relationship of God's redeeming grace to the intrinsic being and powers of our created nature. Neopatristic theologians in the East have stressed the uncreated, divine character of grace that God communicates to us through His uncreated "energies." These theologians, following Saint Gregory Palamas (14th century), hold that the energies of God are distinct from His essence and are the uncreated ground for real participation of created spirits in the divine life. Neothomist theologians in the Catholic West, on the other hand, following Aquinas himself, have stressed the idea that grace has a created aspect, becoming a divine quality infused in the human soul that suffuses the latter not as an extrinsic "supernature" but as a reality that constitutes its very being in relation to God.[109] Taking these positions together, we would want to say that grace is both uncreated, that is, God's gift of Himself, and a reality of our own being, an accident infused into the soul's very substance. At least at the time when Bouyer was most active as a theologian, Neopatristic theologians wondered whether their opponents could maintain the uncreated character of God's grace in postulating the existence of created grace, while the Neothomists wondered

108 Ibid., 191–93.

109 Cf. *Introduction to the Spiritual Life*, 195–99. These pages contain one of Bouyer's best, most straightforward discussions of the dispute between Thomists and Palamists on the theology of grace.

whether their opponents fully grasped the inner transformation of the human person in deification. Bouyer saw the sophiological path as a potential third way in this conflict or debate that might be able to do justice to the concerns of both positions.

The second preliminary point has to do with the need to recognize the narrative and poetic character of Bouyer's thought in order to interpret it adequately. This can be a frustrating dimension of it when dealing with thorny speculative issues, such as that of the relation of grace to nature, which require a more in-depth conceptual development than what he gives. Bouyer provides suggestions, glimpses of possibilities, *aperçus*, that might enlighten us on the theological path forward on the question at hand, but there is a decidedly and intentionally unsystematic character in his exposition. This has some advantages, it must be admitted. In addition to allowing us to savor the paradoxes inherent to the God-world relation, it enables a kind of ecumenical openness in discussion. Neither Neopatristic nor Neothomist concepts are excluded from the dialogue, at least if the latter is taken in its "Augustinian-Thomist" form. If Bouyer is especially intrigued by the possible avenues that sophiological reflection unblocks, he does not fail to take seriously both Neopatristic and Neothomist approaches, the former communicated to him in his friendship with Vladimir Lossky and the latter through his friendship with Guy de Broglie. All of the venerable traditions of Christian thought have a place at the table in the sumptuous theological feast that Bouyer sets before us. He recognizes that dialectical tensions are permissible, as long as they do not lead to oppositional annihilations. None of the three contested traditions—Neopatristic, Neothomistic, or sophiological—is taken to have an absolute monopoly on the speculative understanding of the mystery of grace. It can be frustrating for systematic theologians to accept the kind of open system that Bouyer proposes, but his attempt to clear access to development beyond a theological impasse that unnecessarily contributes to Christian division is instructive for all involved in the debate.

In fact, we think that a single conceptual tool, tied to a particular associative strategy that Bouyer recommends to be utilized in this difficult matter, may be especially fruitful, even if he did not himself develop its fruits in full flower. The conceptual tool is the Thomist differentiation of real and notional relations, and the associative strategy is to see the relation of divine to created Wisdom in connection with that of the Palamist doctrine of the distinction of divine essence and energies. We find this double approach employed at the end of chapter eighteen of *Cosmos* and more fully in *Le Consolateur*. In the former text, Bouyer says that divine Wisdom "such as it exists eternally in God, is . . . in the final analysis nothing but the uncreated whole, the fundamental and primal unity of those divine energies which unfold successively, through the gift of the Spirit, in the course of creation and its history, and which by the same token become distinct from God's immutable essence, but distinct only in a notional sense."[110] In the previous quotation we marked out, Bouyer spoke of Wisdom becoming really distinct from the Word of God in the Eschaton only in order to be perfectly reunited with him for eternity.[111] In the present quotation, he suggests that the divine energies remain ever "notionally" distinct from God's immutable essence, even as they "unfold successively." God's energies never become created, yet awaken created Wisdom in us eschatologically as a real distinction that allows the full personalization of creation by the perfection of our complementary being in the reciprocity of created consciousnesses. These energies seem, on Bouyer's understanding, to indicate the presence of uncreated Wisdom in created Wisdom, even as the latter tends toward real personalization in the eschatological Church.

The theological concept of "notional relation" comes from Aquinas, who distinguished it from "real relation" and applied it especially to the question of whether the divine persons and essence of God are really related.[112] The Angelic Doctor argued that we can draw a distinction between persons and essence in God, but it is only a notional one. We can affirm in our minds that this relation exists, and it clarifies our understanding of both divine plurality and unity, but in God the relation or distinction does not divide the essence. To put it more precisely, the

110 *Cosmos*, 193.

111 See the quotation referenced in n. 106 above: "All creation is ultimately a development of Wisdom from the initial stage when the distinction between it and the Word is purely rational, to the final stage, when it becomes real. . . ."

112 Cf. Thomas Aquinas, *ST* I.28.2.

divine essence and persons are not related in the way of any worldly composition. The divine persons and essence are only notionally distinct, while the persons are really distinct from one another in a relation of real other reference. In *Le Consolateur*, Bouyer points out that Palamists are sometimes befuddled by Western theologians on this issue, because they sense an agreement with them when the Westerners say that the distinction between the divine essence and persons is only rational or notional. How can Western theologians, they wonder, then turn around and disagree with them on the distinction between the essence and energies of God? Bouyer suggests that this befuddlement may indicate that they implicitly understand the distinction between essence and energies in God to be notional and not real.[113]

It is nevertheless clear that Bouyer does not align himself with either a strictly Neothomist or strictly Neopalamist position when he says that the entire work of creation and economy of salvation is through the "divine energies, evidently uncreated, which free themselves in creation, in order to bring into free, distinct being, under the incubation of the Holy Spirit, all the eternal content of the Wisdom, of the great design of the love of God."[114] Wisdom, in its production and destiny, is in a sense only one, he says, "with the Word, the Son where the Father expresses Himself and projects Himself."[115] Likewise, the divine energies are "one with the divine essence common to the Father and the Son," although "aiming at being essentially multiple and limited."[116] Regarding the unity of essence and energies, Bouyer says that the divine energies

> are no less mysteriously distinct from the divine essence in its very unity with them just like the Wisdom of creation—where the free will of the Father and the free counsel of the three who are only one is expressed—is distinct from the Son, from the Word where the Father expresses Himself as His eternal and ineffable Name, while finally coinciding with Him. Uncreated in themselves, they are therefore created in their effects, just

113 *LC*, 430–32.
114 Ibid., 448.
115 Ibid.
116 Ibid.

> like Wisdom, although the ultimate purpose is to bring them back to the uncreated.[117]

In this passage, Bouyer again postulates a homology between the relation of uncreated to created Wisdom and the relation of divine essence to divine energies. He distinguishes "essence talk" from "person talk." In his essence talk, he denominates the distinction between divine essence and energies as notional, and in his person talk he denominates the distinction in God between Wisdom and Word in the same way.[118] So, he indicates, on the one hand, in *Cosmos*, that created Wisdom can be really distinct from the Word, while, in *Le Consolateur*, in the just-cited block quotation, he speaks of both Wisdom and divine energies being created only in their effects, although we take it that he is referring only to *divine* or *uncreated* Wisdom in this latter passage. The divine energies and Wisdom are not collapsed into *created* Wisdom, which is an effect of their operation in the economy through the Father's two hands.[119] Created Wisdom seems to be a kind of "active receptivity" to the action of God working in us through the divine energies, especially in the uncreated grace of the Spirit of Christ, personalized in created persons as created grace, and divine Wisdom seems to contain a kind of "active potency" for the eventual becoming of created Wisdom, which in the end is reunited to the divine life.

Ever linked to the Redeemer, the Immolated Lamb, and to the special, sanctifying grace of the Holy Spirit, God's energies and Wisdom produce in us a transformation, which makes of us a new creation, truly changing us—by creating a new, accidental quality in us?—ever enlivened by the further communication of sanctifying grace.[120] With the Palamists, Bouyer stresses the uncreated character of God's transforming self-communication to the human soul through the energies which create effects in us associated with created Wisdom. With the Thomists, he stresses the transformation that grace effects in us and all of creation in us and through us as the redeemed and saved take on, all together,

117 Ibid.
118 *Cosmos*, 193.
119 *ES*, 403.
120 *IF*, 235.

the character of transfigured personalization in the Son on whom rests the Spirit, especially in the Church of the Last Day. Affirming the real distinction of Wisdom from the Word achieved in the final consummation of creation, Bouyer highlights the ontological density of our new life as real beings in Christ, while endeavoring not to lose sight of the consummation in unity with the Word in the Parousia that we await in hope. If God the Father is not related to us by a "real relation" in creating us, He brings us in Wisdom into the real relations of the *exitus-reditus* of the divine persons by adopting us into His Son and filling us with His beauty in the glory of the Spirit. God really relates to us in filial adoption by virtue of the real relations that constitute His interpersonal being.

This suggestive double approach, utilizing the pairings of real-notional distinction and divine essence-energies in conjunction, may be promising, and it manifests an admirable symmetry. Yet we must admit at this point that we wish Bouyer had developed the implications of his suggestions more fully, with greater care to explore the textual nuances in the debates between Thomists and Palamists. We are not always certain of the coherence of his various, thickly-wooded stanzas where this double approach is deployed. He needed to clarify what he meant by the uncreated energies and their distinction or not from uncreated Wisdom. And, of course, strict Neothomists and Neopalamists are both apt to reject this third way on the theology of grace, so it is difficult to get traction developing the theology of Wisdom in the context of this debate if a more focused, thorough exegesis and exposition is not provided.

The perplexities that Bouyer's thought may elicit in his readers on this question are not so much indicative of a fatal weakness in his own project as of the intrinsic difficulty inherent to recovering in a new context the Athanasian and Augustinian theologies of created Wisdom, compounded by the perplexities perennially elicited by disputes on nature and grace. We think that sophiological theology uniquely holds together the relation of the uncreated with the created, especially with regard to the unity of creation, and does indeed open new avenues of approach beyond age-old impasses, but that Bouyer himself saw this more in a visionary way, and his work needs to be developed on this front.[121]

121 Cf. Stratford Caldecott and Adrian Walker, "The Light of Glory: From *Theosis* to Sophiology," in *Communio* 42 (Summer 2015): 252–64. In this article,

It is our view ultimately that the best approach moving forward in interpreting Bouyer is to associate his theology of Wisdom with the theology of the image and imagination. Describing and extolling the work of the Russian sophiologists, Andrew Louth has simply said that "Wisdom . . . is the face that God turns toward creation and the face that creation turns toward God."[122] The sophiologists, Louth argues, in thinking of the relationship between God and the world in terms of this mutual, co-penetrating, loving gaze, brought the theology of the image to the fore in an important way.

Stratford Caldecott's general understanding of Wisdom puts us especially close to Bouyer and may help to illuminate interpretation on this question in the manner that we are suggesting here.[123] Recognizing the paradox inherent to any serious discussion of this matter, Caldecott explains that Wisdom "both pre-exists the act of creation (in God's foreknowledge), and does not yet exist (in the ever-moving present), and yet is mysteriously present throughout, accompanying the present as a foreshadowing of what will be."[124] This captures the eschatological and historical movement of Wisdom that is so important to Bouyer's thought and helps to understand his deployment of the conceptual apparatus of real and notional relations. The Oratorian's theology of Wisdom seems unique in the way it is able to embrace nature within salvation history understood in terms of its eschatological entelechy, and Caldecott's vision is intentionally close to his. Nature, meaning the whole of creation in its development, is intrinsically ordered to the Uncreated, who aspires to draw all things into perfect unity in the eschatological humanity of Christ. Caldecott connects Wisdom with life, which he understands to be a transcendental of being, at least inasmuch as the whole of being is

---

Caldecott develops Bouyer's sophiological third way precisely with respect to this dispute between Thomists and Palamists. We thank Aaron Williams, a doctoral candidate at the John Paul II Institute for Marriage and the Family, for bringing this article to our attention. This is a suggestive article, but it, too, only represents an initial opening of approach that hardly goes beyond what Bouyer accomplished.

122 Andrew Louth, *Modern Orthodox Thinkers: From the Philokalia to the Present* (Downer's Grove, IL: Intervarsity Press Academic, 2015), 58.

123 Caldecott, *The Radiance of Being*, 273.

124 Ibid., 79.

given being in order to be made fully alive in the Eschaton.[125] Likewise, the Oratorian stressed that only in the Parousia will created Wisdom have the fullness of life in real existence. Created Wisdom is not protologically or essentially distinct from God, in the sense of a real distinction, but only eschatologically and existentially so.[126] Created Wisdom is ordered, as the unity of creation in the reciprocity of finite consciousnesses recapitulated and elevated in Christ, to the perfection of personal communion in the Word of life, without being absorbed into the divine essence in such a way that the unity of God and creation lacks the higher bond of consummated love.

Caldecott, as we saw in chapter one, describes Sophia as "the divine Imagination hypostatized in relation to each of the three Persons in turn."[127] Caldecott's interpretation of Sophia, like that of Bouyer, works in the register of divine-human correlativity, of the image, and with regard to the unity of human experience. Divine Sophia, God's "Imagination," would be, as we suggested, the prototype of the human imagination in its unifying way of seeing and creative mode of acting. As we have said, this will ultimately be the focus that we shall apply in our final, evaluative chapter as it bears on the sophiological motif in the Oratorian's cosmology. Our focus will ultimately be more anthropological or meta-anthropological, centering on created Wisdom. It is with respect to the economy that the unifying core of Bouyer's cosmovision presents itself most fully. Before we get there, much more needs to be said about divine and created Wisdom with regard to really existing creatures: to the angels, to humanity, to Mary in relation to Christ, and to the eschatological Church. It is to these specific topics that we now turn in succession, following the ordering of topics in *Cosmos*, and moving next to the Wisdom of the angels.

125 Ibid., 78.

126 Both Davide Zordan and Bertrand Lesoing argue that Bouyer's eschatological and existential focus differentiate his theology of Wisdom from that of Bulgakov. See Zordan, "De la Sagesse en Théologie," 282–86; Lesoing, *Vers la Plénitude du Christ*, 293–95.

127 Caldecott, *The Radiance of Being*, 273.

CHAPTER 9

The Angels and Cosmic Liturgy

THE SOPHIOLOGICAL TRAJECTORY OF THOUGHT is challenging to modern theology for a variety of reasons, including for the fact that it brings the invisible world of angelic beings to the center of theological consideration in a way that is disruptive of settled modern theological priorities. Bulgakov wrote one of the most important modern monographs on the angels, which Bouyer described as his "essay of angelology *and of Christian cosmology*."[1] For the sophiologists, angelology is at the heart of theological cosmology. Bulgakov's angelology is also a cosmology. Likewise, Bouyer's cosmology is also an angelology, as we see especially in chapter nineteen of *Cosmos*, entitled "The Intelligible World and the Physical World: the Angels," and chapter twenty, "The Fall and Rehabilitation," the chapters whose main themes are our focus in the present chapter. Bouyer knows that there can be no Christian cosmology of any lasting weight or authority that would fail to envision the world in connection with the drama of the invisible persons who are God's created ministers to human beings and to the entirety of creation.[2]

Biblical apocalyptic is suffused with angelology, as are the Church's liturgical texts. Grintchenko firmly asserts in this regard that the "liturgy has best conserved this dimension of the Christian Mystery in its innumerable mentions of the angels, to the point that the very meaning of the Eucharist, from the *sanctus* to the *epiclesis* of the Roman canon, becomes insipid when the angelic dimension disappears."[3] One can rightly say

1 "La Personnalité et L'Oeuvre de Serge Boulgakoff," 140. Emphasis ours.

2 Parts of this chapter are taken from Keith Lemna, "The Angels and Cosmic Liturgy: An Oratorian Angelology," in *Nova et Vetera* 8:4 (2010): 901–21. Reprinted with permission of the publisher.

3 Grintchenko, *Une Approche Théologique du Monde*, 221. On the New Testament basis of biblical angelology, see Bouyer, "The Two Economies of Divine Government: Satan and Christ in the New Testament and Early Christian Tradition," *Letter & Spirit* 5 (2009): 237–62. See also "Le Problème du Mal dans la

that Bouyer's cosmology and angelology helps us to understand liturgy, and his liturgical studies from early in his career inform his understanding of the cosmos as a unity of visible and invisible realms. His Christian cosmic vision is deeply liturgical and sacramental, inserting human history into the center of the drama of creation as a whole, which includes the angelic hosts and the demons who have always sought to disrupt the liturgical service of the angels who are obedient to God. In this chapter, we shall expound main points from the aforementioned chapters of *Cosmos* in the light of the sacramental, liturgical, and monastic bearing of Bouyer's overall thought. There are two main themes in these chapters taken together that we wish especially to explore: 1) the connection of the angels to divine ideas; 2) the angels as causal forces helping to move cosmic history for good or evil. Bouyer draws both of these emphases originally from Newman, and we shall briefly describe in a first section Newman's sacramental principle, which seems to be the decisive theological inspiration informing Bouyer's overall cosmological understanding. We shall further draw out these themes in succeeding sections. In section two, we shall explore his theology of cosmic liturgy, which links the angels with the divine ideas, and in section three provide a response to an important objection that has been made to this linking. In section four, we shall draw on Bouyer's early text on monasticism to spell out more fully the dynamism of his view of angelic and human life in the ascent of cosmic being to the Creator in the midst of cosmic evil. In section five, we shall defend him against objections to his recovery of this ancient monastic vision. In section six, we shall recount his telling of Big History in an angelological light in chapter twenty of *Cosmos*, and in the seventh and final section provide concluding remarks regarding his most conjectural statements on the angels.

NEWMAN'S SACRAMENTAL PRINCIPLE

As we mentioned in chapter one, Newman's *Apologia* and *Parochial and Plain Sermons* made a lasting impression on Bouyer from the time he first read them in his youth. He returns to Newman and especially to passages from the latter text at the end of chapter nineteen of *Cosmos*,

---

Christianisme Antique," *Dieu Vivant* 6 (1946): 15–42.

which is our starting point here, in line with the developmental starting point of Bouyer's overall thought. Our author's cosmology is an expansion of Newman's sacramental principle, enunciated in both the *Apologia* and *Parochial and Plain Sermons*, and he understands the sapiential direction that Bulgakov inspired him to follow as "the crowing of that vision of the world that Newman helped [him] to attain."[4] If Bouyer elsewhere refers to Bulgakov's angelology as a Christian cosmology, as we just saw, in *Cosmos* he refers to Newman's sacramental system as an "angelological cosmology" (*cosmologie angélologique*).[5]

Both of these eminent Oratorians draw our attention to the invisible world of the angels as the deepest dimension of our own world. The invisible world of the angelic persons is, on their shared interpretation, at once beyond our world and within it. They maintain that we are meant to experience it as the hidden essence of our own world through the gift of faith and by our participation in the liturgy. Their angelologies evoke the position shared by Denys the Areopagite and Thomas Aquinas that the angelic hierarchy is vaster in the diverse magnitude of its species than the hierarchy of physical creation itself, which can be understood to be the inverse reflection of the angelic hierarchy.[6] For Newman and Bouyer, all of created reality is personal or reflective of personal existences, uncreated and created. Even the physical world is, as Bouyer says, "but the envelope, the external clothing of a wholly spiritual world."[7] It is, as Newman says, "the skirts of their garments, the waving robes of those whose faces see God."[8]

Newman, early in his ecclesiastical career, powerfully preached about the ultimate reality in our midst of the invisible world of the angels. There were two sermons of his in which he did so in a particularly striking way and that proved formative for Bouyer: "The Powers of Nature," found in volume two of his *Parochial and Plain Sermons*, and

4 *Memoirs*, 78.

5 *Cosmos*, 203.

6 Ibid., 199–200. See Aquinas, *ST* I.50.3. Indeed, Aquinas says that angels exist in "exceeding great number, far beyond all material magnitude." We shall return to this consideration below.

7 *Cosmos*, 195.

8 John Henry Newman, *Apologia Pro Vita Sua*, 28. Newman quotes here from a sermon that he preached on Michaelmas Day in 1831.

"The Invisible World," found in volume four of the same collection. In his *Apologia Pro Vita Sua*, written after his conversion to the Catholic faith, he explained his understanding of the world, which he, in this writing, referred to as his "Sacramental system."[9] Newman saw the world in the light of an analogy between the celestial and terrestrial realms. He admitted in this writing a possible connection between his sacramental view of the world and what he calls the "Berkeleyism" of Anglophone theology, which may have reached him through his reading of Bishop Butler's *Analogy of Religion Natural and Revealed* (1736).[10] Newman's evangelical faith, in connection with his inherent poetic imagination, had given him the capacity, from very early on in life, to see the material universe in a thoroughly religious light. His reading of the Church Fathers confirmed him in this way of seeing. Some portion of the teachings of the Fathers "came like music" to his inward ear,

> as if in response to ideas, which, with little external to encourage them, I had cherished so long. These were based on the mystical or sacramental principle, and spoke of the various Economies or Dispensations of the Eternal. I understood these passages to mean that the exterior world, physical and historical, was but the manifestation to our senses of realities greater than itself. Nature was a parable: Scripture was an allegory: pagan literature, philosophy, and mythology, properly understood, were but a preparation for the Gospel.[11]

Newman saw the whole of creation and history as a *praeparatio evangelica*. The fullness of God's revelation was directly foreshadowed by the teachings of the Jewish prophets and indirectly foreshadowed by those of the pagan prophets, who also were inspired by thoughts beyond their own. It is especially in the school of the Alexandrian masters, Clement

9 Ibid., 18.

10 Louis Dupré, "Newman and the Neoplatonic Tradition in England," in *Newman and the Word*, ed. Terrence Merrigan and Ian T. Kerr (Grand Rapids, MI: Eerdmans, 2000), 137–54, 143–44. Newman claims that he never read Berkeley directly.

11 Newman, *Apologia Pro Vita Sua*, 26–27.

and Origen, that Newman found resonance with this intuitive, poetic vision of the universe and history. This "Christian Platonism" of his comes to the fore in his discussion of how near to his own thinking he found the Alexandrian masters to be in regard to the angels. For both Newman and the Alexandrian theologians, the angels are not only seen as

> the ministers employed by the Creator in the Jewish and Christian dispensations, as we find on the face of Scripture, but as carrying on, as Scripture also implies, the Economy of the Visible World. I considered them as the real causes of motion, light, and life, and of those elementary principles of the physical universe, which, when offered in their development to our senses, suggest to us the notion of cause and effect, and of what are called the laws of nature.[12]

In this same passage, Newman repeats what he said in "The Powers of Nature," the sermon that he preached for Michaelmas Day in 1831: "Every breath of air and ray of light and heat, every beautiful prospect is, as it were, the skirts of their garments, the waving of the robes of those whose faces see God." Though he speaks in the *Apologia* of his angelological cosmology in the past tense, a view that he long cherished, there is no indication that he ever forsook his "Sacramental system," although his Berkeleyism may have been modified.[13]

In his two sermons mentioned above, Newman endeavors to bring his modern congregation to a conscious awakening to the reality of the invisible world. In "The Powers of Nature," he argues that if we are to see the world in its deepest religious significance, we must strive, through faith, to recognize that all created things are in the service of God. The world itself is a revelation of God, and all things have meaning in proportion

12 Ibid., 28.

13 See John Henry Newman, *The Philosophical Notebook, Volume I: General Introduction to the Study of Newman's Philosophy*, ed. Edward J. Sillem (New York: Humanities Press, 1969), 187–88. Sillem says that from 1835 onward Newman distinguished, as he had not done in his earlier thought, matter from sensible phenomena and thus broke from Berkeleyism. We shall return to this question as it pertains to Bouyer in the next chapter.

to their glorification of Him.[14] In "The Invisible World," Newman shows the rationality of the scriptural view of creation. It is no less stunning, he argues, to our quotidian sensibilities to consider the angelic world than it is to consider the myriad unseen worlds that constitute visible nature and even human society. Even the physical and historical worlds, he argues, are constituted by worlds within worlds, a known fact whose consideration should make it less strange for us to acknowledge the existence of angelic realms. The physical world is constituted by animals whose natures we can never fathom, whose activities go largely unseen by us. Indeed, their existence, as we experience it, though we can never fathom the depths of their brute natures, points to a mysterious depth of being at the heart of nature. This depth of being, Newman implies, is a reality that goes far beyond the ability of reductionist science to grasp in its nets. History itself is formed by human societies within societies, whose activities are unknown outside of their respective spheres: of poets, of scientists, of religious men, of scholars, of artists, of artisans. And we live in our respective spheres, going about our daily lives, as if other societies or spheres did not even exist.[15]

Newman teaches that the invisible world of the angels is no less present to us than the worlds within our visible world that go unseen by us but that we know to exist. The world of the angels is always present to us, in our own world, though it will burst forth, into the open, only in the future. Yet, he argues, we can reasonably anticipate this eschatological breaking-in of the invisible world by considering it as analogous to the yearly emergence of the flourishing of springtime in nature, in which life and activity burst forth out of the frozen winter. Just as in the change of seasons from winter to spring the budding of the trees and flowering of the earth transfigures the barren, wintry soil, so the eternal springtime that is to come will break through into our own world. The veil that at present covers the invisible world will be removed. The eternal Kingdom of God, hidden within the world of our direct experience, will shine forth in Christ's Second Coming. "Shine forth, O Lord," Newman prays

14 John Henry Newman, "The Powers of Nature," in *Sermons and Discourses* (New York: Longmans and Green, 1949), 64–71.

15 John Henry Newman, "The Invisible World," in *Sermons and Discourses*, 258–68.

in order to hasten the coming of the eternal springtime, "as when on Thy Nativity Thine Angels visited the shepherds: let Thy glory blossom forth as bloom and foliage on the trees; change with Thy mighty power this visible world into that divine world, which as yet we see not...."[16]

THE LITURGY OF CREATION

Newman's vision of the angelic world is adopted and masterfully developed by Bouyer in a liturgical light in chapter nineteen of *Cosmos*. The French Oratorian shows in this chapter that human and angel exist together inseparably in order to celebrate the uncreated glory of God "through the whole time of creation."[17] Through the shared ministry of angel and human, the plan of divine Wisdom develops to fulfillment in the totality of cosmic history. Bouyer crowns Newman's angelology with a Trinitarian theology of Wisdom that finds a sacramental locus in liturgy. Divine Liturgy is, as Bouyer once said, "the heart of the whole life of the Church: that to which everything is tending, or from which everything results."[18] The Eucharist recapitulates the entire cosmos, which, seen in this liturgical light, is but a reflection of the Trinitarian glory, even though it is marked by struggle and conflict because of the fall of angels and of humanity. Through God's Wisdom and in the light of His glory innumerable created personalities have been given being. The cosmos, reflecting the Trinitarian personality of God, is essentially personal in foundation: "Since God is the quintessentially personal being, the only world He could conceivably create is a world of persons."[19] The Eucharistic *bullitio* of the Trinitarian personality is freely projected by *ebullitio* into the vast hierarchy of created persons and is given a new created center in the sacraments of the Church.

Long before Bouyer completed his trilogies on dogmatic theology he was, as we saw in chapter one, an influential scholar of liturgy and of the Eucharist, and this is how he is still largely known in the United States. His early and enduring studies of the Church's liturgical texts and

16 Ibid., 266.

17 *Cosmos*, 200.

18 *Liturgy and Architecture* (South Bend, IN: University of Notre Dame Press, 1967), 2.

19 *Cosmos*, 194.

rites were foundational for his subsequent theological work.[20] In one of his most famous books as a liturgist, he traced the Eucharistic prayer of the Church in its variations and developments. In the midst of this vast, complicated, and erudite study, as a kind of side remark, he noted that the liturgical texts of the tradition speak of the cosmos in the way Newman did in his preaching on the angels. The Eucharistic texts, like Newman's writings, present the vision of a hierarchy of angels forming a seamless whole with the world of our direct experience.[21] This vision is rooted in the prophetic and apocalyptic traditions of worship in the Old Testament. Expounding the meaning of Isaiah 6 as part of the great prayers of Jewish synagogue liturgy, Bouyer says:

> The higher Angels, the Seraphim, as their name indicates, are themselves products of a mysterious fire which is like a first reflection of the glowing heart of the divine life, and the altar fire and sanctuary lamps act as a reminder of it. This fire recalls the illumination, the transfiguration of all things that is the product of the descent of the *Shekinah*, the divine presence, in the luminous cloud in which it is enveloped. The glory given to God by the Seraphim's singing of the *Qedushah* is the reflection of divine glory returning to its source. But in them it is a conscious reflection expressed in song, just as in God the igneous light is that of the Spirit expressed in the Word.[22]

Humanity, on this interpretation, has the responsibility to join its own song of praise to that of the angels, as all of creation returns together to the Creator. The cosmic dimension of the Eucharist of Christ is understood to be a consummating continuation of this Jewish biblical and liturgical understanding of the lyrical mounting of creation to God. Christ is the fully embodied, personalized manifestation of the *Shekinah*, who

20 See Matthew Levering, *An Introduction to Vatican II as an Ongoing Theological Event* (Washington, DC: The Catholic University of America Press, 2017), 50–80. Levering reads *Sacrosanctum Concilium* in connection with Bouyer's relationship to the text. In the process, he gives the best overview of Bouyer's influence in the liturgical movement currently available in English.

21 *Eucharist: Theology and Spirituality of the Eucharistic Prayer*, 64–68.

22 Ibid., 65.

makes possible, through his expiatory sacrifice on the Cross, the return of all creation to the Father in a liturgy of praise centered in the Church's Eucharist.[23] Bouyer accords great importance to the cosmic symbolism in liturgy in many of his works. The sacramental world of the Church's liturgy, he insists, is never apart from the "real" world. In fact, it renders the world more truly cosmic: "It is the whole world which has to regain from our sacramental experience a transparency to spiritual realities, and our renewed life in it has to tend to reorganize it toward these realities."[24] The material domain of sacramental practice reflects the celestial spheres that transcend humanity. The Church's liturgical symbols are irreducible to merely human constructs. As the Mosaic tabernacle in the Book of Exodus reproduced the celestial sanctuary where the angels give praise to the unnameable God, so the Church's altar of worship has its paradigmatic source in the work and teachings of Christ.[25]

In chapters nineteen and twenty of *Cosmos*, Bouyer invokes wide streams of traditional Christian theology to express the intelligibility of the biblical notion of cosmic liturgy that was present in Jewish tradition and consummated in Christ. He brings to the fore the Christian mystic traditions of Taboric light and the Canticle of Creation as foundational resources for a Christian articulation of cosmology. Saint John's apocalyptic images are given decisive authority by these traditions, for instance, chapters four and five of the Book of Revelation, in which we read about the four living creatures and two dozen ancients who sit round the throne of the Invisible Father and from whom emanates an emerald rainbow that illuminates the cosmos. Bouyer sees these creatures as the angelic guardians of the nations. They sing the Canticle of Creation day and night and play their instruments, which "represent everything that lives and moves in the world."[26]

According to the traditions that carry forward this apocalyptic vision, all of the cosmos was meant to be a choral Eucharist or symphony of

23 Ibid.

24 *Liturgy and Architecture*, 95.

25 *Sophia*, 60. See Exodus 25:40. None of this is to deny the historical realism of Bouyer's thought when it comes to understanding the development of liturgical forms.

26 *Cosmos*, 201.

gratitude and praise to the Creator. It was meant to be a translation into the realm of finitude of the infinite glory of the one, true King. It was first of all the creation of the choirs of angels, the first-born heavenly stewards and ministers of the cosmos.[27]

The theology of music that Saint Francis would bring to expression in his own Canticle of Creation was prefigured by Saint Augustine in the West and by those who followed him.[28] Drawing on this tradition, Bouyer makes striking use of the metaphor of music to describe creation. He sees the angels as a unified choir composed of an immense, harmonious array of tonal components, each tone representative of a person, from which the physical universe itself emerges.[29] All of the particles of the universe, all of its material energies, are, he suggests, moved by and formed through the resonant power of these angelic choirs. But it is particularly in terms of temporal harmony that the musical metaphor has its direct application. By fixing their watchful eyes on the conductor and Creator, the angels keep the temporal measure and rhythm of the universe.[30]

The theology of the Light of Tabor, we learn, was first developed in the Eastern tradition, most especially by Denys the Areopagite, who greatly influenced Western Scholastic theology in the thirteenth century.[31] Saint Bonaventure brought this theology of light to an even fuller articulation. Bouyer follows this tradition in order to express the spatial dimension of creation. He suggests that the world is like a "shimmering white light which breaks down into the countless colors that remain distinctive only by merging imperceptibly into one another."[32] Both of these images, musical and visual, evoke in us a sense that creation is a reflection of the transcendent unity in diversity of the triune God. They enable us to envision materiality as a religious and sacramental reality and to see more readily than is our modern wont that mind and will are ever-present realities in physical nature.

27 Ibid., 197–98.

28 Ibid.

29 Ibid. 198–99, 210.

30 Ibid. This view obviously evokes Tolkien's *Silmarillion*. But Bouyer already developed this angelology in the 1950s, in *The Meaning of the Monastic Life*, before ever having read a word of Tolkien.

31 *Cosmos*, 199–200.

32 Ibid., 209.

Bouyer suggests in chapter twenty of *Cosmos* that the creation of the material world is like a projection into being by God in His sovereign will of His own thoughts communicated through the minds of the angels, just as they were themselves projected into a free and distinct existence through the thought and free will of the Creator.[33] This is an idea very much in line with Newman's preaching that the angels are "the real causes of motion, light, and life" and flows especially from the cosmic theology of Denys the Areopagite. This way of speaking about the angels has been foreign to us in the modern West. It may seem to threaten the integrity of lower levels of secondary causality in their various spheres. Moreover, it might seem to take away from God His unique, creative causality.

In order to allay these concerns, one must have a proper view of the Christian understanding of the hierarchy of creation. Bouyer frequently turns in his writings to Denys's theology of the ecclesiastical and celestial hierarchies to provide needed clarification on this front.[34] He does so in chapter nineteen of *Cosmos*, where he explains that for Denys all of creation is indeed constituted by its interpersonal and hierarchical relationality.[35] It is not, as for the pagan Neoplatonists, a static, compartmentalized hierarchy that returns to its source only by self-elimination. Denys sees that creation extends and communicates in the wondrous diversity of its finitude the eternal *agape* of the divine thearchy.[36] Bouyer follows what he takes to be Denys's theology of the angels according to which the world is a dynamic and ceaselessly intercommunicating hierarchical reality, with each stratum of finite being having its own particular integrity. For Bouyer, this means that each being on a higher level is all that it is, and keeps all that it has, which is itself a divine gift, only in giving itself away. Influence from above and self-completion are not mutually contradictory realities in this view, and it gives a sense of the dynamism of creation, as it sees creation ceaselessly returning to its source in renewed communion. Moreover, the angels do not replace God as the creators of matter. God is understood in this teaching to create

33 Ibid., 208.

34 See especially, *CG*, 277–80.

35 *Cosmos*, 199–200.

36 Ibid. See also *CG*, 277–80.

lower dimensions of being through the higher, and thus it can be said that angels witness, in some mysterious manner, the creation of lower regions of the universe. It is in this sense that we might understand that even though matter is, to use Bouyer's own words, "a kind of projection of angelic thoughts," it is nevertheless a direct creation of the divine will.[37] God, through His divine command, gives the angelic thoughts an autonomous existence, just as he had given the angels, who were once His own thoughts in His eternal Word, an autonomous existence by breathing the life of His Spirit into them.

Denys's theological ruminations on the celestial hierarchy exemplify, on this interpretation, the manner in which the Church Fathers as a whole effected a personalist transformation of the pagan philosophical understanding of divine ideas. As we saw in chapters four and five, Bouyer argues that the great Patristic theologians understood the divine ideas through which all of creation is modeled and created not to be static and lifeless but in some sense vivified, personal presences first of all contained in the divine Word in whom the Father eternally expresses His being.[38] The exemplary ideas through and in which material being was created are in some sense angelic persons given being in the eternally generated Son, who is the "first born of all creation" (Col. 1:15). The eternal ideas tied to creation *in concreto* are not free-floating abstractions, as for Plato, but are free and distinct personal beings. God works in the world in and through the personal, celestial hierarchy that communicates His interpersonal being according to its own capacity.

In Bouyer's own words, recommending a return to this Patristic theology, which he considers to be quite compatible with both Saint Bonaventure and Saint Thomas Aquinas, these spiritual beings, the angels who form the incorporeal world and who constitute the primary cosmos, are "the total and harmonious combination of the individual thoughts which God, in His Wisdom, chose to include in the one thought wherein He recognizes Himself in the person of His Word and Son."[39] He develops this idea in chapter twenty. There he suggests

37 *Cosmos*, 208.
38 Ibid., 196.
39 Ibid., 210.

that the physical world is an ontologically secondary world, a concrete image of the angelic "symphony of light," itself a created image of the uncreated light.[40] The multiplicity of physical forms in this secondary cosmos reflects through the materiality of nature the immense speciation of the angelic hosts. The laws that adapt these physical forms, and the life that fills them, find "an echo in the interplay of melodic changes in which the whole cosmos is involved in a single concert."[41] Each type of body in its microcosmic organization "is like an inverted image of one of those angelic forms in the material mirror that is the web of the cosmos."[42] Materiality, for man, is the "paradoxically translucent opacity" through which the exteriority of his world is open to cosmic being. For the angels, materiality is "the harmony of reciprocal distinctions in which they live."[43]

Is matter, in this way of seeing, a "hardened" or "condensed" expression of angelic praise, with an inherently sacramental significance for both angels and humans? Does Bouyer follow Saint Bonaventure, for whom the angels possess some sort of hylomorphic composition? On the one hand, it seems that he holds matter or materiality to be a gift given to all created spirits, including the angels, by the Spirit of God as an instrument for the symbolic expression of infinite being within the finitude of creation, while, on the other hand, that it is the immanent means by which created spirits communicate themselves to one another and lift their song of praise to the Creator. Matter or materiality would then serve, even when considered solely within the immanent processes of creation and apart from human subjectivity, the function of linguistic expression. It is clear that for Bouyer the material dimension of being is in its very essence an instrument of religious signification. The forms of the world in which materiality subsists are so indicative of angelic presences and reflective of them that the angels might be inferred from this account to be the universals of physical species. Does Bouyer, like John Scotus Eriugena (9th century), understand the angels to be the operative universal causes of material beings?

40 Ibid., 209–10.
41 Ibid., 210.
42 Ibid., 226.
43 Ibid., 210.

ANGELS, IDEAS, AND CREATED WISDOM

This last question gets at a point that has elicited some criticism of Bouyer's cosmology. The French Oratorian argued that although the Church Fathers used Platonic language regarding the intelligible world to give expression to their understanding, they transformed its meaning. If Plato's intelligible or invisible world was one of ideas, he insists, the hidden world of biblical and Christian thought was one of persons. Yet our author does not sufficiently delineate his Patristic references in support of his claims in this regard, and the eminent contemporary Dominican theologian and scholar of the Church Fathers, Juan-Miguel Garrigues, has issued an important critique on this front.[44]

According to Garrigues, there is a common tradition of orthodox Patristic angelology stemming from Saint Irenaeus that gives us a different picture of the relationship of divine ideas to angels than what one finds in Bouyer, who, Garrigues argues, follows Origen's less common and perhaps even heterodox path in this area. Saint Irenaeus and those who are in his line hold that divine ideas are *logoi* (*λογοι*) expressed by God in His Word as the archetypal conceptions of creatures, while for Origen and Bouyer "they are," as Garrigues explains, "spiritual creatures to which man belongs before his sin."[45] The Origenists, he argues, do not take into account the repeated exhortation made by Irenaeus against the Gnostics that creation is made by the Word of God alone and not by the intervention of the angels. This seems, according to Garrigues, to run counter to Bouyer's position—directly quoted by Garrigues and which we quoted in part above—that "the world is a kind of projection of angelic thought into objective existence, just as the angels were the manifestation of the thoughts of God the Creator in a free and distinct existence."[46] According to this interpretation, the Origenists hold, contra the wider tradition that Saint Irenaeus initiated, that the angels constitute a "primordial world," and that the cosmos in its material aspect is only the envelope, vestment, or simple exteriority of a wholly spiritual

44 Juan-Miguel Garrigues, "Le Christ Nouvel Adam, Mais Aussi 'Nouveau Lucifer,'" in *La Théologie de Louis Bouyer*, eds. Bertrand Lesoing, Marie-Hélène Grintchenko, and Patrick Prétot (Paris: Parole et Silence, 2016), 125–39.

45 Ibid., 126.

46 Ibid. See *Cosmos*, 208.

cosmos. Garrigues remarks that Origen and Bouyer are not satisfied to say, as they should with the Book of Job (38:6–7) and the "tradition of the Fathers and Doctors," that the angels have been created, and that the good angels who were divinized from the beginning are confided with the government of the cosmos. Garrigues insists that Origen and Bouyer go too far in holding instead that the angels, as the primordial world of the *Λογιχά*, are themselves the *λογοι* or personal thoughts of God.

Both Bouyer and Garrigues speak of the "Fathers" as if the Patristic writers make up an undifferentiated tradition, with Origen understood by both men to be a questionable outlier on the topic of the angels and redemption, even if Garrigues associates Bouyer's position with that of the Alexandrian theologian. Garrigues is ultimately a proponent of the specifically Western tradition for which the divine ideas are one with the simple essence of God. Garrigues maintains the supposed view of Saint Thomas Aquinas that the divine ideas just are the divine essence as capable of being imitated by creatures.[47] The Eastern tradition of the Church Fathers, whether directly Origenist or not, may be a little less clear on the point than Garrigues suggests. Bouyer's friend Vladimir Lossky differentiated the "Fathers and Doctors" of the East from those of the West in this regard. Lossky held that the Western tradition lost a proper sense of the dynamism and intentional character of the divine ideas by putting them in the essence of God.[48] It was not so for the Eastern Church Fathers, he insists. For them, he argues, the "place" of the divine ideas is not in the essence of God but in the divine energies that come after the essence. The energies are uncreated but express the determinate will of God, and it is in the energies that one finds the divine ideas. Still, the divine ideas are, according to this picture, uncreated. Lossky insists that they cannot be equated with created intellects as in the theology of John Scotus Eriugena. So, while Garrigues may not take due account of the difference between Eastern and Western Church Fathers on whether the divine ideas are in the essence of God, the Neopatristic theologian Lossky is not of the view that spiritual creatures can be identified with the ideas.

47 Cf. John Wippel, *Metaphysical Themes in Thomas Aquinas*, vol. 2 (Washington, DC: The Catholic University of America Press, 2007), 64.

48 Vladimir Lossky, *The Mystical Theology of the Eastern Church* (Crestwood, NY: Saint Vladimir's Seminary Press, 1998), 96.

It would in no way help Bouyer's case in the eyes of either Garrigues or Lossky to point out that it is likely the "modern Origen," Bulgakov himself, who most fully influenced him to associate the intelligible world of ideas with the angels. The Russian theologian says it explicitly: "The holy angels are the hypostatic plan of creation; they are its ideas."[49] Yet it is worth dwelling on this for a moment, as it might in the end help us to give a more sympathetic understanding of Bouyer's position than what Garrigues provides. Bulgakov sought in his monograph on the angels, which was inspired by a near death experience, to establish the ontological implications of the angelic ministry as envisioned in the famous dream of Jacob in which the patriarch beheld the ladder that joins earth with heaven upon which myriads of angels ascend and descend.[50] Bulgakov postulated in his book that the angels are "correlative" in being to humans, and that the two share together in a common cosmic vocation. He saw the world of the angels as the "ideal analogue of the universe in all its parts."[51] He argued that biblical angelology, when understood in its ontological implications, gives true meaning to Platonic idealism. The angels are able to traverse the two realms of created being, ideal and real (material), albeit in a different way than man is able to do as a microcosmic unity of spirit and matter. This is not to say that he denied a sort of pre-existence of ideas in the essential Wisdom of the Word of God. For Bulgakov, the intelligible world of the angels is the first expression of God's eternal Wisdom as refracted through the prism of nothingness, and it is in this sense that he equates the angels with the divine ideas.[52]

It is with respect to the theology of divine and created Wisdom that Bouyer himself endeavors to think the intelligible world of the angels in intimate connection with the Word of God. He says that "in the Son uncreated Wisdom comprehends the form of every creature and all of creation as a unified whole, in the complete fulfillment in history of God's created plan."[53] The angels are first in the essence of the Word as

49 Bulgakov, *The Lamb of God*, 127.
50 Genesis 28:12–13.
51 Bulgakov, *Jacob's Ladder*, 34.
52 Bulgakov, *The Lamb of God*, 126.
53 *Cosmos*, 209.

the primordial created spirits through which God's design for creation is communicated and only then are they freely projected into a distinct existence of their own *ad extra*, although there is no before and after in this sovereign creative act as far as God is concerned. Bouyer's description of Bautain on this issue encapsulates his own understanding: "the ideas harbored in the divine mind are actualized as angels."[54] God's ideas in His Wisdom have an ideal analogue in creation that is never, from the standpoint of eternity, lacking in concrete existence, although this is not to say that the angels, as creatures, are eternal: they exist in the *aevum*. The eternal destiny of these created persons who constitute the plateau of the celestial domain is tied to that of man, and so there is a sense in which for Bouyer as Bulgakov a kind of correlativity obtains between angels and men, although, as we shall see at end of the penultimate section of this chapter, Bouyer conceives of this correlativity differently than Bulgakov. We might invoke Bouyer's employment of the idea of notional and real distinctions on this issue as he does himself with respect to the energies and Wisdom of God. The divine ideas would then be one with the divine Wisdom of God in a notional sense but really distinct from the divine essence in the archetypal created personhood of the angels.

The theology of ideas in both Bulgakov and Bouyer can be associated with the theology of spirit and matter in the Eastern tradition of the Cappadocians that Garrigues does not mention, but with the significant proviso that Bouyer thinks, unlike these Eastern Church Fathers, that "matter is an essential and original part of creation."[55] What Bouyer draws from the Cappadocians is something along the lines of the view that bodies or sensible things are a unity of simple, intelligible qualities in the sense that matter is spiritual in that it is always thought by another, first of all in the mind of God.[56] Denys the Areopagite pushes this understanding to the farthest bounds, and some commentators of his have gone so far as to read him as holding not only that the angels are the divine ideas as

54 Ibid., 139.

55 Ibid., 197. Gregory of Nyssa held that God creates matter only because he foreknows man's fall. Matter prevents our fall from being as total as that of the angels and enables our conversion and redemption.

56 Cf. Lossky, *The Mystical Theology of the Eastern Church*, 103; Hans Urs von Balthasar, *Presence and Thought: An Essay on the Religious Philosophy of Gregory of Nyssa*, trans. Mark Sebanc (San Francisco: Ignatius Press, 1995), 47–55.

exemplary and formal causes of material species, but that they are the efficient cause of matter itself, a position which Bouyer rejects.[57]

We might concede from these considerations that Bouyer's understanding is in fact in some ways closer to a wider Eastern Christian tradition than Garrigues admits, even if the Oratorian does indeed break from the Western Church Fathers. Yet there is another reference point we need to consider briefly, this time the greatest of the Western Church Fathers, which might render this conclusion problematic. It is surprising that neither Bouyer nor Garrigues explicitly brings this reference point to the fore. We refer to Saint Augustine's texts on the connection of the angels to the intelligible world and to created Wisdom in his commentary on Genesis. Attending to these texts, we discover that Bouyer may be a good deal closer to that most seminal of the Western Church Fathers than we initially suspected. We saw in chapter five of our study that Bouyer commended Saint Augustine's theology of created Wisdom. This makes it all the more surprising that he does not expound the Bishop of Hippo on the association of created Wisdom with the angels and the intelligible world. Augustine spoke of the angels as created in light. They are created Wisdom.[58] Some of Bouyer's passages in *Cosmos* that Garrigues thinks to be "Origenist" are in fact quite Augustinian, as we can discern from a couple of important passages from Augustine's aforementioned commentary, *De Genesi ad litteram*:

> And the reason God said, "Let light be made," and light was made (Gen 1:3), was so that what was there in the Word might be here in the actual work. The fashioning of heaven, on the other hand, or the sky, was first in the Word of God in terms of

57 Cf. Dom Denys Rutledge, *Cosmic Theology: The Ecclesiastical Hierarchy of Pseudo-Denys* (Staten Island, NY: Alba House, 1965), 14. Cf. *Sophia*, 154, n. 18. Bouyer says: "The supposition recently recovered by Dom Denys Rutledge, that the angels would be creators . . . is foreign to the whole tradition." Bouyer does not reject Rutledge's book out of hand. See *Cosmos*, 266, n. 48 and n. 52.

58 See Hans Urs von Balthasar, *Man in History*, 14: "the mysterious time-transcending realm of being of the *Confessions* [associated with created Wisdom] becomes in the Genesis commentaries quite clearly the angelic world of pure spirits (spiritual matter)." As we saw in chapter 5, Bouyer knew and commended this text by Balthasar. Yet he does not turn to these texts when expounding his own view.

> begotten Wisdom, then it was made next in the spiritual creation, that is, in the knowledge of the angels, in terms of the wisdom created in them.[59]

Elsewhere in this writing, Augustine says:

> For through the course of all six days, when those things which God was pleased to establish one by one were being established, was not [the angel] receiving first these things in the word of God, in order that they might "become" first in his [the angel's] knowledge, when it was being said: "And thus it has been made" (Gen. 1:3)?[60]

These passages can be supplemented by one from *De Civitate Dei*:

> For when God said, "Let there be light" (Gen. 1:3), and light was created, then, if we are right in interpreting this as including the creation of the Angels, they immediately became partakers of the eternal light, which is the unchanging Wisdom of God, the agent of God's whole creation and the Wisdom we call the only-begotten Son of God, through whom they themselves and all other things were made.[61]

Aimé Soulignac sets forth the meaning of these passages in saying that Augustine "makes the angelic nature the mirror and the privileged witness of creation, to the point of saying that creation is first made in some fashion in the spirit of the angel. . . . The angelic nature becomes in this way the archetype, the consciousness, and the ideal of the condition

59 Quoted by John Gavin, *A Celtic Christology: The Incarnation According to John Scotus Eriugena* (Eugene, OR: Cascade Books, 2014), 20. The following three quotations from Saint Augustine and the one quotation from the Patristic scholar Soulignac are taken directly from Gavin's text, on pp. 19 and 20. Here he quotes Augustine, *De Genesi ad litteram*, I, viii, *xvi*, 23, 20–23; 24, 1–4.

60 Ibid. Quoted from Augustine, *De Genesi ad litteram*, IV, xxi, *xlviii*, 129, 7–10.

61 Ibid. Quoted from Augustine, *De Civitate Dei* XI, ix, 524, 28–29; 525, 1–6.

of creation."[62] Surely this is how Bouyer should be taken when he says that "the world is a kind of projection of angelic thought into objective existence, just as the angels were the manifestation of the thoughts of God the Creator in a free and distinct existence." Bouyer insists, after all, that the angels are not the creators of matter. They bear an archetypal relationship to the rest of creation, in line with Augustine's view, but this is in a secondary sense. It is therefore at least as much Augustinian as Origenist to say, as Bouyer essentially does, that God works through the created light of the angels in order to form the rest of creation. God's direct causal relationship to the whole of creation is in no way denigrated by this position. Instead, the unity of creation in relationship to the Creator is affirmed in the integral wholeness of all its dimensions.

LITURGICAL ASCENT

In chapter twenty of *Cosmos*, Bouyer gives a synopsis of the entire story of creation, highlighting the central mission and role of the angels. He brings out the dynamism of the angelic mission to creation and attempts to elucidate the source of cosmic evil. He had already told this story relatively early in his career in *The Meaning of the Monastic Life*, which we briefly summarized in chapter one, and would do so one last time in his final book *Sophia*. If, he suggests in these writings, cosmic liturgy refers in the first instance to the presence of angelic hosts surrounding the Holy Trinity in worship, the relationship of these beings to the cosmos is not fixed and static. The fall of some of them has rendered their cosmic ministry dramatic and developmental with respect to physical creation. From at least the time of *Cosmos* onward, as well as in *Sophia*, Bouyer relates this dramatic narrative to modern cosmology.

The development of the physical cosmos as we perceive it and know it seems to suggest that created being is something other than a beautiful choir singing praise in unison to the glory of God. The apocalyptic

62 Ibid. Quoted from Aimé Soulignac, "La Connaissance Angelique," in *La Genèse au Sens Littéral*, ed. A. Solignac and P. Agaësse (Paris: Desclée de Brouwer, 1972), 645–53, 653. Gavin shows that Eriugena's understanding of created Wisdom is a development of that of Saint Augustine. He adds succinctly that for Saint Augustine "the hexaemeron took place in the created intellect of the angels" (19).

theology that sees creation in terms of a battle between good and evil forces should seem truer to the mark of direct experience than many intellectuals in the modern age might be willing to admit. Theologians affirm that the cosmos is good in being, and that it was meant by its very creation to be self-diffusive of its own participated, personal goodness in imitation of God's eternal, triune goodness. Common experience tells us that this does not often seem to be a sound description of what is in fact the case. It is not at all self-evident that the ontological dualism in the ancient myths is false. This is why this dualism was so prevalent in the ancient world and continues to exert a popular attraction in the modern age. Dualist mythocosmologies even pervade contemporary culture, particularly as expressions of political ideologies and of movements that associate the liberation of the cosmos with their own causes. Ultimately, we need divine revelation to educate us in another way of seeing creation. Yet pre-biblical myths might seem to be truer to direct experience. The world as we now know it is fractured, divided, beset on all sides by the disintegrating effects of pride, greed, and egoism. How, then, could it have been good in its origin, and how has it come to pass that it appears to be intrinsically disordered in its very being? Only the true myth can set us on the right path in answer to this question, but a renewed interpretive lens is required to see how this can be so.

It is surely not only the fall of man that can account for the situation of sin that has divided creation from the Creator and against itself. Divine revelation can be understood to tell us as much, but our own experience might present a sign that the cosmic scope of evil is inexplicable as a moral force if man is understood to be the lone created spiritual intelligence. Can we explain the suffering of animals which pre-existed the creation of man on this basis? Bouyer sharply comments in chapter twenty of *Cosmos* that on this question "Buddhism shows a lucidity which could be envied by many Christians whose convictions have lost their edge."[63] Cosmic evil, the Oratorian teaches, is tied to the First Adam, but, as the Tradition tells us, it precedes even his creation and fall. The predatory serpent Lucifer is already present in the Garden of Eden seeking to destroy humanity by his deceptions.[64]

63 *Cosmos*, 206.
64 Genesis 3:1–20.

The story of cosmic fall and Redemption is first of all a story of how, as Bouyer puts it in *The Meaning of the Monastic Life*, a text to which we want to attend further at this point because it fills in some details that are lacking in *Cosmos*, "a whole segment of the great mystic rose flowering around the Trinity has become detached and, as it were, torn open."[65] The highest of God's creatures, the first among all the angels, has, out of pride and out of the desire for self-glorification, disturbed the heavenly and cosmic liturgy.[66] Lucifer and his lesser minions have turned away from the divine Word, the eternal image of the Father and the source of all created being, and have directed their love only to themselves. These now-demonic persons have made it their goal to turn the lower hierarchies of creation away from the glorification of the Creator and toward that of their own pitiable and even monstrous being. They have formed, as Bouyer says, "a screen against the spontaneous movement of response which was rising up to the Creator from the most remote strata of creation."[67] The unified liturgy of heavenly and physical creation has been thrust into dissonance, reflected in cycles of destruction that are present in cosmic life.

The Devil and his minions, according to this story, have cast a veil of darkness over the world, obscuring the divine glory at its heart, and the harmonious chorus of heavenly and cosmic praise has turned into the sort of clash of warring factions in the celestial heights with which apocalyptic literature has acquainted us. Bouyer nevertheless insists that it is not within the power of these fallen angels to take full control over the material universe. The French Oratorian holds that it is only by the will of the Creator that the thoughts of the angels could be given autonomous being. God is the Creator of finite spirit, which is in the image of

65 *The Meaning of the Monastic Life*, 30.

66 Bouyer speaks of Lucifer as the highest of the angels. But it should be pointed out that it is not necessarily "common tradition" to hold that Lucifer was the highest of the angels at creation. Nevertheless, Satan is understood to be "the god of this age" in Paul (2 Cor. 4:4) and the "Prince of this world" (John 14:30). This indicates that Satan has a cosmic power that may be a little underestimated by modern Christian theologians. It is as nothing before the power of God. But relative to many other created spiritual powers connected to the world and human beings it is immense.

67 *The Meaning of the Monastic Life*, 30.

the eternal Son, and material being, which is in the image of the angelic images. Matter, as a projection of the angelic thoughts in which created Wisdom resides, is given over to the created ministry of the angels, but because it is not their creature, it is not completely tied to their authority. God, the one and only Creator of material being, has used this created resource—this image of a created image—as the instrument for bringing a new type of spirit into the world. Only this time it is a spirit clothed in flesh, a spirit "who," Bouyer says, "will embrace matter in the ascensional movement of its own creation, and will establish it once more in the cycle of thanksgiving, of the cosmic Eucharist which has been frustrated by Satan."[68] The First Adam was, on this telling, a subsitutionary angel. He was created to be, Bouyer says audaciously, a "new Lucifer," taking his place in the heavenly choir and leading the "remote strata of creation" back to their eternal Source.[69] The First Adam was a potential Redeemer, the potential new master of the earth, whose obedience and faithfulness to the Creator would have reintegrated the world in the praise of love that would have been the essential character of the primordial cosmos.

But, alas, the First Adam failed in his mission, and in his failure the earth has come "under a positive curse."[70] God's eternal will to rescue creation from its fallen condition was not thwarted by this second fall, and from the first instant of the fall of the First Adam, God prepared the world through His Spirit for the final victory of His love in the Incarnation of the Second and Final Adam. God sends His eternal image directly into the world for its salvation. This salvific work is a transfiguring accomplishment of the redemptive vocation of the First Adam. It is its surpassing and unanticipated actualization. The First Adam was called to replace Satan in the choir of the angels. The Second Adam, the Son of Man, though possessing Adamic nature fully, has a more exalted status. Bouyer explains: "the Son of Man, gathering up the whole of mankind in himself and retrieving the whole of creation in that humanity, is henceforth to be identified with the eternal leader of the heavenly choir: with the Word, with the eternal praise of the Father's love."[71]

68 Ibid., 31.
69 Ibid.
70 Ibid., 32.
71 Ibid., 33.

Christ, the Final Adam, as the eternal image of the Father, is infinite self-gift and perfect thanksgiving. He is the eternal, personal Eucharist to the Father in the Spirit, and, by the power of his redemptive Cross, draws humankind and all of creation directly into his perfect filiation. He fulfills Adamic nature through the power of his divine personhood. He does not merely restore the paradisal cosmic or heavenly liturgy. The Christian life, fulfilled in the New Adam, cannot be a simple return to the primeval garden of light and music. In his Ascension, he reunites physical creation in the human being with the heavenly liturgy of the angels. But he does something incomparably greater. He draws all things directly into his own canticle of thanksgiving:

> The cosmic liturgy is not indeed merely restored but reunited to its divine exemplar. Through the incarnation of the Word in humanity, which is itself an incarnation of the created spirit, all things are recapitulated in their divine Model and the choir of spirits is gathered up into the very heart of the Godhead. Christ leads humanity back to the earthly paradise through the Resurrection: through the Ascension he brings it back to the angelic sphere whence the prince of this world had fallen to ruin. Finally, entering right into the heavenly sanctuary, he makes us sit down with him at God's right hand, he makes us, and the whole universe with us, re-enter heaven, taking us with him right to the very heart of the Father from whom all fatherhood proceeds. In the whole Christ, in the heavenly humanity of which Jesus is the head, man, associated with the angels' choir, is initiated into the very canticle of the Word himself.[72]

Christ unites himself to human nature in the hypostatic union and joins us personally to his eternal being, which is one with the Eucharist of the Church. He brings us and the cosmos with us in Wisdom into the very heart of the personal, paternal source of divine and created life. The Eucharistic liturgy of the Church is not only a sharing in the primordial and supracosmic liturgy; it is a "recapitulation" of it in Christ's

72 Ibid., 34.

eternal canticle of thanksgiving to the Father. Through the human being, an angel of substitution, Christ completes, in a transfiguring way, the liturgical mission of both humanity and angels.

The angelology that Bouyer brings forth teaches us a profound truth about our own being, a truth that is in danger of being lost to the extent that strongly demythologizing currents of thought still predominate in Catholic theology. It teaches us that the very reason for our existence is ascent to the light of the divine sanctuary. The Christian life is essentially "ascensional." It is exemplified or lived to its fullest degree in the monastic vocation, which is misunderstood if it is not seen as a life of angelic ascent. The Christian life is a joining in the choirs of heaven, and, even more, it is ordered in its essence to beatific vision. It is a breaking of all ties with the fallen cosmos so that the cosmos itself may be rescued from bondage to the Prince of Darkness and our unbreakable insertion in the cosmos given full actualization in the perfection of cosmic liturgy. In light of Bouyer's angelology, we see that a properly articulated Christian cosmology is essential for us to grasp the very meaning of Christian vocation. Our life must be oriented to the Rising Sun (or Son) of the East, who draws us, as angels of substitution, out of enslavement to sin and egoism and brings us to the true and eternal Garden of the Orient, the otherwise unapproachable temple of his divinity.

HUMANITY AND ANGELS: A RESPONSE TO OBJECTIONS

In spite of the inherent beauty and wisdom in this dramatic narrative, it is unfamiliar to modern humanity, shocking in its seeming otherworldliness to some readers, and was quite controversial even prior to the Second Vatican Council. Bouyer's very manner of expression courts controversy. He speaks of Adamic humanity not only as an "angel of substitution" but of the Eschatological Adam as a "new Lucifer," in that he completes the vocation of the First Adam. The idea of the cosmic importance of the monastic vocation as paradigmatic for all Christians was considered by some to be elitist and unrealistic. Bouyer first told this story in *The Meaning of the Monastic Life*, and its essentials are recounted in chapter twenty of *Cosmos*. Marie-David Weill has given a comprehensive summary and analysis of the controversy that ensued from that first telling. It is important for our purposes in this study to follow her main points, at least those that are pertinent to the cosmological question.[73]

Weill explains that for Bouyer the monk lives a vocation that is exemplary for all Christians, indeed, for all humanity, in that he or she embodies the virtues of contemplation and virginal wholeness. The monk is an eschatological sign, leading the way in the common human vocation to angelic ascent. The monk is joined directly in praise with the angelic choirs and is able to live in contemplation of God through the liturgy of the hours which penetrates the whole of his or her existence. The liturgy of the monastery is a participation in the cosmic liturgy, joining humanity to the myriads and myriads of incorporeal spirits who chant the *Trisagion* or *Tersanctus* in the presence of the Holy Trinity: "Holy God, Holy and Mighty, Holy and Immortal."[74] The virginal purity of the monk has not to do first and foremost with moral purity but with the ardor of a love that belongs wholly to God without diversion or division in a life of praise and contemplation. This contemplation does not recuse the monk from exercising spiritual paternity with respect to the world but is perfective of the love of the monk for human beings.

The objections to such a view are perhaps obvious. In proposing an eschatological and monastic humanism as paradigmatic, does Bouyer not instead present a dehumanizing, disincarnate angelism that counsels flight from the world? In speaking of humans as "angels of substitution," is he not recovering the Gnostic myth according to which man can be saved only by giving up his nature as an incarnate spirit? Can such a tiny elite of Christians, who are oftentimes pessimistic with regard to the world, really be exemplary for all other Christians and truly be regarded as "humanist" in any meaningful sense of the word? As we have seen, Bouyer took the expression "eschatological humanism" from Dom Clément Lialine, a monk of Chevetogne Abbey, and dedicated the book in which he first used it to him.[75] Lialine had a warm appreciation for the book, but he was also critical of aspects of it. In particular, he wondered if Bouyer's eschatological humanism might veer in some pages into a

73 Weill, *L'Humanisme Eschatologique de Louis Bouyer*, 385–402.

74 Cf. Rev. 4:11 and 5:11.

75 *The Meaning of the Monastic Life*, before the contents and preface. Passage already quoted.

"meta-humanism" or false angelism, especially because of Bouyer's use of the Luciferian "myth" in recounting the story of creation and redemption. Weill points out that ultimately Lialine's criticism is directed at certain accents in the book rather than at its foundation.[76]

Without downplaying the importance and justice of certain criticisms, Weill notes that one has to take into account Bouyer's ultimate preference for the eschatological orientation of Eastern Patristic and monastic traditions. Failure to do so will lead, she rightly insists, to grave misunderstanding of his work. Perhaps the most important example of this has to do with the very method of his research. Lialine thought that Bouyer's approach was too myth-centered. Yet, as Weill points out, Bouyer's work was persistently wedded to the biblical "myth" or economy because, like Eastern theologians, he understood that humanity cannot be grasped simply by analyzing it in the present structure of its nature. Weill agrees on this point with Dom Olivier Rousseau, himself a Benedictine monk at Chevetogne, who wrote a positive review of *The Meaning of the Monastic Life*.[77] Rousseau explained that eschatology in the Christian East is treated under a cosmic rather than individualist angle, and that the anthropology of this tradition is based on the view that the human being is an image of God, which implies much more than to say that he or she is a rational animal, as is commonly said in the West. Rousseau (as Weill reports) submits that these premises necessitate the production of a theology that focuses more on the totality of the drama of creation and ultimately on the transfiguration of human nature in Christ than on the presumably fixed condition of our present state. One cannot rest satisfied with a metaphysical theology of the present but must engage the whole of the economy that takes into account "a theology of the primitive state, of the Fall and the Redemption."[78] We are called to ascent and to participation in the glory of God beyond our current condition,

76 Weill, *L'Humanisme Eschatologique de Louis Bouyer*, 388–89. In this whole paragraph, we briefly summarize Weill's own summary of a letter that Dom Lialine wrote to Bouyer on April 13, 1951. This letter is preserved in archives at Chevetogne Abbey.

77 Weill, *L'Humanisme Eschatologique de Louis Bouyer*, 392. Olivier Rousseau, "*Le Sens de la Vie Monastique*: À Propos d'un Ouvrage Récent," *Revue Générale Belge* (Oct. 15, 1953): 957–64.

78 Ibid., 393. Quoted from Rousseau, 959.

in imitation of the angels, and our vocation cannot be understood if our humanity is not connected to this wider whole of the economy. At the heart of this understanding lies the view that "the human nature of Christ divinized by grace, and endowed with the most spiritual qualities" is the eschatological model of our own transfigured humanity.[79]

The ontology of the human being is thus linked intrinsically to the Incarnation, which embraces the vast angelic strata of being, but this interlinking of humanity and the angels was lost in modern Western theology. In *Cosmos* and elsewhere, Bouyer claims that Rupert of Deutz in the twelfth century is the figure who broke away from the earlier tradition according to which the human being is an angel of substitution and so in fact is the ultimate intellectual initiator (well before the Nominalists) of the Western fall away from an integrated cosmology.[80] The Eastern view of the monastic life, summed up no better than in the work of Saint Maximus the Confessor, sees the paradigmatic Christian way as one of angelic ascent. This requires a theodramatic cosmology in order to be understood. Such a form of interpretation is not lacking in premodern Western theology. It is not only present in the Christian East. In the tradition of Western monasticism, monastic life was understood within a larger narrative of man as substitute angel. This story is told, albeit in a differentiated way, by figures such as Saint Augustine, Saint Gregory the Great, and Saint Anselm. According to Saint Augustine, who drew the idea from Origen but transformed it, the human being was created to replace the fallen angels, while for Saint Gregory the Great, he or she exists to complete the ranks of the angels who persevered in fidelity to the will of God from the beginning.[81] Saint Anselm recovered Saint Augustine's view but also postulated that human nature might have been created for itself and not to substitute for or complete the multitude of angelic hierarchies.[82] Rupert developed his critique of the earlier "mythological" narrative on the basis of the door that Anselm opened with this latter postulation.

79 Ibid.

80 See *Cosmos*, 107.

81 See Vojtech Novotny, *Cur Homo? A History of the Thesis Concerning Man as a Replacement for Fallen Angels* (Prague: Karolinum Press, 2014), 137–38.

82 Ibid.

If the earlier tradition is not as univocal on this point as Bouyer makes it seem, it is nevertheless the case that some defenses of Rupert's thought in the twentieth century that were directed against Bouyer's angelology were a little misguided or disingenuous. For instance, the Dominican theologian Marie-Dominique Chenu (1895–1990), among the sternest critics of Bouyer's eschatological humanism, saw Rupert as one of the estimable founders of the modern, secular age precisely because he rejected the old angelology that Bouyer revived.[83] Chenu lauded Rupert for overcoming the false otherworldliness of previous ages and for his concomitant sensitivity to created human nature in its density and intrinsic laws, ordered to its own finite, proportionate ends. According to Chenu, Rupert enabled humanity to grasp its demiurgic *potentia*.[84] Weill notes that it has been shown that Chenu in fact misinterpreted Rupert, who never forsook the monastic ideal and whose entire thought was ordered around the redemptive Incarnation of Christ. He downplayed without absolutely rejecting the motif of the anthropology of angelic substitution, but he nevertheless cannot be taken as a father of modern secularism. He, too, like the great monastic theologians before him, held that our ultimate end is given in a supernatural call to transfiguration in the perfect humanity of the glorified Christ. His theory of the absolute predestination of Christ regardless of whether Adam had sinned is a sign of commitment to the monastic ideal, and in fact shows that he does not prefigure idealizers of the technocratic materialism of a new humanity that seeks perfection according to its autonomous and wholly proportionate capacities. Even though Bouyer himself underestimates Rupert, we cannot but agree with Weill's preference for the Oratorian's position over that of the Dominican Chenu:

> We prefer without contest the perspective of the Oratorian: man is certainly at the heart of the universe, but not as a demiurge tending to replace its Creator. The true exaltation of human nature does not reside in the progress of a technological

83 Weill, *L'Humanisme Eschatologique de Louis Bouyer*, 394–96. See M.D. Chenu, *La Théologie au XII Siècle*, coll. "Études de Philosophie Médiévale," n. 45 (Paris: Vrin, 1957), 59–60.

84 Ibid., 395.

> civilization, but in a theology discovering in man the masterwork of divine grace, illuminated by the incarnation of Christ. In assuming the great tradition of the Patristic East, Father Bouyer presents the transfigured Christ as source and model of all holiness, offering as the premises the humanity fully redeemed and glorified.[85]

Contemporary defenders of Rupert might nevertheless raise an objection that is pertinent to our discussion and not fully addressed by Weill. She suggests that Bouyer's monastic theology keeps us from the questionable view propounded by Chenu that humanity is a demiurge at last fully at home in the regime of modern secularism. In the monastic context that is proper to it, Bouyer's view that the human being is an angel of substitution can still be maintained as a corrective to Chenu's secularist immanentism. Nevertheless, do not the mythic views that Rupert downplayed and Bouyer revives entail the consequence that the human being was not created for his or her own end and is thus a kind of accidental feature of the divine Wisdom?[86] If we revive this sort of anthropology, do we not take away from humanity its own proper ends, its own dignity, and its own intrinsic goodness? Should we not rather say that humanity was created for its own sake and not for the good of the angels or to fill in some sort of gaping cosmic recess that was not originally meant to be there? Rupert's demythologization does not have to be taken to mean that he initiated the demiurgic, anthropocentric turn in theology that Chenu commended and Bouyer deplored, but only that his setting forth of the meaning of human nature in the light of Christ is admirably cognizant of our unique creation in the *imago Dei* as bearers of the Incarnate Word. In response to this potential objection, we should take note again, as in chapter two of our study, that Bouyer's dramatic construal of human nature and personhood in the vast fresco of the liturgy of the angels is connected to that of Saint Maximus the Confessor according to

85 Ibid., 396.

86 Novotny, *Cur Homo? A History of the Thesis Concerning Man as a Replacement for Fallen Angels*, 137–48. Novotny defends Rupert against certain criticisms that have indirectly threatened his view in that they call the apparent anthropocentrism of Vatican II's *Gaudium et Spes* into question.

which the human being is a "workshop of creation." In this understanding, the good of our nature is intertwined with an intrinsic vocation that is interconnected with the entire condition of the cosmos into which we were born. This view affirms, and rightly, we think, that humanity was created both for its own sake *and* for the good of the whole of creation. We carry a mission of responsibility before an already-fallen cosmos, but this does not have to entail that we were an afterthought in the mind of God, given being only on condition that the angels sinned. It can be taken to mean instead that we were given being with the vocation we bear before the whole of creation in accordance with the condition of the cosmos as it was, as already fallen. The concreteness and existential character of Bouyer's thought is highlighted in this recognition. It is no degradation of the good of our "being-for-its-own-sake" to qualify our nature as inherently redemptive. We are, after all, at once and inseparably individual, social, and cosmic, and our actualization cannot be rightly described if any of these particular dimensions of our being is abstracted from the others or from the totality of significance that accords to our attainment of divine likeness in Christ.

EVOLUTION AND THE ANGELS

One of the most remarkable aspects of Bouyer's cosmology is the role he gives to the angels and demons as central protagonists and antagonists in the drama of Big History, with its evolutionary valences as detected by modern scientific inquiry. He affirms both evolution and the cosmic presence and ministry of the angels. Few theologians have understood or admitted, as he did, that, if one is going to stay in line simultaneously with classical theology in its monastic roots and with the scientific understanding of the cosmos, then the story of cosmic development will have to be told differently in its particulars than either classical theology or modern evolutionary theorists have done. Ultimately, the truth or validity of Bouyer's development of the traditional cosmic story is not dependent upon whether what he says is exactly the same as authorities in the Tradition before him. We are in a situation where particulars of the old story must be put in a new context, because the Fathers and Doctors did not know that the physical universe is billions of years old and that biological species have come and gone, suffering and dying in extinction before the human race was brought into being. At the same time, particulars of the new story must also be altered in order to make allowance for the effective presence in the cosmos of created spirits in their vast and variegated potencies. Evolution requires something other than a Neodarwinist interpretation.

Did any other Western theologian of stature in the twentieth century connect the dots on these points in the way Bouyer did, at least in broad outline? In chapter twenty of *Cosmos*, our author recounts the story of creation, cognizant of the need to integrate this just-described double alteration.[87] Remarkably, this enables him to synthesize two different traditions of angelology from the early Church, which he does not himself acknowledge. Thomas Aquinas recounted these in his *Treatise on Separate Substances*, and it is a bit surprising that the Neothomist Garrigues did not mention this in his critical presentation on Bouyer's angelology. According to the first tradition, which Saint Augustine upheld and Aquinas favored, the angels, although associated with the first light of creation, were created along with corporeal things. According to the second tradition, whose orthodoxy Aquinas acknowledges even if he does not prefer it, the angels pre-existed the physical creation for ages and ages. He quotes Saint Jerome (327–420), who seemed to prefer the Eastern view: "Six thousand years of our time are not yet completed and how many eternities, how many times, how many origins of ages are we to think first existed in which the Angels, Thrones, and Dominations and the other orders served God without the succession of and measurement of time and did God's bidding!"[88] Drawing these two traditions together in an evolutionary perspective, as Bouyer does in chapter twenty of *Cosmos*, we could say that before the existence of humanity with its relatively limited history the angels existed for "many origins of ages" but not without or in separation from the material cosmos with which they were created. They have been, as Augustine thought, linked to the material cosmos from the beginning, but they pre-existed the creation of human being in connection with a vast spatio-temporal expanse of which Saint Jerome had no knowledge.

87 See especially *Cosmos*, 212–15.

88 See Thomas Aquinas, *Tractatus de Substantiis Separatis*, xii, 96. The quotation is from Saint Jerome, *Commentarium in Epistola ad Titum*, I (PL 26:560a).

Bouyer affirms, first, that angels have an archetypal relationship to physical species, and, second, that the evolution of physical species is a reality.[89] There is much that is implied in the first affirmation to which Bouyer, perhaps out of commendable prudence, did not call direct attention. Stratford Caldecott, who assents to Bouyer's angelological recasting of the story of creation and the Fall, quotes an anonymous monk from Holy Trinity Monastery that vividly attests to the sort of tradition of thought to which Bouyer gives much credence and that is concordant with potential developments of the Augustinian view that we discussed above:

> None of the individuals of a species can account for the nature of the species as such. In order to explain the nature or form of any species, we are required to posit a cause which transcends the individuals of the species. Such a cause must contain the form of the species within it in a higher way without being itself part of the species. These are the spiritual substances or angels. The angels contain within themselves in a purely spiritual way the forms of things in the material world.... The causality of the angel of sparrows on the young sparrow is more intimate than even the sparrow's own mother. When the mother dies the sparrow continues to exist; but without the angel there would be no sparrows at all.[90]

Each physical species, on this view, has its own angel, and the angel is a direct causal presence in the life and death of each individual of the species. This develops Bouyer's already-quoted conjecture that the bodies of physical species are an inverse image of the angelic forms. Caldecott refers to the angelic ideas, in a way that is very much evocative of Bouyer, as the "morphogenetic fields" of each species, present at "the beginning of each evolutionary chain, needed to explain the emergence of new forms in nature."[91]

89 *Cosmos*, 208–13.

90 Caldecott, *The Radiance of Being*, 55. Caldecott does not provide further bibliographical information.

91 Ibid., 56.

The situation is complicated by the Fall. Angels have archetypal, causal impact on physical creation, but this cannot be limited to the angels who clung from their first creation to the will of the Creator. The warp and woof of the cosmos is tangled by the presence of fallen angels who conspire against the divine Wisdom. This is Bouyer's view in chapter twenty of *Cosmos*. He holds that a cosmic, physical fall issued from the Fall of the first created spirits and that this brought death into the world. Modern cosmology, on this postulation, should not force us to reject the link between sin and death. Death cannot be reduced to a necessary passivity directly willed from the beginning by the Creator. But the human fall cannot explain adequately the presence of death in the cosmos prior to the existence of humanity.[92]

According to Bouyer's suggestion, the angels who fell made a mess of creation, but, because opposed to divine Wisdom, were not the created spirits who had the upper hand in moving the cosmic process along. Bouyer teaches that the divinized angels were able to integrate "death itself... into the development of life, which we see constantly surviving worldly death, and even feeding on it."[93] Remarkably, he sees the effects of their ministry in all of the advances of biological life, including the development of sexuality and the physical generation through which species are propagated in a world haunted by death. In making this bold (or outlandish?) suggestion, does Bouyer follow those Eastern Church Fathers for whom sexuality or sexual differentiation exists only for the sake of procreation in a fallen world? In fact, he does not. He thinks that sexual difference exists as essential to God's sovereign will and Wisdom for humanity from the very beginning, irrespective of the Fall, but that it takes on the function of physical propagation in the way that we know it only as a result of God's reparative action through the ministry of the angels.[94] He embraces what we might call a modified Eastern Christian view. For Bouyer, the biological world is what it is even in its manner of sexual expression as a result of sin and its consequences. However, unlike these Eastern Church Fathers, he does not hold that sexual complementarity and union exist as such

92 *Cosmos*, 212

93 Ibid.

94 Ibid., 212–13, 226–27.

only as a concession for the human fall.[95] Creation is essentially nuptial in his view and the nuptiality of Adam and Eve was meant to heal an already divided creation, taking the form that it did because creation was as it was. Human sexuality has a positive connotation, but it is inscribed in the concrete vocation of man (male and female) as workshop of a created order disrupted by a primordial fall preceding the creation of man (male and female).

The work of the faithful angels in accordance with the divine will allowed life to develop and flourish in the cosmos. Bouyer says:

> Supported and nourished by the creative grace to which the angels' fidelity contributed, their struggle was to bring about the consistent triumph in the world of the forces of life over those of death. It was also to achieve the ascent of life, a steady development that seems to be the essence of the entire history of the physical world—in spite of entropy, the deterioration which appears inevitably to accompany creative energies.[96]

The emergence of the human being "would be the first stage of the reassertion of the forces of life in the world,"[97] but God's image was first projected from the first moment of creation by the angels onto the "materiality" of the cosmos. As with Saint Gregory of Nyssa, so with Bouyer, "materiality," although distinct from spirit, is already in a sense "spiritual" or qualitative because mind, both uncreated and created, always projects its image into it. Man is nevertheless, Bouyer affirms, uniquely an incarnate spirit, the first potential redeemer of a world that was already in disarray before he was ever brought into being through the direct power of God. This sets the stage for the "evolution" of the human species and culture that is made ready by the Father's two hands to receive the Incarnation of Christ. It is in chapter twenty of *Cosmos* that Bouyer speaks of the kenoses of God in relation to creation that we have already described. These kenoses shape the entire course

95 We shall explore this point in much greater detail in chapter 11.
96 *Cosmos*, 213.
97 Ibid.

of human history "which culminates in the supreme manifestation of uncreated love as the gift of self which is the divine essence."[98]

At the end of chapter twenty, Bouyer directly addresses the question of how we should understand the nature and dignity of humanity in relation to that of the angels. Which type of created spirit, angelic or human, holds ontological superiority? Bulgakov, we learn, argued that man is relatively superior to the angels, because he is both spiritual and incarnate and therefore possesses a plenitude of being that is lacking in the angels, as he is able to provide the flesh for the Incarnation, which the angels could never do.[99] Bouyer argues to the contrary that because the angels are intimately connected to the being and destiny of the entire cosmos and are called with man to participate in the life of God through the Incarnate Son, they do not exist in a condition of relative lack, in the way that Bulgakov and others have suggested. Both angels and man, he insists, possess the same grace of God even though our situations mirror each other in relationship to the physical world.[100] Bouyer holds that with the exception of the "Virgin Mary and of the mystical body of Christ, the Church militant in union with Christ," the angels have a greater capacity to receive grace and respond to it and so "remain relatively superior to mankind, in spite of the completion which men contribute to the angels, and in spite of the consummation which the Savior's incarnation brings to the whole cosmos."[101]

Does this last assertion contradict *Gaudium et Spes* from Vatican II, according to which "all things on earth should be related to man as their center and crown?"[102] Bouyer certainly undermines the anthropocentrism of modern theology with this statement in a way that we think should rejoice certain proponents of "Deep Ecology."[103] But he in fact does not

98 Ibid. 215.

99 Ibid., 216.

100 Ibid., 217.

101 Ibid.

102 *Gaudium et Spes*, I, 12.

103 Surely we would relate much more respectfully and humbly to the beings of the physical world that we consider as being beneath us if we recognized the presence behind them of invisible powers, much more powerful than us, who communicate the formal elements of being to them. This is one of Newman's main points in the aforementioned sermon "The Powers of Nature."

downplay the unique ontological dignity of humanity. Rather, he associates it with humanity's supernatural elevation in the Mystical Body of Christ and in Mary. He holds that Mary is the pinnacle of human personhood in the supernatural order who elevates it to an even higher plateau than the angels by virtue of being the Mother of God (*Theotokos*), and that her unique "maternity of grace" is communicated to the whole Church.[104]

CONCLUDING REFLECTIONS

Bouyer does not tell the story of the angels in the way that he does in *Cosmos* and elsewhere first and foremost as an apologetic defense of the doctrine of the angels. His primary aim is theological: to give the angels their due as cosmic presences biblically revealed and also to illuminate thereby the human creature's own ontological dignity in a cosmos that is unified in the divine Wisdom. He realizes that theologians have largely, timidly avoided taking modern cosmology into account in light of angelology. Whatever the particular views of the Fathers and Doctors, it is undeniable that all held the angels to be ministers to the whole of creation. Their shared view on this broad but not insignificant point was faithful to the sapiential and apocalyptic heart of Christ's own teachings. It does not have to be forsaken today, but in order to find new life as something more than an historical monument it has to be successfully integrated into our Big History or our Big History into it. Bouyer is unique in calling attention to this point, but it may remain true today that any theologian who would be bold enough to develop his insights risks marginalization. On the other hand, in a day when a sitting pontiff speaks openly about the Devil and the Archangels, now may be a particularly propitious time to recover the monastic cosmology that Bouyer sets before us. If the Devil and the Angels exist, and have existed in relation to the physical cosmos from the very beginning (as the Western tradition held) or before it (as the Eastern tradition held), and are cosmic ministers, as the whole orthodox tradition of Christian thought has maintained, then they bear relation to its very evolution. How are we to understand this? We think Bouyer generally sets us on a good course in this manner and shall explore the issue a bit further in our next chapter.

104 We shall explore this issue in depth in chapter 11.

However, to be true to Bouyer, we have to recognize that in the end he maintained caution and deference to the symbol in telling this story of cosmic history. In *The Church of God*, he says that the Word of God describes the angels "only through images that instruct us more by deficiencies we sense than by anything positive."[105] In *Sophia*, he described his conjectural narrative of the angels in evolution as "*paramythia*" in the sense meant by Plato.[106] In the penultimate section of chapter twenty of *Cosmos*, entitled "Symbol and Reality," he again asserts the need to maintain caution on these matters, and that we must respect symbolic intelligence in this area. If, as some suggested at the time of the publication of *The Meaning of the Monastic Life*, Bouyer's story seemed a bit too neat and tidy in marking the stages of cosmic development, we should be aware that he recognized the distance that exists between what he tries to describe and his power to describe it. His dramatic narrative is best read as an attempt to help elicit anew in us a sense of wonder in the glory of the Creator. His angelological cosmology indeed calls to mind the aforementioned maxim shared by both Denys the Areopagite and Thomas Aquinas that "the angels, even inasmuch as they are immaterial substances, exist in exceeding great number, far beyond all material multitude."[107] There is no scientific or philosophical reason to deny it. In fact, Bouyer once directly said that "in a world [our world] whose material dimensions have been revealed to be ever more vast than we thought, it is more believable than ever that man is not the only emergence of the spirit, or the highest or the first."[108] If we evacuate the cosmos of angelic beings, we are inclined to fill it in with other personal beings of great power, with UFOs, space aliens, or beings from other dimensions. This has been noticed by theologians at least since the time of Karl Barth (1886–1968). Is there not a glimmer of truth in conspiratorial histories of the world that view human

105 *CG*, 582.

106 *Sophia*, 159–60. See *Phaedo*, 83a. See also *Cosmos*, 216: "But these things can be spoken of only in images. . . ."

107 Thomas Aquinas, *ST* I.50.3. We quoted this passage in n. 5 above. Saint Thomas says this in explicit connection with the view of Denys the Areopagite and the *Celestial Hierarchy* xiv. They both agree that "the more perfect some things are, in so much greater excess are they created by God."

108 *CG*, 583.

progress as guided by spiritual beings from other planets? Thinking ourselves unable to claim reasonably that angelic ministers have guided our cosmic and cultural evolution, perhaps we are left with a genuine explanatory lacuna, and extraterrestrial humanoids are thus understandably invoked to fill in this lacuna in the popular mind, which rightly suspects that mainstream academicians have discrepancies of plot in the stories they tell. The universe is billions of years old, with billions of galaxies, many of which contain hundreds of billions of stars—and this speaks only of its macroscopic magnitudes! The immensity of the physical cosmos above, below, and within us boggles the mind. It is incomprehensible. Can we not imagine that our cosmos is spiritually suffused with finite intellects and freedoms as vast as or even vaster in its differentiated, spiritual expanse than what materiality shows forth in its incomprehensible magnitudes? And do we not have to say, in order to be true to what is certainly the common mind of the Fathers and Doctors, that the angels were present with the physical cosmos at the very least from the beginning? It is just that now the "beginning" of the physical cosmos is understood to reach back to the many "ages" that Saint Jerome thought were the province of the angels before physical creation with its spatio-temporal dimensions was brought into being. Certainly, it is not impossible for God to create a cosmos that is as expansive in its spiritual diversity as it is in the horizon of its materiality. If we simultaneously affirm both the material and spiritual grandeur of the cosmos in imitation of God's own being, our sense of wonder in the incomparable greatness of the glory of the Creator can only be magnified. If Bouyer does not himself directly offer the reflections we have just made, his angelology points out the conditions for telling a coherent Big History in the purview of biblical revelation and helps to awaken in us contemplative wonder at the infinite reaches of God's eternal glory as Creator and Redeemer in its conjoined material and spiritual scope. It can contribute to a renewed liturgical way of seeing. The French Oratorian thus gives new life with his angelology to the doxological cosmology at the heart of Christ's own teaching.

CHAPTER 10

Human Corporeity and the Mystical Body

THE FOREGOING DISCUSSION OF ANGELOLOGY brings to the surface fundamental questions on the nature of the relation between matter and spirit that are crucial for the theological recovery of the cosmos. It inevitably spurs the challenge of whether such a view can be considered at all credible given what modern scientific cosmology tells us about the world. Much that is said throughout *Cosmos* addresses this challenge tangentially or indirectly. The very defense of the legitimacy or validity of mythopoetic thinking and imagination in this book is already intended to meet this challenge. In chapter twenty-one of *Cosmos*, entitled "Created Spirit, Matter, and Corporeity," our author more directly responds to the questions and challenges that arise from this topic. He develops in a suggestive manner certain philosophical positions that he thinks it is important to take seriously if the Big History of creation as told in the modern age is to maintain viable connection to the orthodox Christian Tradition.

In this present chapter, we shall focus on some of the theological and philosophical points raised in the twenty-first chapter of *Cosmos*. We shall explore in successive sections the Oratorian's history of the concept of matter, his philosophy of spirit and *qualia*, his recovery of Christian "microcosmism" in light of modern science, his theology of sin and cosmic evil, and his theology of the New Adam and the Church as macrocosm or macro-*anthropos* (words he does not himself use). The central issue we shall ultimately focus on, in the midst of many twists and turns that we think it necessary to follow in order to draw out adequately the implications and connections in Bouyer's reflections, has to do with the unique corporeity of human personhood. We shall see that a qualitative philosophy of materiality in line with the Cappadocians and Berkeley is the philosophical support given to this understanding. We shall demonstrate that for our author the meaning of materiality can be found only in the hypostatic union of divinity and humanity in Christ and in affirmation of the meaning of his Body, the Church, as the mystical principle of the whole of creation.

THE HISTORY OF MATTER

We live in an age of rampant materialism, but the "matter" on which we presume to base our lives is an extremely elusive being. There are different understandings of it that correspond to different types of materialism, whether mechanist materialism, dialectical materialism, or the practical materialism of everyday consumerism. The concept of matter is in fact so elusive to our absolute comprehension that a significant tradition of modern thought has raised the question of whether matter even can be said to exist at all. Is not the concept of matter the cause of a great deal of mental and spiritual confusion? Would it not be better, in order to allay this confusion, just to do away with it altogether? This latter proposal was defended by A.A. Luce (1882–1977), the foremost scholar and disciple of Bishop Berkeley in the twentieth century and a philosopher with whom our author maintained a cordial correspondence. Bouyer read with serious attention some of Luce's books that offered this proposal and, as we have seen, took seriously Berkeley's matter-denying challenge to philosophical materialism, even if he did not himself fully embrace the theses of either Berkeley or Luce.

In one of his books that Bouyer read, *Sense without Matter or Direct Perception*, Luce provides an etymology of the concept of matter and argues that the postulation of the existence of a subsistent matter unavailable to perception is deleterious to human flourishing. Such a postulation leads and has led, he argues, to confusion of thought and to skepticism about the possibility for our senses to put us in contact with a really existing world. His argument might be surprising for many Catholic theologians and philosophers, for Luce does not wed denial of the existence of self-subsistent matter to transcendentalist or representationalist epistemologies or to subjectivist idealism. He is not anti-realist. Just the opposite is true. He argues that immaterialism is the only sure path to a cogent, hard-headed realism. Sensible realities, he urges, are not founded on a material substructure that can never appear. There is no good reason to postulate the existence of this substructure. Luce shows that realism and idealism—although not *subjectivist* idealism—are not opposed to one another. This is, he avers, the profound and abiding truth of Bishop Berkeley's system. Luce proposed, in line with his philosophical master, that the only sure path beyond sophism and skepticism is to recognize the ideal character of sensible being.[1]

The work that Luce does in this book is a foundational source for Bouyer's brief but highly suggestive presentation of the history of the concept of matter in chapter twenty-one of *Cosmos*. Bouyer prefaces this history by acknowledging this book's importance. He even admits that "without always agreeing, we do not lay claim to any substantially improved results [than what Luce obtained]."[2] Bouyer certainly adds extra data beyond what Luce provides. For one thing, while Luce skips over the Middle Ages entirely, Bouyer begins his brief treatment of the meaning of matter by referencing the dispute between Saint Bonaventure and Saint Thomas Aquinas on the materiality of the angels.[3] Aquinas, as is known, held that angels were purely spiritual beings and without matter, while Bonaventure held that matter was one of the principles of angelic individuation, even though he recognized that the angels were incorporeal. It may not be true to say, as Bouyer does, that the theologians differed on this score because they had different understandings of matter.[4] Instead, they differed on a broader issue regarding the constitution of individuality and personhood, leading to different ways of understanding "form" as much as "matter."[5] This does not undermine but indeed supports Bouyer's claim in chapter twenty-one that both matter and spirit are recalcitrant to comprehensive definition or conceptualization.

In the Western tradition, the two most important philosophers of antiquity, Plato and Aristotle, were also the two most decisive thinkers concerning the question of whether we can speak of a really existing material substratum of being. Bouyer next traces in chapter twenty-one the views of both philosophers on the topic.[6] He disagrees with Renaissance

1 Luce, *Sense without Matter or Direct Perception*, 1–11.

2 *Cosmos*, 218.

3 Ibid.

4 Ibid.

5 Cf. Étienne Gilson, *The Philosophy of Saint Bonaventure*, trans. Dom Illtyd Trethowan (New York, NY: Desclée and Co., 1965), 226–30.

6 *Cosmos*, 218–21.

Platonists and their descendants for whom Plato was the first philosopher to postulate a sharp distinction between spirit and matter as metaphysical principles. Plato, in the *Phaedo*, does, Bouyer argues, distinguish an active *nous* that gives us true knowledge of permanent ideas from sensory or sensible realities that only give us opinions (*doxa* or *δόξα*).[7] In the *Timaeus*, he speaks of *hyle* (*ὕλη*), but he does not mean by this an abstract concept of matter. Bouyer maintains that the word retains its original meaning as "wood" or "foliage" in this text, and that Plato contrasts ideas not with a subsistent and univocal matter but with an undefined receptacle that is more akin to the concept of space than of matter. Bouyer quotes from *Timaeus*:

> [The] mother and receptacle of all created and visible and in any way sensible things is not to be termed earth, or air, or fire, or water, or any of their compounds or any of the elements from which they are derived, but is an invisible and formless being which receives all things and in some mysterious way partakes of the intelligible and is most incomprehensible.[8]

Aristotle took this teaching on the receptacle as the basis for his final clarification of the meaning of matter. A statement from the *Metaphysics* is evocative of Plato's *Timaeus*, and Bouyer quotes it in full:

> By "matter" I mean that which in itself is not stated as being the whatness of something, nor a quantity, nor any of the other senses of "being." For there is something of which each of these is a predicate, whose being is other than that of each of the predicates; for all the others are predicates of a substance, while a substance is a predicate of matter. Thus, this last is in itself neither a whatness nor a quantity nor any of the others. . . .[9]

This statement, Bouyer informs us, gives us the last of three stages in Aristotle's ruminations on matter. At this final stage, we learn, matter is

7 Ibid., 219.
8 Ibid., 220. *Timaeus*, 51a.
9 *Cosmos*, 220. *Metaphysics*, 1029a, 20.

understood to be the suppositum that underlies all visible or tangible existence, lacking qualitative being in itself. Matter cannot exist independently but underlies change. Luce had described these three stages in his aforementioned book, and Bouyer follows him directly.[10] In the first stage, according to both authors, matter was accorded a relative reality in keeping with its relationship to the various supporting metaphysical constituents of the world. Taking the example of a knife, one can see that matter is construed differently based on one's vantage point. With respect to the knife taken as a whole, the cutting edge is the form and the steel is the matter. From an alternative vantage point, steel is form to the particles that constitute it. In the second stage of Aristotle's thought, matter is understood to exist *in posse* or as possibility in relation to action. This, Bouyer adds, stems from the terminological and mythical distinction between feminine passivity and masculine activity.[11] In the final stage, summarized in the quotation above, we get to the Aristotelian understanding summarized by the Scholastic theologians according to whom matter is *nec quid, nec quantum, nec quale*— neither a thing, nor a quantity, nor a quality.

Luce argues that none of these views is helpful in enabling us to grasp the constitution of the sensible plenum before us that gives us to live in a blooming, buzzing world of qualitative life and multiplicity. Bouyer adds that ancient philosophers coming after Plato and Aristotle were themselves not terribly helpful in clarifying the meaning of matter.[12] The Stoics, he teaches, conflated spirit and matter. In *Sophia*, he compares their view to that of modern panpsychists, such as the Teilhardians, whose definition of matter Bouyer rejects.[13] For Plotinus, we learn in *Cosmos*, matter constitutes a limit and is essentially indistinguishable from nothingness. Bouyer differs from Luce in interpreting the Wisdom of Solomon on the meaning of matter. Luce argues that this biblical book construes matter in a philosophical way that is very much in line with Aristotle, and Anglicans, he suggests, are to be relieved that it is not taken

10 Luce, *Sense Without Matter or Direct Perception*, 140–49, especially 145.

11 *Cosmos*, 220–21. In making this suggestion, Bouyer goes beyond Luce's explicit text.

12 Ibid., 220.

13 *Sophia*, 158–59. We shall discuss this further below.

by their communion as a canonical text.[14] Bouyer, on the other hand, says that the meaning of *hyle* in this text is not to be taken in a philosophical sense. Wisdom texts do not juxtapose spirit and matter but heaven and earth, the former the abode of God and His angels, the latter the abode of flesh and blood beings, of man moved by a created soul (*nephesh*) given life by the Spirit of God (*ruach*).[15]

Grintchenko points out that Bouyer's history or etymology of the concepts of "matter" and "spirit" in *Cosmos* is intended to show that matter is no less mysterious a reality than spirit.[16] This is an argument for the existence of spirit on the basis of analogy and mystery, evocative of Bishop Joseph Butler's apologetical strategy in *The Analogy of Religion Natural and Revealed* to prove that God directly intervenes in the cosmos. Butler argued against the Deists of his day that the intervention of God in the world in miracles, prophecies, and ultimately in the Incarnation and Resurrection of Christ is no more mysterious, no more a violation of our sense of what is rational and orderly, than the riddles of creation itself. Science, properly understood, he argued, is as accustomed to mystery as orthodox Christian faith. There is an analogy between mystery as present in the created order and mystery as present in the order of redemption. This type of argument goes back at least as far as the Alexandrian Church Fathers, and Butler begins his famous book with a quotation from Origen that encapsulates it: "He who believes the Scripture to have proceeded from Him who is the Author of Nature, may well expect to find the same sort of difficulties in it, as are found in the constitution of Nature." Newman was deeply influenced by Butler's *Analogy*, and in chapter twenty-one of *Cosmos* the analogy of mystery, in this case between matter and spirit, is invoked as part of Bouyer's own development of the "Sacramental system" that we discussed in our previous chapter. However, Bouyer's exposition of the elusive concepts "spirit" and "matter" in *Cosmos* has a directly metaphysical end in view and not just an apologetical one. It signals, if not an absolute preference for, at least a desire to recover as part of our heritage in Christian thinking on the cosmos, a way of understanding matter that has not been commonplace among mainstream theologians in the Western tradition.

14 Luce, *Sense Without Matter or Direct Perception*, 131–32.
15 *Cosmos*, 220–21.
16 Grintchenko, *Une Approche Théologique du Monde*, 243.

THE REALITY OF SPIRIT AND *QUALIA*

If attempts to pin down the definition of matter have the unexpected outcome of giving rise to a sense of mystery, this is, as we just indicated, no less true of the definition of spirit, to which Bouyer next turns his attention in chapter twenty-one of *Cosmos*.[17] The word "spirit" derives from the Latin *spiritus* and draws together the meanings of the Greek *nous* and *pneuma*. It is a principle both of intellect and life. Bouyer shows that the biblical distinction between *ruach* and *nephesh* was adopted by the Church Fathers and enabled them to develop a theologically-enriched understanding of spirit, going beyond pagan philosophy. Divine Spirit (*ruach*) communicates life to the human soul (*nephesh*).[18] Saint Augustine's vastly influential understanding of spirit exemplifies this assimilative endeavor, elevating pagan wisdom in the light of biblical revelation. Bouyer teaches that the Bishop of Hippo understood created spirit, whether angelic or human, to be an image of God, combining the Greek qualities of life and intelligence with the biblical understanding of the heart as the seat of God's self-communication to humanity. Created spirit is, we learn, *capax Dei*, an openness to grace in God's outpouring of love in the person of the Holy Spirit. Spirit is understood in terms of its theological end in a way that could not be true in pagan philosophy. The metaphysical definition of spirit was now tied to a wider soteriological understanding of the origin of both spirit and matter.[19]

At this point in his exposition in *Cosmos*, Bouyer shifts attention to Saint Gregory of Nyssa and to Berkeley, who share similar understandings in this regard.[20] He very quickly moves through positions that are common to the two and that he takes to be of great relevance for theologians, philosophers, and scientists in our own day. Consideration of these positions enables us to draw together some dimensions of our earlier exposition as well as some considerations from Bouyer's wider writings to comprehend better his total understanding of the meaning of spirit and matter as well as their interaction in the unity

17 *Cosmos*, 221.
18 Ibid.
19 Ibid.
20 Ibid., 221–22.

of the human person in the midst of salvation history.

In one sense, Bouyer teaches, "matter" designates only an abstraction, in that it can be turned by human conceptual manipulation into a self-subsistent, separate substance only by eliciting grave contradictions. The Oratorian notes that Saint Gregory of Nyssa and Saint Basil of Caesarea, over a thousand years before Berkeley, argued that one cannot conceive of matter in isolation from spirit without separating out parts of our total experience from its living whole, which alone gives them meaning.[21] Berkeley's axiom *esse est percipi* is, Bouyer insists, best understood in this light.[22] Luce clarifies more fully than Bouyer that Berkeley meant by this famous expression that bodies in the world are constituted by the total experiential context of personal existence and are ontologically grounded in *qualia* or ideas that exist first in the mind of God.[23] The human mind does not construct these ideas by the power of its transcendental subjectivity. The inner categories of mind or its forms of sensibility do not create them. The *qualia* that compose sense realities do not go out of existence when the human mind ceases to alight upon them in a finite act of sense-perception, because they always exist within the divine intellect.

It is accurate to say that Berkeley's thought is a form of realism in the mode of what Louis Dupré has described as "spiritual empiricism."[24]

21 Ibid. Bouyer says: "More specifically, as St. Gregory of Nyssa was the first to discern and to assert, it is impossible to conceive matter except by separating from the human spirit some of the elements of its experience, although they are endowed with consistency and sense only in the context of its living unity."

22 Ibid., 222, 132–34, 146.

23 See Luce, *Sense without Matter*, 162–64. Luce quotes from Berkeley's *Principles*, Sect. 6, here: "Such I take this important one to be, to wit, that all the choir of heaven and furniture of the earth, in a word all those bodies which compose the mighty frame of the world, have not any subsistence without a mind, that their being is to be perceived or known; that consequently so long as they are not actually perceived by me, or do not exist in my mind or that of any other created spirit, they must either have no existence at all, or else subsist in the mind of some eternal spirit: it being perfectly unintelligible and involving all the absurdity of abstraction, to attribute to any single part of them an existence independent of a spirit."

24 See Louis Dupré, "Newman and the Neoplatonic Tradition in England," in *Newman and the Word*, ed. Terrence Merrigan and Ian T. Ker (Louvain: Eerdmans Press, 2000), 137–54, 137. The expression is Dupré's. It is interesting to

He sees the cosmos as essentially spiritual, with its inverted roots firmly planted in the heavens above, in the infinite divine mind. His spiritual empiricism teaches that our experience really reaches out to realities, to beings outside the human mind, which are not constructed in their concrete existence or formal differentiation by human subjectivity. The qualitative richness and diversity of creation is not a product of human mental additions applied to sensory data that are rooted in a plenum that is undifferentiated, absolutely noumenal, and antithetical to the delights of perception. As Bouyer describes it, finding deeper sources for Berkeley's insights in the writings of the Church Fathers and medieval theologians, the material world or the world given to our intellect through the senses really exists, but it exists as a medium of communication for spiritual beings. It is intrinsically meaningful as ordered by and to the perfecting intercommunion of created freedoms.[25]

The position of the Cappadocians is very much akin to what we have just described. A certain instinctive, uninformed reaction to their view, a particularly modern anti-Hellenism, might immediately raise the suspicion that they harbored a latent, world-denying Gnosticism or Manichaeism. Did the Cappadocians hate matter so much that they questioned its very existence? Just the opposite is true. They called into question the existence of matter as a self-subsistent metaphysical principle precisely because they so strenuously rejected Manichaeism. The Manichaeans, after all, thought that matter was the uncreated source of evil in the cosmos. The Cappadocians strenuously insisted against this aberrant view that the world, even in all that is material in it, comes from immaterial thoughts in the divine mind. Their immaterialism was part and parcel of a rejection of ontological dualism. Matter, they insisted, does not exist eternally beside and outside of God, having its own principle of origin. Material objects are combinations of divine thoughts. All that is tangible, limited, and visible in the world comes first from the intangible, unlimited, invisible Spirit of God:

---

note that the inspiration for this article by Dupré was the chapter from Bouyer's bachelor's thesis on Newman's thought, "Newman and English Platonism," that was translated into English for the first issue of *Monastic Studies* in 1963.

25 *Cosmos*, 222.

> Being wholly mighty, by his wise and powerful will he [God] forcefully brought together all that matter consists of for the completion of beings, [namely,] lightness [and] heaviness, density [and] perviousness, softness [and] hardness, humidity [and] dryness, coolness [and] hotness, colour, shape, contour and extension. Taken one by one all these are mere thoughts and concepts; none of these constitutes matter of its own but when they reciprocally converge they become matter.[26]

A. A. Luce argued that we could remain true to the thought of Berkeley and still admit the existence of matter if we define matter in connection with the sensible whole of direct perception. "Matter," he said, "may be defined . . . a permanent Possibility of sensation."[27] This is as far as he is willing to go in defense of matter. He thought that by turning matter into a noumenal being subsisting in its own right and unavailable to perception we have created a monster, all the more terrifying in its capacity to devour us in that it is never perceived. It carries out its consumption of the world invisibly and silently. What is real, Luce strenuously argues, is not the metaphysical matter that is presumed to be the substrate of the world of sense, but the world of sense itself. What the tradition of Western philosophy going back at least as far as Galileo held to be secondary or merely epiphenomenal qualities are in fact alone what give us access to the cosmos as it really is. Direct perception of *qualia* is not an illusion formed by mental operations on the unperceivable objects of sense. The scent of a rose or the color of the sky exemplify the wider reality in which the presumably "primary *qualia*," chemical properties, wavelengths of light, and so forth, have reality and meaning. Matter exists only in relation to and for the purpose of the actualization of spirit. Bouyer, in this general line of opinion, says "matter, or more accurately, the entirety of so-called material realities, exists only as a

26 Gregory of Nyssa, *Apology* 7 (*PG* 44:69c). Quoted by Doru Costache, "Making Sense of the World: Theology and Science in St. Gregory of Nyssa's *An Apology for the Hexaemeron*," *Phronema* 28.2 (2013): 1–28, at 19. All parenthetical additions belong to Costache. We have removed some Greek text that she includes in parentheses in giving this quotation.

27 Luce, *Sense Without Matter or Direct Perception*, 6–7.

common content given to the created spirits by the uncreated Spirit from whom they and matter both derive."[28]

The distinction that Bouyer makes here between "matter" and "the entirety of so-called material realities," or between the "material" and "materiality" (as he will do later in this passage) is an important one, and one that is in fact embraced by some philosophers of a decidedly different cast of mind than his own.[29] Some of Alfred North Whitehead's contemporary followers have called into question just as vigorously as Berkeley and Whitehead himself the usefulness of the concept of matter, although they are far indeed from Luce's "immaterialism" or the wider "spiritual empiricism" that influenced Bouyer, with its indebtedness to the exemplarism of Sacred Scripture and the Church Fathers.[30] In his recent Gifford Lectures, Bruno Latour (b. 1947), one of these contemporary devotees of Whitehead, implored theologians to defend "materiality" while urging them to rebuke the vulgar materialism that stems from the reification of the concept of matter.[31] Bouyer is himself a defender of "materiality" or of "the entirety of so-called material realities," but he recognizes, in a way that the Whiteheadians do not seem to do, that this defense requires a full embrace of the divine Wisdom which carries, in God's eternal transcendence, the *logoi* of all creatures, the divine ideas which are paradigmatic for the meaning of the world in its materiality.

In his own recovery of Berkeley's thought, John Milbank (b. 1952) has emphasized that for the early-modern Anglican bishop language and symbol are irreducible to a noumenal material substratum. He commends Berkeley for developing a metaphysics without substance for which the things of nature are "composed of unsupported sensory qualities," making it possible for him to initiate a development in Western thought for

28 *Cosmos*, 222.

29 Ibid: "the material world [*le monde matérial*], or rather the world considered in its materiality [*le monde envisagé dans sa matérialité*], i.e., in its sensible existence, is but a shared language among spirits." We highlight that in this quotation from *Cosmos* Bouyer precisely distinguishes the material world or "matter" from the world in its "materiality" or sensible existence.

30 Cf. Didier Debaise, *Nature as Event*, trans. Michael Halewood (Durham, NC: Duke University Press, 2017), 3–8.

31 Bruno Latour, *Facing Gaia: Eight Lectures on the New Climactic Regime*, trans. Catherine Porter (Medford, MA: Polity Press, 2017), 70–72.

which the world could be recognized entirely as an "implicatory network of signs which encompasses our whole practical inhabitation of the world."[32] He, too, argues that the Cappadocians prefigured Berkeley's understanding of matter, and he quotes Saint Basil of Caesarea in a passage that is very much in line with what we have just expounded from Saint Gregory of Nyssa:

> Do not let us seek for any nature devoid of qualities by the condition of its existence but let us know that all the phenomena with which we see it clothed regard the condition of its existence and complete its essence. Try to take away by reason each of the qualities which it possesses, and you will arrive at nothing. Take away black, white, weight, density, the qualities which concern taste, in one word all that which we see in it, and the substance vanishes.[33]

Milbank argues that the Cappadocians in fact deny the existence of not only material substance but immaterial substance. There is, for them, no finite substratum that can ground the reality of a thing. All things are sustained in being only in the mind of God by the direct power of God's creative will. This means that nature is not self-sufficient. It is fundamentally relational and has no inner subsistence, whether material or spiritual. It is therefore incomprehensible.

The cosmos is fundamentally a language, as both the Cappadocians and Berkeley saw it. Bouyer associates Saint Bonaventure with this view as well.[34] Yet he does not interpret this strand of thought, with which he associates his own position, in quite as extreme a manner as Milbank. Certainly, the Oratorian's understanding of worldly being is deeply relational. However, given the nuptial ontology underlying his developed thought, which affirms the ontological depth of creation in its intrinsic relationality, there is no ground to accuse him justifiably of disregarding the *meaning* of signification in a world replete with expressive signs. In

32 Milbank, *The Word Made Strange*, 98.

33 Ibid. Quoted from Basil of Caesarea, *The Hexaemeron*, in *Nicene and Post-Nicene Fathers*, vol. III (Oxford: Jakes Parker, 1895), 8.

34 *Cosmos*, 222.

commending aspects of Berkeley's thought, he does not de-substantialize natural beings. He does not explicate relation in quite this extreme a manner. Nevertheless, in chapter twenty-one of *Cosmos*, he explicitly aligns for once with Saint Bonaventure, who, through the logic of analogy, understood the cosmos as "a system of images or signs . . . the language spoken to man by the Creator."[35] The Seraphic Doctor understood creation to be a book in which is inscribed the brilliant meaning of the creative Trinity. The particular beings of creation are, for him, words communicating the divine Wisdom, shadows or vestiges of God, receiving the intelligible ray of the divine light and participating in the divine analogy.[36] In *Sophia* more so than in *Cosmos*, Bouyer explains what he understands this sapiential analogy of language to mean and develops it in his own way as essential to his personal assimilation of the sacramental Christian cosmovision. "All the things, all the objects of this world that we call 'physical'. . . must be considered as an expression, a common language of the first born spirits, of which we can even say that in a sense it constitutes a body with them, even if they are not properly speaking bodies as in our case."[37] We find here again the Augustinian theologoumenon according to which God's intelligible light and Wisdom is communicated to the lowest reaches of the hierarchy of being through the intellects of the angels. All of creation is together in some sense a unified body, but it is not yet the body of the eschatological Church.

We saw in the last chapter that this does not mean that the angels are the efficient cause of matter or its creators. If the physical world emanates from angelic spiritual beings, it is because God directly willed it so. As the first repositories of the light of created Wisdom, the angelic intellects have a mysterious, archetypal relationality to the beings of the physical world, but they do not have the power to create these beings.

35 Gilson, *The Philosophy of Saint Bonaventure*, 195. See *IF*, 258–60. We say "for once" because Bouyer generally neglects the Franciscan tradition. In the pages of *IF* referenced here he expresses distaste for Bonaventure's "paradisaic philosophy," that is, its lack of realism in the face of the concrete world of suffering and sin. Yet the theology of language and analogy articulated here shows that this distaste was not an absolute or permanent disposition.

36 Gilson, *The Philosophy of Saint Bonaventure*, 197.

37 *Sophia*, 155. This passage seems to indicate a preference for Bonaventure's metaphysics of the angels.

This Bouyer insists upon. Yet the physical world is a language, not only an expression of the Wisdom of God in the characters of its myriads and myriads of *logoi*, but a means through which the vast panoply of angelic intellects hold a common intercourse among themselves and with us.[38] Bouyer suggests that Job 38:7 should indeed be understood in this sense, which is doubtlessly more Augustinian and Bonaventurian than Garrigues would admit: "the morning stars sang together and all the heavenly beings shouted for joy." The world, Bouyer says, "in its materiality, i.e., in its sensible existence, is but a shared language among spirits."[39] The physical world emanates by the power of the sovereign will of God from the angelic intellects, just as, in the reverse, the human intellect emerges from our own specific materiality or corporeity, which is inseparable from the cosmos at large.

HUMAN BEING AND MODERN SCIENCE

By this last sentence, we mean to indicate that for Bouyer human beings possess a singular corporeity, a special type of bodiliness uniquely informed by a spiritual soul. The Oratorian holds that the very materiality of the cosmos takes on new meaning in humanity in and through its unique manner of being in the world. This is the ultimate theological point in chapter twenty-one of *Cosmos*, and, in bringing it to the fore, Bouyer explicitly recovers the ancient Patristic idea that humanity is a microcosm recapitulated in Christ.[40] Bouyer not only recovers the theme of the microcosm but develops it, or at least suggests how it might be developed in the context of modern scientific cosmology. He assesses the meaning of the corporeity of the human in the train of evolutionary recognition that the cosmos has increased in complexity through the aeons of physical history leading to the emergence of the

38 *Sophia*, 155.

39 *Cosmos*, 222. See n. 29 above.

40 See William Norris Clarke, "Living on the Edge: The Human Person as 'Frontier Being' and Microcosm," in *The Creative Retrieval of Saint Thomas Aquinas: Essays in Thomistic Philosophy* (New York: Fordham University Press, 2009), 132–51. Clarke gives a good summary of the theme as it is found both in the Eastern and Western traditions. He notes that the Western Augustinian tradition has tended to downplay the theme. He has himself recovered it in his writings as part of his larger recovery of Thomist metaphysics.

human person. He sees the emergence of the human in space-time as the potential factor of recreation of cosmicity in a world plunged into chaos by the fall of angels.[41]

The very vision of matter and spirit that Bouyer brings forth is evocative not only of philosophical "spiritual empiricism" but of certain ways of construing matter that have come out of debates on the implications of quantum mechanics in the twentieth century. Bouyer references again in chapter twenty-one of *Cosmos* Tolkien's friend R.G. Collingwood. We saw above that he interprets Aristotle to have maintained a certain level of agnosticism as to the meaning of matter. This point of interpretation follows Collingwood directly.[42] Collingwood suggested that what modern scientists call "matter" is what both Plato and Aristotle meant by "form."[43] Aristotle thought that we could not get a clear definition of matter, because the only thing of which we can have a clear concept is form.[44] What we know of matter is in reality form incorporated in matter. Bouyer takes this to mean that matter exists in order to be in-formed. This understanding aligns with Aristotle's ultimate definition of matter as *potentia*.[45] If Bouyer privileges the Cappadocians on the question of the meaning of matter and so seems to think the concept "materiality" to be rather useful, he in no way rebukes Aristotle's definition of matter in so doing. As Collingwood says, matter for Aristotle "is the limiting case or vanishing point at the negative end of the process of nature."[46] It cannot be banished from the cosmos, because cosmic development can occur only if there is some limit aspect or potentiality to be actualized. Unlike the Cappadocians, Aristotle cannot think of matter as intrinsically ordered by the eternal and ideal providence of divine, creative Wisdom, but Aristotle's understanding of matter is not without its place in Bouyer's cosmology. The ultimate definition of both spirit and matter that we find in Bouyer centers on the human person,

41 *Cosmos*, 223.

42 Collingwood, *The Idea of Nature*, 91–92.

43 Ibid. See *Cosmos*, 218. Bouyer references Collingwood on this page.

44 Collingwood, *The Idea of Nature*, 91.

45 Cf. *Dictionary of Theology*, 302–3. On p. 302, Bouyer says explicitly: "Philosophically, matter could be defined only as pure potentiality with regard to the forms."

46 Collingwood, *The Idea of Nature*, 92.

and the human person, as microcosm, is, for Bouyer, a hylomorphic unity, the privileged mediating point or *pontifex* uniting spirit or form and matter, the latter constituting, as for Aristotle, a limit. Yet Bouyer recommends a certain poetic caution in assimilating this understanding, as is evidenced by his tentative deployment of the concepts of spirit and matter in his definition of the person: "the person, as we experience it, i.e., the human person, is a living whole whose center is what we imprecisely call 'spirit,' and whose periphery is what we, even more imprecisely, call 'matter.'"[47] He seems to hold that the boundary between spirit and matter is not as fixed, rigid, or clear in the human person as some philosophical and theological traditions have maintained.

Modern physics, in quantum mechanics, may seem to have arrived at this vanishing point, this limit or exteriority of created being that we nominate "matter" which never really exists without being involved in the interchange or exchanges of forms and spirits, of information and *qualia*, of corporeity and materiality.[48] Bouyer is not the only theologian to have suggested, as he does in chapter twenty-one, that there may be conceptual similarities connecting the ancient philosophers and theologians with modern physics.[49] The work of a triumvirate of twentieth-century French philosophers of science—the already-mentioned Raymond Ruyer, Olivier Costa de Beauregard, and Aimé Michel—is especially, albeit rather surreptitiously, linked to Bouyer's cosmology on this point. Two of these figures, Michel and Beauregard, wrote for *France Catholique*,

47 *Cosmos*, 222.

48 See especially, Smith, *The Quantum Enigma*, 77–94. Smith says that what modern physics has in fact accomplished is to have discovered by empirical inquiry the realm of matter marked by quantity: *materia quantitate signata*. Smith develops in a corrective manner certain insights from Werner Heisenberg (1901–1976). Heisenberg defines elementary particles as "possibility for being." This is so because the only description that we can give of them is a probability function. This aligns with the recognition that these particles consist of energy. See Heisenberg, *Physics and Philosophy: The Revolution of Modern Science* (New York: Harper and Brothers Publishers, 1958), 70–71.

49 Recall our exposition, in chapter 7, of pp. 145–46 of *Cosmos*. Cf. Joshua Schooping, "Touching the Mind of God: Patristic Christian Thought on the Nature of Matter," *Zygon* 50.3 (September 2015): 583–603. Without, of course, any reference to Bouyer, Schooping himself connects the Cappadocians to Berkeley and to the theoretical physicist David Bohm on the concept of matter.

a popular journal for which Bouyer made many contributions of his own.[50] The third, Ruyer, was the author of a best-selling book in France, *La Gnose de Princeton*, on the societal implications of twentieth-century revolutions in physics.[51] This book, which Bouyer seems to have read, has never been translated into English but created somewhat of a stir among French readers. It was a serious work of philosophy of science, but in its external trappings it was a ruse. Ruyer pretended in the book to be an acquaintance of a group of eminent American physicists and astronomers, modern "Gnostics," who secretly held to a new form of religious knowledge that they claimed took its basis in modern physics. These scientists were presented as holding that philosophical materialism has been undermined by recent advances in physics. They were said to have influenced and admitted into their inner circle certain eminent Christian leaders, whom Ruyer never names. Ruyer's most intimate friends were always in on the ruse, and they knew the reason why he perpetrated it. Ruyer was a provincial philosopher at the *Lycée Poincaré*, in Nancy, France, without much of a name or popularity in his writings. The point of his deception was to get across to a larger audience than would have been otherwise interested in a rustic philosopher's work his own serious reflections on matter, spirit, and the meaning of modern scientific discoveries. The French public would much more readily drink in the secret views of American physicists at the forefront of global scientific and technological advance than those of a provincial philosopher from their own land.

Aimé Michel, in his review of the book in *France Catholique*, went along with the game.[52] He promoted the book as an inventory of

50 See https://www.france-catholique.fr/-Articles-du-R-P-Louis-Bouyer-parus-dans-France-Catholique-de-1957-a-1987-.html. This link contains the list of articles that Bouyer wrote for the print edition of *France Catholique* over many years.

51 Already cited in chapter 7. For an overview of the cultural impact of this book, see Aimé Michel's review, "La Gnose de Princeton," in *France Catholique* 207, no. 1487 (June 13, 1975). Online at https://www.france-catholique.fr/LA-GNOSE-DE-PRINCETON.html. See the important footnotes to the online version provided by Jean-Pierre Rospars. Ruyer exercised an influence on several seminal twentieth-century French philosophers, including Gilles Deleuze and Maurice-Merleau Ponty. Bouyer references Ruyer and this book in *IF*, 51 and 60.

52 Ibid.

revolutionary implications flowing from the new physics, signaling the end of mechanist materialism. Ruyer, Michel, and Beauregard were linked together by sharing this understanding of the twentieth-century revolution, and the latter two were influenced by Ruyer's "neo-finalism" or return to the emphasis on final causality in interpreting the development of the cosmos.[53] They each held that only a spiritual interpretation of the cosmos can make sense of it as science now knows it to be. The neo-finalism of these thinkers represents a direction of philosophical thought with which Bouyer's *Cosmos* can be reasonably associated, most directly so with respect to the work of Beauregard, to whom *Cosmos* is dedicated. This association comes out perhaps most of all in chapter twenty-one of *Cosmos*, where Beauregard is referenced for the second time in the book.[54] It is of interpretative utility to detail Bouyer's connection with this great French physicist and student of Louis de Broglie, while expounding some essential dimensions of his thought.

We have mentioned that there are two brief letters that Bouyer wrote to Beauregard that can be found in the latter's archives. In the first, written on November 27, 1964, Bouyer commends Beauregard for his book on time, *Le Second Principe de la Science du Temps*, and raises a couple of questions or points for further consideration.[55] Beauregard possessed, in a manner appropriate to his status, a thorough command of the nuances of the mathematical formalism through which his discipline is practiced and developed.[56] This disciplinary command comes across in even his popular writings, which are difficult for the educated lay person to grasp in full. Bouyer acknowledges upfront in his first letter that he is himself

53 See Raymond Ruyer, *Neo-Finalism*, trans. Aloysha Edlebi (Minneapolis, MN: University of Minnesota Press, 2016). First published in France in 1952, this book is considered to be Ruyer's *magnum opus*.

54 *Cosmos*, 271, n. 37. Bouyer refers the reader to the two books of Beauregard that we have already referenced and shall briefly expound here, *Le Second Principe de la Science du Temps* and *La Physique Moderne et le Pouvoirs de l'Esprit*.

55 The stationary for both letters contains the address: "Villa Montmorency, 8 Avenue de Montmorency, Paris, XVI." The letters are not organized in Beauregard's archives by a reference system.

56 See especially Olivier Costa de Beauregard, *Time, The Physical Magnitude* (Boston, MA: D. Reidel Publishing Company, 1987). This book, one of two books by Beauregard in English, shows forth his command of the subject matter to the English-language reader.

an amateur and autodidact in these matters and so cannot draw from his reading of Beauregard's book all of its fruits. He recognizes nevertheless that a book like Beauregard's is "infinitely precious" for theologians or philosophers who take an interest in cosmology. Beauregard's book indeed enables one "to see how the problems of freedom, creation, and of the spirit can be posed in the context of contemporary science." Moreover, the book helps the theologian or philosopher to distinguish better what is essential in their "respective meditations from mythological shadows with which one tends always more or less to confound them."

There are nevertheless two questions that Bouyer thinks it is important to raise. The first has to do with a perceived lacuna in the book. Bouyer is surprised that Beauregard mentions only by vague allusion arguments coming from Louis de Broglie and other French scholars of the day according to which the determinism of classical physics was on the way of being overcome by contemporary physics. Regarding the second question, Bouyer professes to be astonished that Beauregard situates "exclusively in quantum indeterminacy the place of possible insertion of freedom." This is astonishing to Bouyer because it appears to him that Beauregard's "general reasoning postulates an indeterminacy even vaster, first wherever natural laws appear as statistical laws, and finally as an inevitable correlation of all information." Tangentially, Bouyer professes surprise at Beauregard's rehabilitation of the philosophy of Henri Bergson, particularly because of the latter's confusion in the face of Einstein's theory of general relativity.[57]

The second letter that Bouyer wrote to Beauregard was on November 23, 1971. This letter shows that the two scholars of modern science must have maintained occasional contact with one another over the years. Beauregard wrote a letter to Bouyer dated October third of that year, to which Bouyer was able to give a response only in November. The Oratorian professes in this letter to be in profound agreement, "even more than ever, if possible," with the orientation of Beauregard's most recent

57 On this last point, Bergson entered into discussion with Einstein and published a book as the fruit of this discussion, *Durée et Simultanéité* (1922), which he refused to allow to be republished, having later recognized the inadequacy of his understanding of the mathematics behind the debate. See Leszek Kołakowski, *Bergson* (New York: Oxford University Press, 1985), 72.

publications. He tells Beauregard that the latter's research is unsurpassed in importance given the "present debacle" (could he be referring to the cultural revolution of 1968 and its materialist aftermath?), and he assures Beauregard that there are several young researchers in France whose pursuits move along the lines of Beauregard's work and who could benefit from his experience and counsel in directing their future investigations. This second letter suggests a growing accord between the two men. Perhaps it is no surprise that when the French-language edition of the prestigious international journal *Communio* was launched in 1975, Beauregard was called upon, at Bouyer's behest, to serve on its editorial committee. Beauregard in fact served in Bouyer's place. The Oratorian thought that his own criticisms of the postconciliar ecclesial establishment in France after the Council, found in a book like *The Decomposition of Catholicism*, had rendered him a notorious figure and would end up making the journal unnecessarily a target for French progressivist Catholics, who were at the time the main leaders of the Church in France.[58]

It appears that Beauregard's interpretation of the philosophical implications of modern physics influenced Bouyer by the time of his writing of *Cosmos*. His dedication of the book to Beauregard can hardly have been an empty gesture. It surely has significance for the interpretation of the book. In order to plumb a little more deeply into this issue, it is necessary to expound a couple of points in Beauregard's philosophy of physics that seem to us to be pertinent in order to understand some philosophical points in Bouyer's cosmology, particularly in chapter twenty-one of *Cosmos*.

The question of the meaning of time is the paramount concern in the French physicist's research. Standard theories in physics in the modern age postulated that the flow of cosmic time is irreversible, moving from the past to the future through the edge of the present, and that this can be deduced from statistical theory. Beauregard argued, to the contrary, that the unidirectional flow of time is a fact of experience and not a necessity of mathematical or statistical deduction. He distinguished a "law of fact," to which the irreversibility of time belongs, from a "law of right," for which the reversibility of time is not ruled out. The arrow of

58 Jean Duchesne, one of the founding editors of the French-language edition of *Communio*, has related this to us.

irreversible time is connected by fact to the global increase of entropy, that is, to the waning of energy in the cosmos and the waxing of disorder. The reversibility of time, on the other hand, a law of right, would be associated with decreasing entropy and disorder as well as with the increase of energy.[59]

Beauregard argues that a new calculus of probabilities, associated with quantum mechanics and relativity physics, has forced us to confront the fact of our inability to deduce the irreversibility of time on the basis of sound mathematical premises. The principle according to which time is irreversible in its movement from past to future through the present entails that time is dissymmetrical. Only the present, on this view that Beauregard disputes, can be held to exist. The past no longer exists and the future does not yet exist. We live in the present on the edge of time, between past and future. We can see into the past but cannot act on it. We can act on the future but cannot see into it. In acknowledging the law of right for the reversibility of time, Beauregard argues that in the larger cosmic context of the arrow of time, ever darting toward the future, there may be regions of the cosmos that flow in the reverse direction. He references the paradox of Einstein, Podolsky, and Rosen to instantiate this claim.[60] Following a thought experiment that Einstein and his colleagues devised in order to demonstrate what they took to be the absurd ontological implications of quantum mechanics, it has been experimentally verified—against Einstein's hope for maintaining what he considered to be an intelligible construal of the structure of the physical universe—that two particles launched separately from a common source on separate trajectories toward spatially divergent endpoints may instantaneously affect each other's physical properties upon arriving at that endpoint. Even though separated by vast distances, indeed, as vast as the physicist's imagination can conceive, the particles seem to act on each other in the flash of a moment. They cannot be said to have impacted one another on their separate trajectories through a common cone of space. If they were to have influenced each other in the present instant in their separate locations, that causal influence would have to have been communicated across space at a velocity quicker than the

59 Beauregard, *Le Second Principe de la Science du Temps*, 103–24.

60 Cf. Beauregard, *La Physique Moderne et les Pouvoirs de l'Espirit*, 53–62.

speed of light, which is impossible. Beauregard explains this result on the basis of "retrocausation."[61] The two particles are connected to one another, he argues, by virtue of sharing a common link in the temporal magnitude: they can "zig-zag" in time in movement between past and future, even reversing the arrow of time.[62] For Beauregard, this zig-zag embodies the flow of information in the cosmos and is dependent upon the existence of conscious observers, for instance, the scientists who have experimentally verified the paradox.[63]

Basing himself in part on Ruyer's work, Beauregard argues that the presence of information in the cosmos, specifically in the micro-sources that do not follow the universal time arrow of common experience, is both a communication of knowledge and a source of organization disrupting the increase of entropy and disorder. Information is allied with reverse entropy or "negentropy," through which cosmic realities increase in order.[64] Negentropy is associated with willful volition. There can be a movement from negentropy to information, and this characterizes observation or growth in knowledge. Inversely, there can be movement from information to negentropy, and this characterizes processes of action or organization. Beauregard sees cybernetics as a way of recovering the Aristotelian-Thomist understanding according to which form includes

61 See Aimé Michel, "Sur le Seuil de la Nouvelle Physique," in *France Catholique* 294, no. 1613 (November 11, 1977). Available online at https://www.france-catholique.fr/SUR-LE-SEUIL-DE-LA-NOUVELLE-PHYSIQUE.html. Once again, see the important footnotes added by Jean-Pierre Rospars, on March 24, 2014, especially n. 3. Beauregard held that there is a "zig-zag" in space-time in the movement of the particles, communicating information between them at once in the future and in the past. Rospars explains that this is not the most common explanation in present physics to account for the E-P-R phenomenon. Current physics prefers an explanation that links the correlated pair of particles in a context that transcends space-time. Rospars points out that Beauregard was nevertheless one of the first physicists to grasp the importance of the E-P-R paradox, whose strange implications are nowadays acknowledged by everyone. As for Beauregard's attempt to explain the paradox, Rospars says: "But who can say what the future reserves for this conception of the reversibility of time that he defends?"

62 Ibid.

63 Beauregard, *La Physique Moderne et les Pouvoirs de l'Espirit*, 63–71.

64 Beauregard, *Le Second Principe de la Science du Temps*, 61–72.

both the communication of knowledge and the power to organize.[65] If the larger cosmic arrow of time points in the direction of dissipation and disorder, "mini-sources" of decreasing entropy and increasing order are present everywhere. These include "cosmological generation of particles, atoms, molecules, organic molecules; and astronomical building of stars, planets, and life on planets . . . producing phylogenesis, ontogenesis, plant, animal, and human activity."[66]

The paradox of the presence of information and negentropy in a cosmos in which the time arrow seems to point in only one direction, namely, toward death, is an underlying consideration in the final chapters of *Cosmos*, especially in chapter twenty-one. There is another dimension of Beauregard's interpretation that is significant in helping us to understand Bouyer's association with the great philosopher of science, having to do with his manner of connecting cosmology with psychology. Beauregard speaks of the universal presence of the subconscious mind in matter.[67] With Bergson, he holds that only the hyper-aware consciousness of the individual, concentrating on the present shape of life, is aware of the dissymmetry of time, the difference between past, present, and future. Relativity theory in fact teaches us that space-time and matter do not unfold objectively in an advance from past to future. Matter, space, and time exist together enfolded in a kind of block.[68] The material universe in its fourth dimension, time, has a thickness or extent that embraces the totality of the periods of the cosmos. The subconscious is spread out along the duration of this fourth dimension of space-time and thus across the whole block of cosmic matter that is inseparable from it. Beauregard differs from Bergson in that the latter did not extend the subconscious into the future, while Beauregard thinks that by doing so we may be able better to understand empirically verified phenomena such as pre-cognition and psychokinesis. Both men distinguish the clear consciousness of everyday, individual, empirical existence from the cosmic depths of the subconscious mind. It is with respect to the former that time advances. Time is the progress of living psyches which are clearly conscious and

65 Ibid., 76–78.
66 Beauregard, *Time, The Physical Magnitude*, 162.
67 Beauregard, *La Physique Moderne et les Pouvoirs de l'Espirit*, 77–83.
68 Beauregard, *Le Second Principe de la Science du Temps*, 109–10.

attentive to life. It is not matter, separate from living organisms, which advances in time, but living beings adapting to life. Their shared progress constitutes time's arrow.[69]

The paradox of the presence of entropic decrease and the communication of information, both in its organizing power and as exchange of knowledge, surely provides one of the links with scientific cosmology that Bouyer thinks it is important to establish for theologians who take cosmology seriously. In the vast domain of information exchange that Bouyer took as noteworthy in Beauregard's book on time, he sees entailed the presence of indeterminacy and freedom. In *Cosmos*, he says explicitly that modern physics teaches us that information exchange in its human transmission signals that intellectual consciousnesses possess a limitless potential to act on and modify the world.[70]

Information ultimately has a spiritual source, both uncreated and created. Aimé Michel argues on the basis of information theory as understood by Ruyer and Beauregard that spirit is present in the world from the very beginning. It is both within and outside of the physical processes of the world. It is Teilhardian to speak of physical processes as if they possess a "within," but Michel distinguishes the view that he shares with Ruyer and Beauregard from that of Teilhard, who

> has too easily accepted the materialist image of our origins. He [Teilhard] places his Omega Point at the end of time, when everything indicates that the entire history of the universe is underpinned by a thought, that there does not exist a second in this abyss measured by the centuries and the millennia which does not have its spiritual significance.... For Teilhard, thought emerges from matter. And certainly it is true that if we stick to what the spirit of man can see, there is more of thought in a fish than in a worm ... and finally in man than in all the terrestrial animal world. But by wanting to limit the interpretation of this immense ascent to what the mind of man can perceive of it, Teilhard falls into the most insidious form of the vicious circle....[71]

69 Ibid., 134–36.

70 *Cosmos*, 225.

71 Aimé Michel, "La Gnose de Princeton," in *France Catholique* 207, no.

If Bouyer's intellectual direction is rigorously eschatological in orientation, he is nevertheless certain of the presence, from the very beginning, of God's thought in all things or of all things in the thought that God has of them in His Wisdom and Word. Bouyer adds to this picture an overt and developed theological affirmation of the universal immanence of angelic intellects throughout space-time. In the Oratorian's cosmology, the fight within creation for negentropy and information to overcome the standard arrow of time, with its degradation of energy and exhaustion of life, is indicative of the presence in creation from the very beginning of these purely spiritual creatures. The angels are, as we have seen, the keepers of the measure and rhythm of time. This measuring function cannot be unaffected by a fall within their ranks. Time as we know it, interwoven with our history, must bear the mark of this collapse of angelic freedom. This view is somewhat along the lines of Bautain's postulation that we mentioned in chapter six, but Bouyer does not directly postulate that the existence of time as such is dependent upon the angelic fall.

The Dominican theologian Benedict Ashley (1915–2013), who read with appreciation Bouyer on the angels, mounted an argument for the existence of angelic spirits on the ground that negentropy in the physical cosmos could have no other sufficient cause.[72] He thus gave a modern twist to the ancient philosophical argument for their existence. According to the Greek astronomers of antiquity, the heavenly bodies that we perceive exist within perpetually rotating celestial spheres. The philosophers realized that it was impossible to account for the endless source of energy that keeps the spheres turning without postulating the existence of separated intellects to move them. On this basis, Aristotle argued to the existence of a Prime Mover who maintains the unity of the universe by directing all of the other separate intellects to work together in unison.[73] Ashley, with a little more philosophical concern and precision than we find in Bouyer on this point, updated this argument with

1487 (June 13, 1975), available at https://www.france-catholique.fr/LA-GNOSE-DE-PRINCETON.html. Quoted in n. 3 by Rospars.

72 Benedict Ashley, *Theologies of the Body* (Braintree, MA: Pope John XXIII Medical-Moral Research and Education Center, 1985), 649–52. Ashley does not himself use the term "negentropy," but he describes the reality that the word encapsulates.

73 Aristotle, *Metaphysics*, Book XII (Lambda).

reference to the mystery of evolution in an entropic cosmos. He argued that the counter-current to entropy that enables the growth of increasing complexity in unity and consciousness is not explicable exclusively by reference to natural laws. It involves an immense history of sequential occurrences constructed by a constellation of natural, contingent forces that lack the direction of a known, undergirding natural law. Christian theists, he contends, tend to jump too quickly to the conclusion that the guided direction of this process is explicable solely by recourse to the existence of the divine mind to create and guide it. This solution is simplistic, he argues, because evolution, though apparently guided, is not a smooth, continuous progress of perfection in speciation. It is conflictual, agonistic, dialectical. It moves by revolutions sometimes, by ruptures or saltations. The guiding movement of this process is not necessarily explicable by reference solely to the infinite mind of the One God. Lesser intelligences, filling an ontological gap in creation between God's infinite being-beyond-being and human being, might also be at work in cosmic evolution, and these might work in contradiction to one another, some in unity with God and some at cross-purposes to the divine Wisdom and will. Ashley suggests that cosmic evil is ultimately indecipherable if we fail to affirm the existence of these separate intelligences.[74]

It may be that both entropic increase and negentropy have created, spiritual sources driving their contradictory advance. Stratford Caldecott suggested as much. He held that both angels and demons affect the physical cosmos in their relation to it and to one another. He clearly drew directly from Bouyer on this point. The increase of entropy in the cosmos is, he suggested, the first sign of the malignant presence of the demons. The violence of nature, its plagues and pestilences, its earthquakes and extinctions, might have a causal source in the conflicting interactions of these spiritual beings. There are indeed physical laws that account for these seeming flaws in nature. Yet that does not have to be taken to mean that the influence of drama and history in physical process is negligible. Caldecott says that "the whole order of nature that we observe and which is defined by these laws is influenced by the separation of the evil spirits from the good, which together form the spiritual order

74 Ashley, *Theologies of the Body*, 651.

on which the physical is based or in which it participates."[75] If, as Bulgakov reminds us, "Everything in the world is preserved by angels, and everything has its angel and its correlation in the angelic world," there is another side of the coin.[76] The fallen angels also have an exemplary status, or, perhaps it is better to say, a mission that they have chosen for themselves opposed to the archetypes of divine Wisdom, and this too is reflected in the constitution of cosmic laws and in the very constitution of space-time as we experience it.

Toward the end of chapter twenty-one of *Cosmos*, Bouyer returns again to his narrative of the Fall and redemption and puts the issues at stake in light of considerations that are deeply concordant with all that we have expounded so far in this section. Each body in the evolutionary world, he says, "taking and changing shape in spite of the forces of death weighing on it, reaches, through this evolution, ever higher stages of corporeal being, increasingly integrated in their very complexity, in a universe apparently doomed to degradation, and is in fact the germination of a new universe."[77] Each body in the universe is, as we saw in the last chapter, if not in number then in kind, "an inverted image of one of those angelic forms in the material mirror that is the web of the cosmos."[78] Created spirit does not first emerge at the endpoint of a long evolutionary chain. It is there from the very beginning. Bouyer seems to be saying in this chapter that the material body comes into existence in the concrete order of being as an intrinsically redemptive and salvific reality. It is on the material plane that the cosmicity of creation will be reestablished, ultimately only through the Paschal Mystery of Christ. This is a radically anti-Gnostic point of view. Bouyer teaches that matter not only is *not* the source of evil in the cosmos but the very instrument through which God "rebuilds" the cosmic Temple in and through his Son to make reparation for the initial act of cosmic evil perpetrated by purely spiritual beings.

There are, according to Bouyer, two "saving evolutions" through which God has providentially guided cosmic history.[79] The advance of time is

75 Caldecott, *The Radiance of Being*, 242.
76 Ibid. Quoted from Bulgakov, *Jacob's Ladder*, 24.
77 *Cosmos*, 223.
78 Ibid., 227.
79 Ibid.

a movement through these evolutions, which are discrete stages of the unified course of salvation history from the physical body of the cosmos to the Mystical Body of Christ. The first has to do with the growth of complexity that brought forth a corporeity fitting for the emergence of a new level of freedom within material being itself in the creation of humanity. There is a progressive ascendency of life in the cosmos, not without setbacks, conflicts, and tragedy. This advance takes place through kairotic cosmic events. First, plant life emerges and then animal life. Both are indicative of the work of the angels. The sexual, reproductive vitality of plant life "represents the first sign of the animation elicited by the angels' descent along Jacob's ladder, which initiates the rehabilitation of a world held until then in night and death by the fall of Lucifer, its prince."[80] The further advance of cosmic vitality with the emergence of animal mobility "is linked to an embryonic consciousness which is like a muted reflection of the angels' superconsciousness and provides an outline of what the human world was to become."[81]

This first saving evolution leads to the development of human corporeity. Adam could have cultivated through his preternatural bodiliness "the paradise-like garden which the universe would have become around him."[82] He could have banished death and attained immortality in the fullness of his unique corporeity if he had lived in accordance with divine Wisdom. Because of his sin, death gained ascendency over humanity, as it had already done over the physical cosmos with the fall of the angels. This fall sets in motion a second saving evolution, leading the development of human corporeity to the point of its renovated perfection in the womb of Mary, through and in which the divine Word assumes the body animated by a human soul that will become the body of all humanity renewed and reconciled in him.[83] It is ultimately the development of the human corporeity of the Mystical Body of Christ from the New Adam which drives back in a definitive manner the entropy of creation.

80 Ibid., 224.
81 Ibid.
82 Ibid., 223.
83 Ibid.

SIN AND COSMIC EVIL

We should pause to comment on the importance of Bouyer's vision of the drama of materiality or corporeity, especially as this can help us to grasp better the mystery of cosmic evil and animal suffering that we have broached in these last two chapters. The link between the fall of created freedoms and animal suffering has not always been, we think, sufficiently highlighted by the theological tradition. Teilhard de Chardin's challenge to traditional Catholic doctrinal positions was based precisely on this point, albeit not with respect to the fall of the angels but to the fall of the First Adam.[84] Teilhard disputed the Thomist theologians of his day who limited the scope of the deleterious effects of the actions of the First Adam and, by consequence, in his view, diminished the breadth of the reconciling work of the Second Adam. These Thomist theologians, without compensating for their position by proposing that the sin of the angels accounts for the existence of suffering and death in the animal kingdom, denied that the sin of the First Adam had cosmic consequences that were all-pervasive. They held that Adam's sin was local in its effect, having an impact on humanity on Earth alone. Teilhard argued that such a view made a travesty of the Pauline anthropology expressed in Romans (5:12, 6:23), according to which the sin of the First Adam brought death as such into the cosmos, communicating its wages even to non-human animals. Teilhard thought that the Pauline view of the First Adam's universality could be taken literally, and make sense, in a geocentric cosmos. It could not, however, be made to convey intelligibility on the plane of literal interpretation of Scripture in the cosmos such as we now know it. Suffering and death preceded by aeons the emergence of humanity. How could human sin have been responsible for this situation? According to Teilhard, the whole body of traditional Christian doctrine was developed on the basis of Saint Paul's understanding of the universal effects of the sin of the First Adam. Traditional doctrine is therefore inextricably tied to geocentric cosmology. Teilhard thought that all of Catholic doctrine would have to be rethought

84 See especially Pierre Teilhard de Chardin, "Note on Some Possible Historical Representations of Original Sin," in *Christianity and Evolution*, trans. René Hague (New York: Harcourt Brace Jovanovich, 1971), 45–55. This book is a collection of essays by Teilhard on speculative theology.

in light of post-Copernican and post-Darwinian scientific advances. His attempt at reformulating the doctrine began with the postulation that death, suffering, and cosmic evil are "necessary passivities" that inevitably arise in the creation of an evolutionary cosmos, and that "original sin" is a particular manifestation of a law inherent to the universal process of becoming. More precisely, it expresses "in an instantaneous and localized act, the perennial and universal law of imperfection which operates in mankind in *virtue* of its being '*in fieri*.'"[85] Cosmic evil is a necessity inherent to the movement of creation from multiplicity toward unification in Christ, who is the Omega Point of the common process of universal history.

Teilhard's question was appropriately targeted. It marked one of the most crucial stages, perhaps *the* most crucial stage, in the battle between progressive and conservative theological opinion in the Church of the twentieth century, and it has yet to find theological resolution. Few grasped the impact of this question as directly as he. Few understood the link between cosmology and theological doctrine as clearly. His question is one of the factors setting the context for Bouyer's *Cosmos*, but the Oratorian gives a very different answer to the question, whose great significance he nevertheless clearly appreciated. He does not think that traditional doctrine needs to be redone. Bouyer's position maintains, in line with Saint Paul, the universal effects of sin, but, unlike Teilhard, he begins with the angels and not with the First Adam, as we have seen. He recognizes in the fact of animal suffering a sign of something catastrophic and calamitous, not merely a necessary passivity or an inevitable imperfection as a stage in the immanent course of cosmic becoming. The paleontological record as discovered by modern science should perhaps attune us better to animal suffering as tragedy. The horrors of biological destruction have been well detailed. Is it enough to say, as some theologians are content to do, that animal suffering need not be linked to sin, because it follows from the good, natural order of the world as set forth by the Creator? Some have held that it is in the very nature of certain animals to be predators and of others to be prey, as if such an assertion is enough to put the matter to rest. We thus do not

85 Teilhard de Chardin, *Christianity and Evolution*, 51.

need to link suffering in the animal kingdom with sin, and we might speak very poetically and beautifully of the unfolding of creation from the beginning according to its own inner physical laws, which includes animal suffering as a sacrifice needed to bring about cosmic maturation. In this respect, the Thomist position that Teilhard disputed and the Jesuit paleontologist's own position in fact exist in concord. Both positions hold that God allows creation to be creation, which means, as they understand it, that it is able to unfold itself by its own powers, to mature in the midst of entropy and degradation in accordance with the laws that He has set for it, by virtue of the drama of predation and extinction that necessarily accompanies the maturation of biological life. We ourselves think that Bouyer is correct to dispute this position, which downplays the horror of the cosmic violence that characterizes so many of the natural processes on earth and has led to the suffering and death of untold myriads of organic beings even to the point of unleashing much mass extinction. Is this really the direct, sovereign will of the triune God of perfect Wisdom and mercy? Did He delight in the dramatic display of untold ages in the evolutionary process where nature so often devoured itself? This seems to us to necessitate a manner of construing divine providence that is inadequate to the Christian understanding of God and His desire to reconcile all things in Christ (cf. Col. 1:20; Rom. 8:15). In contrast to this sort of view, the voice of Saint Isaac the Syrian stands out in the Tradition. Speaking of the cosmic scope of human intercessory prayer, he leaves us with profound words that help us to understand that it is not only human persons and human nature but the cosmos as such that is under slavery to sin and its violent, death-dealing consequences. All of creation is in need of redemption, and, as an act of charity, we should pray for it:

> What is a charitable heart? It is a heart which is burning with charity for the whole creation, for men, for the birds, for the beasts, for the demons—for all creatures. He who has such a heart cannot see or call to mind a creature without his eyes becoming filled with tears by reason of the immense compassion which seizes his heart; a heart which is softened and can no longer bear to see or learn from others of any suffering, even the

> smallest pain, being inflicted upon a creature. This is why such a man never ceases to pray also for the animals. . . .[86]

Bouyer does not follow the line of thought here that would counsel hope for universal salvation, even for the demons. And he does not speculate on the nature of salvation in the animal kingdom. Will animals be saved? If so, how? As individuals or as species, or perhaps as a memory inscribed in the re-created cosmos of transfigured corporeity? The French Oratorian does not address these questions. He respects the mysterious breadth contained in the promise of redemption as found especially in the Pauline writings and passes over the particular possibilities of cosmic renewal in silence. But he does not rest content in his explicit affirmations with the view that the animal kingdom was made to bear suffering by an express act of God's antecedent will. Suffering and death are incorporated into God's Wisdom given the presentiality of creation, in all of its acts, even in sin, to God's eternal "foreknowledge." Materiality and corporeity bring about a new cosmos in the way that they do in the tragic order of creation such as it is. Animal suffering is, as a fact of the concrete order of being, part and parcel of the movement of developing cosmic corporeity. The physical universe did indeed develop or "mature" through the suffering and death of organic species. Perhaps we may indeed speak of the whole of nature as "cruciform," as some theologians do in our day, for whom death in nature is the inevitable path to resurrection. If we were to affirm this expression, it should not be taken to mean that suffering and death do not represent in their initial manifestation—even if rectifying in the way that they may be of the actual order of creation—the outcome of a calamity or violation at the origin of cosmic being. Could not physical creation have been otherwise? Could it not have been integrated peaceably and non-redemptively into the angelic choirs without having to presume the fall of portions of the latter as a precondition for its existence? Perhaps there could have been maturation of physical being without original sin and a fall. Bouyer leaves us no reason to think that it could not have been so. It is problematic to speak of the cruciform ordering of

86 Quoted by Lossky, *Mystical Theology*, 111. Taken from Isaac the Syrian, *Mystic Treatises*, ed. A. J. Wensinck (1923), 341.

nature as if it were a necessity inherent to biological existence.[87] Such a view tends to encourage genuinely Manichaean cosmogonies and soteriologies for which escape from finitude emerges as the only plausible desideratum associated with the passion of hope.[88]

ADAM AND THE CHURCH AS MACROCOSM

Giving such prominence to the consequences of angelic sin, does not Bouyer diminish those of the sin of the First Adam? This is not necessarily so. Bouyer insistently emphasizes that the process of what he calls, with Teilhard, "hominization," brought a decisive, new factor into play in the cosmos that enhanced the power and altered the direction of corporeity in the midst of angelic dissolution manifested in the universal increase of entropy.[89] As we saw, he holds that the human fall brought a "positive curse" on the world. The endpoint of hominization in its first stage, with the creation of Adam and Eve, was the concrete personalization of a universal potency that could have been actualized in either of two directions: either to reconcile all of creation to its Creator and to itself or to unleash a further cosmic calamity. It is the latter that was initially realized. Bouyer does not downplay the cosmic impact of the original sin of the First Adam and Eve, but he does set the possibilities inherent to their being in the wider context of the pre-existing drama of creation.[90]

Bouyer pauses briefly at the end of chapter twenty-one to reflect on the ontology of the human being as microcosm in the midst of a larger

87 Cf. Celia Deane-Drummond, *Wonder and Wisdom: Conversations in Science, Spirituality, and Theology* (Philadelphia: Templeton Foundation Press, 2006), 119. We agree with Drummond's position, which is concordant with that of Bouyer on this point, that "Crucified Wisdom does not mean that the cross was an inevitable aspect or principle of creation. . . ." The Immolated Lamb, "Crucified Wisdom," is the principle of creation for Bouyer, but this is not by way of divine or cosmic necessity.

88 Ibid., 172. Suffering does not have to be taken to be a presence inevitably built into the process of the cosmos.

89 *Cosmos*, 223–25. The expression "saving evolution" is Bouyer's own. See the bottom of p. 223 for "hominization" (*hominisation*).

90 Cf. *Catechism of the Catholic Church* 401. In this most recent summary of Catholic doctrine issued by the papal magisterium the consequences of Adam's sin with regard to death are limited to human history. Because of his sin: "Death makes its entrance into human history." CCC 391–95 makes clear that the fall of the angels preceded the fall of Adam.

consideration of the movement of grace in creation through which physical bodies are integrated into the Mystical Body of Christ. At the same time, he both evokes and directly invokes the writings of Beauregard in suggesting ways that modern science can help us to deepen our understanding of what it means to speak of the human microcosm. This brings us full circle with regard to the themes that we have explored in this chapter.

It is necessary to provide a bit of background in speaking about the human being as microcosm. As we noted in our introduction, when the Church Fathers confronted this ancient idea they did not always see it as readily compatible with God's revelation of humanity as created in His image and likeness. The Cappadocians, and those following them, held that it does not safeguard the unique dignity of the human being to speak of him or her as a "microcosm" or a little world in a big one. Saint Gregory of Nyssa in fact scorned the expression: "People said, Man is a microcosm ... and thinking to elevate human nature with this grandiloquent title, they did not notice that they had honored man with the characteristics of the mosquito and the mouse."[91] Saint Gregory Nazianzen said, as we noted in the introduction, that God created man in his image and likeness as a "macrocosm" or great world in the small one.[92] Saint Maximus the Confessor gave this anthropology the status of central motif in his seminal, brilliant synthesis of the theology of creation and redemption.

This theoanthropocosmic vision, common in the Christian East, is at the foundation of Bouyer's theology of creation. Can we still have this view of things in the modern context? Does not the Copernican revolution in physics challenge the Patristic vision and preclude the very possibility of making it our own? At the end of chapter twenty-one, Bouyer underlines the naïveté of this supposed challenge. In fact, he suggests, in line with insights drawn from Beauregard, that developments in physics seem to open the possibility that the metaphysical scope of *anthropos* as suggested in the writings of the Church Fathers can be expanded beyond even what they foresaw. Bouyer succinctly proposes three responses to the challenge.

91 Gregory of Nyssa, *De Hominis Opificio*, *PG* 44:177d–180a.
92 Gregory of Nazianzus, *Oratio* 38, *On Theophany*, ch. 11, *PG* 94:921a.

First, he insists that there is confusion in the view that time exists or has meaning in the absence of human consciousness.[93] Given the immense duration of time, should we not be overwhelmed by the relative insignificance of the measly span of human duration? Much like Beauregard, Bouyer seems to understand time to exist as an advancing frontier only in human consciousness and its reciprocal exchanges, in its concrete bodiliness, in a world already fallen. He holds that we extrapolate the time of pre-human origins in an act that is inevitably coextensive with our own consciousness. Time, Beauregard showed, is deployed all at once and possesses its irreversible arrow in dissymmetrical movement to the future only in relation to the human body and its consciousness. We think that this helps to explain Bouyer's statement that Einsteinian science shows that "without such consciousness time is but an imaginary variable: it simply appears in equations as a fourth dimension of the space where the human body was in gestation, so to speak, but into which historical time, the field of our own freedom, had not entered."[94] Bouyer does not deny the duration of astronomical and geological time but shows that this duration has differentiation and real meaning only for embodied human consciousnesses in their common historical advance. This does not have to lead to a negative response to his question in chapter one of whether human history has a field that extends to the whole cosmos, because "the human body was [always] in gestation" in the block of space-time that Beauregard describes.

Second, Bouyer maintains that there is confusion regarding the anthropological significance of the immensity of space. He assents to Pascal's view that it is not the Copernican universe, so much as the Ptolemaic one, that "decentralizes" humanity. Between macroscopic and microscopic infinities the human being's place is much more central than in the premodern cosmography for which the earth has a centrality in a world of diminished spatial extent.[95]

As to the third point, the Oratorian maintains that it is a gross confusion that leads some to compare the quantitative greatness of the cosmos with the spiritual greatness of humanity, especially in the order of

93 *Cosmos*, 224.
94 Ibid., 224.
95 Ibid., 224–25.

charity. Bouyer draws this idea also from Pascal, who distinguished, in contrast to Descartes, a realm of charity that is incomparably magnified in dignity beyond the physical magnitude of the scientists.[96] Related to this point is the basic observation about the power and extent of human consciousness that we have already discussed. Bouyer invokes Leibniz's monadology in this context. Human intellectual consciousness, he urges, comprehends the totality of material beings in itself without losing any of its own being. It not only comprehends the totality of material things but unifies the whole. Bouyer suggests that each consciousness constitutes a world in itself and has its own specific way of unifying the world. The body that is informed by the intellectual soul has its own kind of corporeity with the power to reach to the farthest bounds of creation:

> We may add that the human body itself, the body of each human individual in mankind—through the roots sensible consciousness gives it in the entire cosmic reality, which both penetrates us through these senses and is entirely accessible to them—actually extends to the limits of the universe. In fact, the latest findings of science indicate not only that there is theoretically no limit to our capacity to act upon the world, but also that any information has the effect of modifying whatever it conveys to us.[97]

We have seen that this interpretation of information as communication of knowledge and formative power flows directly from Beauregard, and it is at this point that Bouyer directly references his friend's work.

This returns us to the first section of this chapter. The Cappadocian and Berkeleyan understanding of materiality that Bouyer embraces is a dynamic one that enables us to see that there are different degrees of corporeity or materiality. The corporeity or materiality of the human body is singular, because it is the irreplaceable means through which human spiritual consciousness is extended to the limits of the cosmos, actualizing thereby the subconscious mind with which physical nature can be associated in an understanding of matter that sees the cosmos,

96 Ibid., 225. Bouyer references Pascal's *Pensées*, in the Chevalier edition (La Pléiade Collection), 1341.

97 *Cosmos*, 225.

even in its noumenal depths, as ordered to expression or relationship through spiritual communication. There is a uniquely human way of seeing, potentially recapitulative, through which the materiality of the cosmos can be elevated onto a new plane in transfiguration. When the First Adam renounced his mission, this could only have had calamitous cosmic consequences. Bouyer does not diminish the Pauline anthropology in this regard. Cosmic death was still linked to human sin, in that the potential renovation of the cosmos was delayed. As things stand in the one, concrete order of salvation history, it is only in the Incarnation of Christ in human flesh that this renovation can now take place. All of what Bouyer talks about in chapter twenty-one of *Cosmos* concerns a potentiality or capacity possessed by human nature that can be actualized only in Christ. We return to a quotation that we explored in an earlier chapter:

> Our bodies have no limit except those of their own sensations and impressions. And these have no frontiers but those of the universe itself. Contrary to ancient physics, which imagined bodies as being mutually defined by their reciprocal exteriority, modern physics sees them as mutually permeable systems that are defined, not according to an apparent exteriority, but according to a variety of perspectives, a set of coordinates or rather a formula of unique coordination containing the same elements that make up what is common to all that exists—something like the Leibnizian monads which, each in its own way, encompass the whole universe but have form, are a single universe together.[98]

This quotation is from Bouyer's text on Christology, where he makes clear that human potentiality is actualized only in the hypostatic union of Christ, as all of creation is reconciled in unity by the global extension of his own being in the Eucharistic Body of the Church. This is his own Body, and he is its head. Describing the Cappadocian understanding of matter, Vladimir Lossky says that for the Cappadocians the world is a single body, through which material elements pass from one body to another.[99] Bouyer's development of this idea stresses the interiority or

98 *ES*, 397.
99 Lossky, *The Mystical Theology of the Eastern Church*, 103.

permeability of material bodies in their reciprocal exchanges. This is a fact of life in the cosmos as uncovered by modern science. Individual bodies do not have to be understood to lose their limit, their individuality, in these exchanges. There is indeed a "within" of things that is shared, as the above quotation indicates. The theologian should affirm that it is only in the Body of Christ that this permeability of one thing to another is actualized in perfecting communion without division or degradation. Matter can be transfigured. In fact, this is why it exists. The illuminated body of the transfigured Christ on Mount Tabor is paradigmatic, and the Mystical Body of Christ is the ultimate instrument of the transfiguration of cosmic being. Bouyer would surely agree with Lossky, who said: "The history of the world is a history of the Church which is the mystical foundation of the world."[100] Bouyer himself said almost the same thing, albeit with a Western twist: "The Catholic Church is indeed . . . the very principle of creation."[101] Lossky and Bouyer are simpatico in this understanding, except with one considerable proviso beyond the fact that Bouyer is a Catholic theologian: the latter construes the Church as the "mystical foundation" or more simply the "principle" of the world, because it will be the actualization of created Wisdom. He sees the descent to earth of the Heavenly Jerusalem as the apocalypse of Wisdom, the final, existential fulfillment of created Wisdom as the Bride of the Lamb.

100 Ibid., 111.
101 *CG*, 592.

CHAPTER 11

The Eschatological and Nuptial Cosmos

COSMOS ENDS WITH AN OVERVIEW OF THE meaning of the cosmos as revealed in the conjoined eschatological and nuptial disclosures of the final, apocalyptic book of Sacred Scripture. Bouyer teaches, on the basis of John's Apocalypse, that the driving purpose embedded within the metaphysical sinews of creation from the very beginning was to be united in nuptial communion with the Immolated Lamb of God in and through the eschatological Church. The meaning of creation is unveiled only with the definitive apocalypse of Wisdom in the Parousia. In this chapter, we shall explore the themes explicit and implicit in chapter twenty-two of *Cosmos* that join together, in light of John's Apocalypse, nuptial theology/anthropology and apocalyptic disclosure of the sophianic entelechy of all finite being and beings. This will require developing, beyond what we just explored in the previous chapter, Bouyer's theology of the Church as the mystical principle of creation. Once again, it will be necessary to draw on other texts by Bouyer to understand the meaning of his succinct and dense passages in *Cosmos*. We shall explore further with Bouyer the Church in its maternal and spousal characteristics. This involves seeing the foundation of the Church in Mary's motherhood, a theological topic which we shall develop in a first section of the chapter. In a second section, we shall set forth what Bouyer means by speaking of the predestination of the Church in Mary and in a third as the Bride of the Lamb flowing from divine predestination in Mary. In a fourth and concluding section, we shall relate these themes to the aesthetic cast of Bouyer's thought, centering on the glorification of created Wisdom in the immolated humanity of the spotless Lamb of God.

THE MOTHER OF GOD AND THE CHURCH

In the final chapter of *Cosmos*, entitled "The Nuptials of the Word and of Wisdom," Bouyer's Christian theoanthropocosmic vision is brought to display through utilization of Johannine apocalyptic imagery connected with Pauline and Old Testament sapiential theology. The chapter again explores the story of creation and redemption, this time with respect to a more fully developed theology of the eschatological Church in connection with Wisdom than is explicitly present in earlier chapters of the book. It begins in the first two of six brief sections with consideration of human consciousness in relation to the divine Word and presents initially a sort of confluence of the motif of ecclesiastical hierarchy in Denys with certain modern, partial recoveries of the way the link between consciousness and the world was understood by the Areopagite and by those who followed in his footsteps.[1] Bouyer sums up at the beginning of the chapter ideas that we have already discussed, but he puts them now in the full light of the apocalyptic understanding of the eschatological Church as the Bride of the Lamb, whose establishment is made possible in history through the fiat of Mary, by her definitive "yes" to the will of God, her acceptance of the gift of virginal motherhood, and the extension of her "maternal grace" to all the faithful who are reborn in Christ through the Church.[2]

In describing the essential characteristics of human consciousness in the first section of the chapter, Bouyer delineates its structure as rational, free, and personal.[3] By distinguishing in this list the word "personal," he means to say that human persons are not only constituted by rationality and by freedom oriented toward some indeterminate good but by a specific call to enter into interpersonal relationship with God and to one another in the perfection of charity. The chapter moves to consideration of the articulated grammar of that call. Bouyer holds that the ultimate fulfillment of human personalization will come only

1 *Cosmos*, 226.

2 Ibid., 228–30. Cf. *CG*, 585. What Bouyer describes of Mary and the Church in the final chapters of his great book on the Church is essential to understand the implicit theme in the final chapter of Cosmos, as we shall see further in what follows here.

3 *Cosmos*, 226.

when we are made a "Temple of the Spirit"[4] in our bodies and are given, as he says elsewhere, full "participation in the very life of the Divine Persons . . . in the eschatological Church."[5] This will be a perfection of the image of God in us, which can be attained only by entering, as he says in *Cosmos*, "into the interpersonal relationships which constitute the divine personality in whose image every created personality has been fashioned."[6]

Bouyer insists that because we are beings in the world, bodily and corporeal, our rational consciousness is intertwined with animal consciousness or "the consciousness of a body, of one of those specific organisms within the cosmos which, from a certain viewpoint, espouse it in its entirety."[7] As we have seen in earlier chapters, Bouyer holds that each body or kind of body images an angelic form, and the angelic forms and the physical cosmos that mirrors them establish together the one, intricate web of cosmic being. In chapter twenty-two of *Cosmos*, he clarifies that the spiritual dimension of our being exists with and for the material dimension of it and indeed emerges from it through a special gift of the Holy Spirit.[8] This harmony of relations in the vast cosmic web of interrelated beings has been thrown out of balance by the fall of angels and later of humanity, and a drama has been inaugurated that affects human consciousness at its very roots. The thrust of Bouyer's presentation in the final chapter of *Cosmos* shows that human consciousness cannot be entirely understood as a static structure analyzable apart from human bodily insertion in the drama and tension of all finite beings, who exist together in a relational cosmos that has become

4 Ibid., 230.

5 *CG*, 530. This text from Bouyer's book on the Church sheds light on the final chapter of *Cosmos*. We shall expound pertinent portions of this text in the current section to help draw out and develop certain themes that are a little under the surface in *Cosmos*.

6 *Cosmos*, 226. We think that when Bouyer refers to the "interpersonal relationships that constitute the divine personality" he is saying something to the effect that the relationships of the persons do not refer simply to separate relational terms in "I-Thou" polarity but to a communal "We." The "divine personality" is relationally constituted: the relations are not abolished by some higher-level unity.

7 Ibid.

8 Ibid., 223.

agonistic but is called to liberation in Christ from its ensnarement to the empty promises of the god of this world.

Each corporeal image of God, each human person, is, Bouyer urges, a definite personality called to freedom in the Holy Spirit in a determinate place within the material network of creation.[9] Each of us is nevertheless free by virtue of our creation in the image of the divine Word. We are natively receptive to all the forms and degrees of being, and it is through this capacity as *imago Dei* that the grace of the Spirit of God can work in each of us and all of us together to render us *capax infinitum*.[10] In line with nineteenth- and twentieth-century theologians who recovered and detailed the anthropological meaning of the nuptial imagery in Sacred Scripture, Bouyer understands that the human image of God is not limited to an individual in isolation, subsisting with his or her own personal faculty of reason and will or God-given mission irrespective of his or her sexed, bodily alterity and complementarity.[11] Bouyer's anthropology of the *imago Dei* is resolutely nuptial and Marian. He sees Adam and Eve as fallen types of the New Adam, Christ, and the New Eve, Mary, and that the First Adam and the First Eve were meant to image the divine personality together, existing in a fundamental communion of one flesh (cf. Gn. 1:27). Their existence together was

9 Ibid., 226–27.

10 Ibid., 226. Interestingly, Bouyer says in the relevant footnote that for Saint Thomas, "this is certainly the point at which grace can penetrate human nature; it is also what makes our nature both capable and spontaneously desirous of receiving grace" (see *Cosmos*, 271, n. 3). This may seem to support Henri de Lubac's famous argument that Aquinas's statements regarding a "natural desire for beatific vision" need to be emphasized in order to interpret his thought adequately. Cf. Aquinas, *SCG*, Bk. 3, ch. 57: "Every intellect naturally desires a vision of the divine substance"—"*Omnis intellectus naturaliter desiderat divinae substantiae visionem*." We suggest that this would have to be seen in accordance with Bouyer's early preference for the tripartite constitution of the soul as found in the tradition of Saint Paul.

11 See Fergus Kerr, *Twentieth-Century Catholic Theologians: From Neo-scholasticism to Nuptial Mysticism* (Malden, MA: Blackwell, 2007). As the title indicates, Kerr describes the turn to nuptial mysticism in twentieth-century theology, which he criticizes. Although he explores different, major theologians in different, respective chapters, he does not devote a chapter to Bouyer. Nevertheless, this theme of "nuptial mysticism" is central to Bouyer's work, as our present chapter shall indicate.

"proto-sacramental."[12] Bouyer develops in chapter twenty-two of *Cosmos* Saint Paul's analogy in Ephesians of the natural union in marriage of male and female to God's relation to the Church as Bridegroom to Bride. The vocation of humanity to be reconciler of creation is concretely embodied in the nuptial relationality of man and woman and their complementary being, body, soul, and spirit.[13] Bouyer suggests in this chapter that Eve, in her feminine perfection as both mother and spouse, would have brought together the whole of creation in an immanent unity returning to its source, while Adam, in his masculinity, was to be a sign of conjunction of humanity with the Uncreated, with God, who alone could actualize the unity of creation contained in the first Eve.[14] In the obedience of faith, the shared work of Adam and Eve would have built the world into a single Temple proclaiming the glory of God. Although we cannot know precisely how they would have accomplished this task, Adam and Eve would nevertheless have been together the agents of "spiritualization of their own bodies and of the entire material universe, and therefore the instruments of universal divinization."[15]

As is usual in Bouyer's work, narrative summary in the final chapter of *Cosmos* dominates speculative, ontological explication. At different points in this book, as we have already seen, he tells the story of salvation history in its cosmic scope, but at each point he does so with different

12 This expression comes from John Paul II but describes well Bouyer's position. See Grintchenko, *Une Approche Théologique du Monde*, 250. See also Christian-Noël Bouwé, *L'Union Conjugale et le Sens du Sacré: La Sacramentalité du Mariage dans la Théologie de Louis Bouyer* (Paris: Les Éditions du Cerf, 2016), 284–351.

13 *Cosmos*, 227.

14 Ibid. The parallels between Bouyer's thought on this issue and those of John Paul II are clear. Grintchenko notes a parallel especially in regard to the way each theologian sees the unity of male and female as a sacrament intended by God's sovereign will irrespective of the Fall (of man): "A primordial sacrament is constituted in this dimension understood as sign which transmits efficaciously in the visible world the invisible mystery hidden in God from all eternity. And this is the mystery of Truth and Love, the mystery of divine life in which man really participates. In the history of man, it is original innocence which opens this participation." John Paul II, *Homme et Femme Il Les Créa. Une Spiritualité du Corps* (Paris: Les Éditions du Cerf, 2010), 105, quoted by Grintchenko, *Une Approche Théologique du Monde*, 250.

15 *Cosmos*, 227.

emphases, highlights, or accents. In chapter twenty-two, his emphasis is decidedly Marian and ecclesiological. Mary's cosmic motherhood is a culminating theological motif in his monograph on cosmology, although it is hardly drawn out in full in this text, which presumes much of his earlier work and foreshadows some of his last writings, especially his very last writing, *Sophia*.[16] The reader should keep in mind in exploring chapter twenty-two his exposition of Baader and Bautain from earlier in the book.[17] These two Catholic theologians were among the rare modern practitioners of their craft up to their time to show that Marian theology entails a nuptial theology of the whole creation, including physical nature. This common sophiological understanding forms a single picture with biblical angelology, and a central consideration in this final chapter of *Cosmos* is Mary's divine mission in relation to that of the angels, although this relation was in fact spelled out more thoroughly in his earlier text on ecclesiology.[18]

Bouyer insisted in this latter writing that the Church cannot be understood without reference to its vast, cosmic interlinking with the celestial hierarchy in its totality. The Church is not only human but cosmic in scope. Following Denys, he holds that ecclesiology is essential to a fully integrated cosmology. There is, he insists, a cosmic Church, and its first saints are angels.[19] They are themselves bearers of the motherhood of grace through which God's agapeic self-communication is promulgated in creation. The angels are the ninety-nine faithful sheep of whom Christ spoke in the parable of the Lost Sheep recorded in the Gospels of Matthew (18:12–35) and Luke (15:3–7).[20] Bouyer remarks that God does not in fact leave them behind to go in search of the lost sheep, that is, fallen humanity. Instead, He utilizes them as co-workers in this search in order effectively to instill His Wisdom in the world.[21] The Holy Eucharist, which founds the body of the Church, is first of all a gathering of the angels, who are, as Bouyer says, "the first to have

16 *Sophia*, 161–76.
17 *Cosmos*, 136–39.
18 *CG*, 582–90.
19 Ibid., 582.
20 Ibid., 583.
21 Ibid.

known the love of God and to respond to it in praise."[22] The faithful angels share in God's saving work of loving self-communication in the Holy Spirit through which God brings about a new creation from within the old, establishing the elect of God who have co-operated with His grace as "a new, holy personality in the life of the Church."[23] They are the first to instantiate the "motherhood of grace," the share that creation has both in the divine life and in God's redemptive mission to the world.[24]

In chapter twenty-two of *Cosmos*, Bouyer recounts the primary, kairotic stages or events through which God in His Holy Spirit heals and elevates human consciousness, lifting it up from the misfortune that attends its fallen condition to the pleroma of its re-creation in the holiness of Mary.[25] The angels, he suggests, were the ministers through which God first disclosed His preparatory revelation, in an initial stage restoring human consciousness to recognize the divine law immanent in creation.[26] The patriarchs of the Old Covenant were called to detachment from the cities of the ancient world, which they knew to be under the sway of the Devil, and a tradition of prophetic inspiration was opened over time in their hereditary line, awakening humanity to hope for deliverance from its exile from paradise, until at last, through and beyond the great biblical Wisdom tradition, the anticipatory figurations of the Suffering Servant and Son of Man were made manifest to prophetic consciousness.[27] At length, the Virgin Mother was born. She recapitulated in her human personhood the entire trajectory of salvation history that preceded her. She is the full "flower and fruit of the holiness and the motherhood of grace of Israel."[28] She demonstrates the power of human personhood to contribute to its own divinization.[29]

22 Ibid.
23 Ibid., 577.
24 Ibid., 584–85.
25 *Cosmos*, 228–32.
26 Ibid., 228.
27 Ibid., 228–29.
28 *CG*, 588. On this whole paragraph, see *Cosmos*, 227–31.
29 Cf. Weill, *L'Humanisme Eschatologique de Louis Bouyer*, 203. Weill says: "In Mary, humanity cooperates not only in its own salvation, but also, what is

This story recounts the preparatory development that made possible the historical birth of the body of the Church, the very Body of Christ, in Mary. The historical and cosmic development through which the entelechy of creation is actualized in the personalization of created Wisdom as the perfect Spouse of the Bridegroom has a maternal dimension from the beginning, which is realized by Mary in history.[30] Through her divine motherhood the Church is able to be mother to all the faithful.[31] The faithful angels themselves, as we just indicated, participate in this maternity of grace in and through which God's agapeic glory is made present in creation and the new personality of regenerated humanity is born. But the angels could never have been God's biological mother and therefore mother to all the faithful. Mary, as *Theotokos*, thus surpasses even the first created spirits in the order of holiness.[32]

The final chapter of *Cosmos* indicates the link between Mary's motherhood and the maternal grace of the Church but does not develop the theme in full. This was already done in *The Church of God*. From this latter text we learn that grace has a maternal character and is a gift of the Spirit of God, who is in some sense feminine, as the Syrian Church Fathers insisted in noting the gender of the Hebraic expression *Ruach Adonai*.[33] God transcends sexual characteristics in His eternal triune being but "reveals Himself in the motherhood that this Spirit communicates to creatures in raising them up to the Creator."[34] The angelic choirs, though lower than Mary in the order of holiness, nevertheless constitute the primordial existence of the ecclesiastical hierarchy in the first divine communication of motherhood, as they cooperate with God to draw humanity by the love of the Holy Spirit into the cosmic Eucharist.[35] The prophetic and apostolic ministries actualized in the visible Church are a share in the celestial ministry of service of the angels, which is itself a participatory imitation of the thearchic *agape* of the triune God. There is a practical consequence to this: in order to be a

---

even more astonishing, in its divinization, in its filial adoption."

30 *Cosmos*, 230–31.

31 Weill, *L'Humanisme Eschatologique de Louis Bouyer*, 203.

32 *CG*, 582.

33 Ibid., 584.

34 Ibid.

35 Ibid., 585.

person of Christian action one has first to be a person of contemplation and prayer, in imitation of the angelic persons.[36] It is in this way alone that God's paternal *agape* can begin to take possession of the whole of human being in the latter's maternal capacity.[37]

In the absolutely unique human personality of John the Baptist, the ecclesial ministry of humanity was directly joined with that of the angels. The Baptist was, as Bouyer teaches, himself a "messenger," an angel in the flesh, set apart to wander the desert as the greatest "Friend of the Bridegroom," totally possessed by the Holy Spirit and sent to purify "the temple of the world by emptying it of all idols, through the example of a life freed from earthly detachments."[38] However, it is in the person of the Most Blessed Virgin Mary, Bouyer insists, that the motherhood of grace is universalized concretely in its full effectiveness. Christ has communicated to her uniquely his own holiness in plenitude, without equal in relation to any other created person, angelic or human. In the supernatural order of the Holy Spirit, she is uniquely Mother, the Mother of those reborn in her Son, and the motherhood of the Church is realized in her because it has found in her its personal anticipation within the economy of salvation. As Bouyer teaches in *The Church of God*, she possesses "the supreme created holiness, a unique communication of the holiness of Christ to her, who is not only our Mother, as the saints or angels could never be, but first of all *his* Mother."[39] She is the personal consummation in freedom of a vast and specific cosmic and historical movement led by the Holy Spirit that is progressively accomplished throughout the history of the People of God. The very maternal depth of creation, ubiquitously linked to human personality in the latter's unconscious mind, is consummated in Mary, by her fiat to the will of God. "Like the Church," Bouyer explains, "Mary comes from the earth, from the desert, which flowers again beneath the showers of the sky, and yet comes down from God as the gift of grace incorporated in mankind, in fallen creatures in the process of being saved."[40] Mary is the herald of "the supreme moment of human and cosmic history, in which

36 Ibid.
37 Ibid.
38 *Cosmos*, 229.
39 *CG*, 588. The emphasis is Bouyer's own.
40 Ibid.

the saving Word is fully heard by perfect faith, its supreme creation, and elicits that response which gave birth not only to all those saved in grace, but first of all, to the Savior himself."[41] Through her fiat humanity and the cosmos are associated in a renewed, immanent unity.

The motherhood of Mary is a real spiritual maternity that perfects biological motherhood in creation. As Bouyer says in his dense little book *Woman in the Church*, "the mystery of woman, throughout the Bible and Church tradition, is presented as the final mystery of creation, especially redeemed creation, saved and made divine by the Incarnation of God in flesh which he took from a woman."[42] Following Saint Athanasius, Bouyer argues in this book and elsewhere that fatherhood is proper to God alone, and that it is possessed by human fathers only in a transitory and imperfect way.[43] This is especially so, he insists, the more spiritually elevated this fatherhood becomes, as in the spiritual fatherhood of the priest, which is a gift of the Holy Spirit in the supernatural order.[44] It is otherwise with motherhood, in which the final vocation of the creature is disclosed in its immanent perfection. "Every human being finds itself," as Jean Duchesne explains, "facing the eternal Father in a feminine situation."[45] The spiritual actualization of motherhood is the highest perfection of the capacities inherent to the creature *qua* creature.[46]

Bouyer spells this out most fully in *Woman in the Church* with respect to the anthropological and theological significance of virginal

41 Ibid.

42 *Woman in the Church*, trans. Marilyn Teichert (San Francisco: Ignatius Press, 1979), 28.

43 Ibid., 32–34. Saint Athanasius interprets Matthew 23:9 on the point: "And call no man your father on earth, for you have one Father, who is in heaven." Bouyer frequently invokes Saint Athanasius on this matter. Texts that he cites include: *First Discourse Against the Arians*, 18–19 and 21; *First Epistle to Serapion*, 17; *Second Discourse Against the Arians*, 29–31. See Weill, *L'Humanisme Eschatologique de Louis Bouyer*, 185.

44 *Woman in the Church*, 31–32.

45 Duchesne, *Louis Bouyer*, 90. It is of interest to note in assessing Bouyer's wider thought that he had intended to publish a book on marriage, which he had mapped out but was never able to write, due to age and sickness. Duchesne says that Bouyer himself recognized that his life's work would thus remain incomplete (see *Louis Bouyer*, 112).

46 Ibid., 90. Duchesne says: "Femininity is in consequence the proper human condition, as paternity is that of God."

motherhood. Mary, he argues in this book, perfects created virginity, which is much more a property of the female than of the male, because in her it embraces "all future humanity, both masculine and feminine, for it will never come into being if not by an interior development of feminine being, which the male does nothing but set in motion."[47] There are two specific reasons why virginity belongs properly to the female and not the male. The first is that the exercise "of sexuality inscribes itself irreversibly and intimately in the body of the woman."[48] The second is that it is "the woman who brings in her flesh the child to birth."[49] Hans Urs von Balthasar explains:

> His [Bouyer's] basic affirmation is that, in the sexual realm, woman is the full explication of the dignity bestowed on the creature of being a second causality alongside, in and through God. Because of this, furthermore, woman enjoys the role of being the world's comprehensive answer to God.[50]

This affirmation is truest of the woman in her virginal integrity. Bouyer holds that the virginal woman images by virtue of her encompassing potency the virginal actuality or *actus purus* of God the invisible Father, who transcends, as the antitype of virginal motherhood, both the masculine and the feminine as they exist in creation.[51]

47 Hans Urs von Balthasar, "Epilogue," *Woman in the Church*, 34. Balthasar wrote the epilogue (pp. 113–21) to this book in defense of its main theses, which are compactly stated. In what follows, in this section, whenever we refer to this epilogue, we shall mark it as we have done here. Balthasar notes that he defended a position similar to Bouyer's, albeit from a different angle, in his book in defense of the Petrine primacy. See *The Office of Peter and the Structure of the Church*, trans. Andrée Emery (San Francisco: Ignatius Press, 1986), 180–225. Balthasar argues here that the Marian principle of the Church encompasses all other dimensions of it, including the Petrine.

48 Quoted from Bouyer by Weill, *L'Humanisme Eschatologique de Louis Bouyer*, 183. Weill says further: "From this fact, the feminine virginity, associated in all cultures with the idea of a certain integrity, is representative of a dignity, of a particular prestige, even in strongly eroticized societies." The two reasons that we list here are drawn from Weill.

49 Ibid., 184.

50 Balthasar, "Epilogue," *Woman in the Church*, 114.

51 Cf. *Woman in the Church*, 77.

This anthropology does not have to entail, as some critics of Bouyer insist, that the dignity of man the male is undermined, although the emphasis is certainly a little one-sided. Bouyer ultimately recognizes that man the male (*vir*) and man the female reveal the complementary bearing of human personalities as differentiated but coequal in the respective activities through which they are actualized. The feminist challenge in the contemporary period, and reactions against it, can both be dehumanizing to the extent that they do not acknowledge that all creatures are absolutely dependent on God and that created femininity is especially symbolic of finite being in the plenitude of its ontological interiority.

Balthasar explains that in the natural sexual relation between male and female "the woman enjoys the inward role of bearing, which is more perduring, while the man provides an external, episodic function: he merely *represents* a primal, creative principle which he himself can never be."[52] Bouyer directly invokes, in this context, Aquinas's hylomorphism, for which the intellectual soul is the substantial form of the body. Though neither he nor Aquinas speak in terms of a "masculine soul" or a "feminine soul," all that Bouyer says might entail this distinction: "And remembering that the soul of man [*l'homme*, not *vir*] is simply the substantial form of his body, we must expect that his physical being will reveal and define his metaphysical being itself."[53] Man the male, Bouyer shows, realizes himself outside of himself in the act of procreation, by way of necessary detachment from what he has procreated.[54] The being of the woman, by contrast, is actualized in self-surrender to what she carries in herself and which germinates from the nourishment provided by her own substance.[55] In order to become perfect in her own motherhood, she has to consent to separation from her child, so that it may flourish in growth and maturation in autonomy from direct biological connection to her. This is very difficult for her, while man the male accepts detachment from his child as a matter of

52 Balthasar, "Epilogue," 114. Balthasar agrees with, defends, and develops this position.

53 Bouyer, *Woman in the Church*, 53.

54 Ibid., 52–53.

55 Ibid., 54.

course. The woman, by virtue of her specific biological and therefore spiritual constitution, cannot as easily consent to this separation.[56]

The human male and female have, as a result, different corresponding tendencies in their conscious habituation to the world, to one another, and to their own selves. Man the male does not come to adult completion of his consciousness by nourishing another person from within his own being. It is only by way of intentional relationality to an exterior other that he can exercise paternity. This exercise is always effected in a kind of incomplete, transitory, and borrowed manner, and his consciousness may, as a result, flit about in an unstable way between a wholly objectivist exteriority and a subjectivism that too easily veers into solipsism.[57] The consciousness of the woman, by contrast, is actualized through the formation of another being within her own. She is thus, as Bouyer says, "natively adapted to empathy, to that sympathy with the object which is not conceivable except in a subject such as the feminine subject is, for whom the object does not appear from the outset as exterior."[58] Femininity in both its maternal and spousal dimensions instantiates perfection of human personal being in a manifestation of consciousness that is more attuned to empathic receptivity and, we might say, to *intellectus* or even to "imagination" as Bouyer uses the term.[59]

The idea of the human microcosm is given a nuptial twist in this wider anthropology of the relation between male and female. Bouyer sees in the virginal purity of the woman "the whole of created human reality with all its inexhaustible richness, but first of all in its organic unity . . . initially present in the microcosm of her body as a shadowy perception which awakens her spirit."[60] In her spousal and procreative encounter with

56 Ibid. See Weill's discussion, *L'Humanisme Eschatologique de Louis Bouyer*, 205–10.

57 Balthasar, "Epilogue," *Woman in the Church*, 114–15. On this page, Balthasar explains this same point: "The role of the man consequently acquires a peculiarly open bi-polarity where woman's role exhibits a closure: as a representative of the Creator God, the man is more than himself, and yet, at the same time, as a mere transmitter who can as such *only* represent, he is also less than himself."

58 *Woman in the Church*, 56.

59 Ibid., 65. Thus the woman is more naturally religious than the man. This point has often been recognized, but Bouyer endeavors to explain on the level of anthropological ontology why this is so.

60 Ibid., 62.

man the male a portion of her limitless potentiality is actualized, but precisely because it is only a portion of her germinal, feminine being, it does not reveal the full breadth of her microcosmic capacity.[61] Only in the enfleshment of God's Word in the womb of Mary is this potentiality actualized in totality.[62] Divine motherhood in Mary is thus not a reductive, fragmented, purely earthly motherhood. It is the consecration of motherhood in absolute perfection. Christ possesses the male prototype of perfected humanity as the only true representative of the eternal Father's paternity, and his relational subsistence as eternally begotten by the Father is translated into the flesh he has assumed in and through the womb of Mary.[63] In him, the second person of the Trinity, a divine person, human nature is raised beyond itself, because it is assumed into his hypostatic subsistence. In Mary, a created person, human personhood is realized in its integral fullness. The respective modes of human perfection in Christ and Mary are distinct, and we should not downplay the fact that, strictly speaking, in following the Council of Chalcedon, Christ himself is not a human person. In Mary alone we can see "all that grace was able to make of a creature, of human nature, while still leaving it in its order as a created being."[64] Nuptial distinction and communion are consecrated by attentiveness to this particular, exemplary individual person.[65]

The preparatory stages in the emergence of the body of the Virgin Mary constitute a vast history, as vast in spatio-temporal expanse as the physical cosmos itself, including all of human history and pre-history

61 Ibid.

62 Ibid. 62–63.

63 See Balthasar, "Epilogue," *Woman in the Church*, 115–16. Thus for Bouyer as for Balthasar the Son must eternally possess a kind of receptivity to the Father. He knows in himself eternally the active receptivity of the creature exemplified in the perfection of maternity.

64 *Mary: Seat of Wisdom*, viii.

65 See Zordan, *Connaissance et Mystère*, 669; Weill, *L'Humanisme Eschatologique de Louis Bouyer*, 95–96. Zordan and Weill both argue that Bouyer's thesis in this regard was unique for its time. The thesis was at the heart of his book on Mary, which we shall discuss in the next section. Bouyer presents Marian doctrine as inseparable from the most basic theological questions. See also Michael Heintz, "Mariology as Theological Anthropology: Louis Bouyer on Mary, Seat of Wisdom," in *Mary on the Eve of the Second Vatican Council*, ed. John C. Cavadini and Danielle M. Peters (South Bend, IN: University of Notre Dame Press, 2017), 204–26.

in its train. This brings us back again to Bouyerian reflections that we explored from *The Eternal Son* and briefly discussed both in chapter five and in the preceding chapter. Christ, we learned, actualizes the potential cosmic and human universality that is present in each human person but that remains latent without his intervention in the flesh of historical humanity.[66] Only enhypostatization in Christ's human nature, without division or confusion with his divine nature in his divine personhood, could actualize the recapitulative potency of human being. In the last chapter, we explored more fully the first of two suggestions that Bouyer makes for how we might meet the challenge of the question of whether the concrete universality of Christ does not pose a threat to his individuality, regarding the permeability of material being. Here it is pertinent to discuss further the second suggestion in this unified response, in which Bouyer stresses the interplay of the concrete universality of exceptional religious personalities in the totality of history who tended always, under the pressure of the Holy Spirit, to make ready the Incarnation of Christ.[67]

These individuals, Bouyer argues, possess a uniquely powerful qualitative universality, characterized by the perfection of interiority and empathy.[68] The historical figures of religious importance to which Bouyer refers represent the full progress of a specific cultural history in its total, qualitative meaning. This reference is ultimately to the history of the formation and development of the People of God, whose exemplary figures embody the common experience of the whole of the People throughout its history, leading up to the most exceptional human person of all, the Blessed Virgin Mary. This particular history is "at the heart of our human history, shattered by sin."[69] Christ himself is thus linked to the whole of the history that precedes him, as he infinitely surpasses in perfection the creative religious personalities who prepared the way for his coming. He could not be the surpassing inheritor of this total history in its concrete perfection as set apart in salvation history if he did not receive his humanity in and through the gift of the divine maternity that is given to the one who is the full flowering of Israel's unique motherhood.

66 *ES*, 393–99.
67 Ibid.
68 Ibid., 397–38.
69 Ibid., 398.

"All creation," Bouyer says in the final pages of *Cosmos*, "tends toward man, all mankind tends toward Christ, and, in turn, Christ, as he has revealed himself to us, tends to unite with all mankind, and through it with the universe."[70] In creation, God's Wisdom unfolds by the power of the Holy Spirit, making ready not only the One who will reunite creation in himself but the one whose supremely human actualization of receptivity enables his definitive entrance into the very flesh of the world. Mary's maternity, which sums up the totality of preparations that belong to the People of God, is extended to all God's people in the subsequent history of the Church. There is a "prolonged . . . maternity of grace" in the life of the Church in all those who are born anew in baptism into the death and Resurrection of Christ. Bouyer says that our acts of faith and hope, insofar as they are carried out in charity, link "our humanity to this bringing to birth that involves all creation."[71] At the center of the world lies the history of the people of God, and so, as Vladimir Lossky said, "the history of the world is a history of the Church."[72] Perhaps it is truer in Bouyer's case to speak of a pre-history of the Church intrinsically connected to its historical and eschatological actualization. It is in the context of his apocalyptic, Marian, nuptial vision of creation that Bouyer defines the meaning of cosmic space-time in the final chapter of *Cosmos*: "Time is but the mysterious transit in which the created freedoms signify their consent to the uncreated liberty, in a process of Love calling love."[73] Mary's fiat is emblematic of this consent. She is, like her Son, a figure of recapitulation. She sums up the healing elevation of created freedom that marks the history of the People of God as well as the microcosmic capacity of human personhood that is uniquely present in the virginal female.

PREDESTINATION IN MARY AND THE CHURCH

Consideration of this supernatural anthropology of Mary's motherhood sheds light on the apocalyptic theme of the Church as the Bride of the Lamb in Revelation. The Paschal Mystery, the most decisive divine action in our history, is not possible without Mary's perfect human consent

70 *Cosmos*, 231.
71 *ES*, 405.
72 Lossky, *The Mystical Theology of the Eastern Church*, 111.
73 *Cosmos*, 231.

to the divine will and has no purpose unless creation is brought into perfected communion with her Son at the end of time. Bouyer says in *Cosmos* that the "Nuptials of the Lamb are the goal of history, just as the principle of history is in the Lamb's immolation."[74] The ultimate *telos* of the cosmos is the nuptial union of the eschatological Church with Christ the Bridegroom, and this is rendered possible only by the sacrifice of Christ on Golgotha, which is, in turn, capacitated by Mary's "yes" to God's concrete demands for her as perfect Mother throughout the course of the historical mission of her Son.

The theme of predestination in divine Wisdom is entailed in this articulation of divine nuptials and sacrifice. Predestination is an implicit emphasis in chapter twenty-two of *Cosmos* and provides the opportunity for us to address in greater depth Bouyer's theology of predestination, whose focus is as much Marian as Christological. Bouyer insists that our predestination is not only in Christ but in Mary. He draws out this point more fully in his book on Mary than anywhere else, but what he says in this first book of his dogmatic trilogies was maintained in essence all throughout the trilogies and is an underlying theme in *Cosmos* with important implications for cosmology.[75]

Saint Paul, in Ephesians 3:10, speaks of the Church as the bearer of the *polypoikilos* (*πολυποίκιλος*) Wisdom of God. Bouyer translates this expression to mean Wisdom "shimmering with numberless lights."[76] His third-to-last section in chapter twenty-two of *Cosmos* is given the title "Ultimate Revelation of the *Polypoikilē* Wisdom" (an expression that inspired the title for our book). According to Saint Paul in this passage, the Wisdom of God is revealed through the Church to the authorities and rulers in the heavenly places. Bouyer understands this Wisdom to be centered on the Mystery of the Cross. "Wisdom," Bouyer says, "is

74 Ibid. We saw in the previous chapter that he referred to the Church as well as the "principle" of creation. Here he is making a rather loose distinction. These differences in expression are not contradictions. After all, the Church is the Body of the Immolated Lamb.

75 *MSW*, 103–30. These pages contain an explanation of the Catholic doctrine of the Immaculate Conception and the meaning of God's predestination of humanity's salvation in its light. This will be further discussed below.

76 *CG*, 591; see *Cosmos*, 231.

destined to be both bride and mother" of the Lamb.[77] This actualization of Wisdom will be the final gift of the Holy Spirit to creation. The completed Church at the end of time is the Body of the Son that God received from the womb of Mary in history. It will be the final and complete incarnation of Wisdom as the Bride of the Lamb. In the eschatological Church, Christ will come to fullness in his Body and the "pre-existent" design of God concretely realized. The eschatological Church is the Body of Christ brought to plenitude as the Bride who exists with the Trinity in full personal otherness consummated in a perfect union.[78] In the nuptial ecclesial motif, much more so than that of the Body of Christ, we distinguish Christ from the Church. The metaphor of the Bride of the Lamb entails an ontology of other-reference and relation. Even so, the Church has Christ as its head and will be perfectly fulfilled in him in the end.[79]

Mary, the Mother of the Church, is, for Bouyer, the Seat of Wisdom at the center of history. God's eternal plan for the completion of creation in the Church runs through the human person of the Virgin Mary as its central nodal point. Mary is thus inextricably tied to the Church and her predestination to holiness, which is signified in the Western doctrine of the Immaculate Conception, is not simply a glorification of her motherhood in its individuality. Predestination, Bouyer insists, should not be conceived in an individualist fashion, as was often the case in Western theology from the time of the sixteenth century. The object of predestination, as we suggested in chapter five, is not individuals taken in isolation from one another but Christ in the totality of his mission, his Church, and all the members whom he brings together to form his one Body.[80] The Blessed Virgin Mary is "preeminently the exemplar of pre-destination, as being the link between Christ and the Church."[81] To

77 Ibid.

78 *CG*, 592.

79 Cf. *CG*, 528. Neither the unity of Christ and his Church nor the radical distinction between them should be downplayed. Indeed, Bouyer says that it is even truer at the end of history that the Church will not only be perfectly one with Christ but as distinct from him "as one person from another." This refers to the personalization of created Wisdom in the eschatological Church as the eternal goal of history.

80 See *MSW*, 108.

81 Ibid.

speak of the predestination of Christ requires recognizing the predestination of Mary and through her the fullness of the Body of the Lord. These realities necessarily intertwine in the unified cosmos of the Western Church's mystical theology. It is with respect to the figure of the Virgin Mary that Bouyer seeks to overcome distortions common to the Christian mind regarding God's omniscience and omnipotence, human freedom, and the relationship of nature to grace. Mary full of grace is the exemplar of human personhood perfected in freedom. Mary takes center stage. The peak of human maturation in freedom is her response to the will of God.

In *Mary: Seat of Wisdom*, Bouyer explores the vexing topic of divine predestination in light of the question of whether human salvation requires an explicit "yes" to the Incarnate Lord in encounter with him. He presents his own Marian account of predestination as a third way that overcomes two common, distorted positions. The first of these is the rigorist, Jansenist option of early modernity, which condemned pre- or extra-Christian humanity *en bloc*, because it lacked direct access to the person of Christ. This view held redemptive grace to be entirely absent as an actual reality in history prior to the coming of Christ or outside the visible bounds of the Church.[82] The second is the latitudinarian option that was prevalent already when *Mary: Seat of Wisdom* was written in the 1950s but became especially widespread in the Church after the Second Vatican Council. It proposed the existence of an "anonymous Christianity" pre-existing the Church and existing in our own day outside of its visible bounds. Rightly seeking to combat the narrowness and rigidity of the Jansenist view, proponents of this latter option have sometimes tended to hold that those who have never encountered Christ have equal access to the life of grace as, or even greater access than, those to whom he has made himself explicitly known.[83] Against the first view, Bouyer stresses

82 Ibid., 105.

83 Ibid., 105–6. See also *IF*, 78–79. In these pages from *The Invisible Father*, Bouyer roundly criticizes Rahner's version of Transcendental Thomism and its idea of the "anonymous Christian." Bouyer criticizes Rahner for collapsing the orders of nature and grace. He says that "an examination of the theological anthropology with which Rahner is increasingly replacing theology as a whole makes one suspect that his system is taking him to a goal exactly opposite the one he intended to reach . . . his conception of humanity and of God himself, has without realizing it plunged God, Christ, and the Christian supernatural into

that it has to be acknowledged that God indeed dispenses preparatory grace and holiness. Against the second view, he stresses that it is necessary for the believing Christian to align with biblical revelation and the indisputable unanimity of the Church's tradition, which together hold that God's communication of grace in its preparatory stages is directly connected to Christ and his so-called "categorical" or explicit self-revelation.[84]

The latter, transcendentalist option has tended to diminish the doctrine of predestination, which its proponents associate with Jansenism. Bouyer suggests that both Jansenist and transcendentalist positions have construed predestination in too individualist a fashion. Following Saint Paul in his Letter to the Ephesians (1:3–14), Bouyer stresses that the predestination of the individual should not be our first affirmation but that of Christ and the Church associated with him, and only then, each person in the Church.[85] This, he argues, gives a unified vision of history and links divine as well as human freedom to the absolute decision of God in His Wisdom. However, Bouyer thinks that we must also give attention to God's special predestination of Mary. God in His Word and Wisdom is the source of all history as well as all the determinations of created liberties. Human freedom is entirely a gift whose source is God's conjoined thought and will. Mary, the Seat of Wisdom at the center of history, shows forth the glorification of human freedom in its perfect response to the divine design for creation. To be "full of grace" or fully responsive to the call of God, letting God do God's work in us, is to be fully human.[86] It is of eminent significance to the doctrine of predestination to recognize that God always willed Mary's fiat.

We have seen that Bouyer understands predestination in line with Aquinas's doctrine of "presentiality": all creation, past, present, and future, is seen by God in His eternal present in its actual existence. God's "foreknowledge" of our future free actions thus includes not so much the abstract possibility of our sin but its actuality as present to

---

the pure subjectivity of our human nature, naturalizing all these realities by the very fact of relieving us of our own proper human nature."

84 *MSW*, 106–7. Bouyer does not use the expression "categorical revelation." It comes from Karl Rahner and is juxtaposed to "transcendental revelation."

85 Ibid., 112–14.

86 Cf. *MSW*, 113.

His all-encompassing vision. It includes as well His eternally decreed and actualized response to our sin, summed up in the Paschal Mystery. God does not will our sin by the direct power of His sovereign will, but He does directly will His decisive response to it. This ultimately means that from the very beginning "there is no part of human history that is unconnected to Christ."[87] Grintchenko correctly sees that it is thus for Bouyer that there can be no opposition between nature and the supernatural.[88] The fullness of integral nature resides in the one who possesses the plenitude of grace in full connection to Christ as his Mother. All cosmic history was moved by the divine mind, according to God's eternal Wisdom, working in corrective conjunction with the first created spirits to prepare the way for her perfect reception of this grace, which was a preparatory grace joining her to the totality of the humanity that preceded her all the way back to its primordial origin in the First Eve.[89]

One of the major, crushing impacts of Nominalism on the whole of modern theology has to do with its manner of conceiving the relation of God's sovereign freedom to human freedom. Bouyer consistently and vehemently rejected the oppositional dialectic that Nominalism propagated in this regard.[90] In Nominalist currents of thought, God's creative causality was placed on the same plane of potency as that of creatures. God's causal influence on creation was understood to be competitive with the causality of finite beings, as if God were simply a highest being in a chain of beings in the same order of causal efficacy.[91] A notion of general, causal concurrence (*concursus generalis*) prevailed, according to which in a free human action God is causally present only in a part of it and certainly not in the decisive response of the creature on the level of choice. God and the human being split the work of free human action between them, in the way that two human beings do in human work in order to accomplish the goal of a common endeavor. One person does part of the work, and the other person does the remainder. For instance,

87 Ibid., 108.

88 Grintchenko, *Une Approche Théologique du Monde*, 264–65.

89 *MSW*, 123.

90 Cf. *Cosmos*, 187–88. We explored this, in part, in chapter 8.

91 See Aaron Riches, "Christology and *Duplex Hominis Beatitudo*: Re-sketching the Supernatural Again," in *International Journal of Systematic Theology* 14 (January 2012): 44–69.

in lifting a table, two human beings split the burden of the action. Neither does the whole of the lifting. In the line of authentic Thomist thought, to which Bouyer is wedded on this and many issues, the common work of God and the human being is not split apart in this way. In truly free human action, God does all the work. At the same time, He makes the work wholly our own. So, there is a sense in which we do all the work in free human action that is genuinely free, but one must be careful: God's causal efficacy is total and primary.[92]

This latter explanation of freedom gives us a properly synergistic understanding of divine causality. It entails that the more fully God is at work in us causing our freedom, the freer and more "self-possessed" we are. Bouyer emphasizes in *Mary: Seat of Wisdom* that in free human actions God's freedom is in a sense projected outside of Himself into that which could "not-be."[93] All good—and therefore truly free—human actions are "entirely from God, being the act of God before being our own."[94] Only evil actions, which are falsely autonomous and a violation of our freedom, can ever be said to be fully our own in the sense of having no other ultimate source than our own will detaching itself from the divine will. In these actions, we identify ourselves with the nothing, from which, in one sense, we were created. The re-creative grace of God is a second work *ad extra* by which God draws us not only from nothingness but from our self-willed identification with the nothing. He saves us from, as Bouyer says, "this self of nothingness that has, so to speak, been contra-created by our sin."[95] God in His Wisdom takes up our sinful condition of alienation by a work of transfiguring "condescendence" (*συγκατάβασις*) through which creation will be made resplendent with the glory of divine holiness shining from within.[96]

92 Cf. *MSW*, 109. We think this description, which we draw from Riches, helps us to understand a quotation found in *Mary: Seat of Wisdom*: "Whatever exists does so only inasmuch as it is contained within the pre-existing Wisdom. Nothing will ever exist except what will be given existence by the sovereign decree which expresses, or rather, constitutes, this Word . . . for it is by virtue of God's own freedom as affirmed by his Word, that the freedom of man has been, as it were, projected outside of God."

93 *MSW*, 109.

94 Ibid., 110.

95 Ibid.

96 Ibid., 111.

The gradualness of God's redemptive work is a sign of His faithfulness to His own good creation and the integrity of its natural being. Though Bouyer stresses the unity of the supernatural and the natural, he nevertheless takes care not to collapse the two orders of being onto a single level. He holds that God's intervention in history for our salvation works within the tension that exists in us as a result of the self-determining choice for the "nothing" that our first parents actualized.[97] God works with us from the starting point of our being as determined by the history of human decision-making that stems from this fateful first exercise of human freedom. God does not choose to wipe us clean without our own full consent to Him in our humanity as it really is in its actual historical constitution. In order for humanity's freedom to be healed and elevated as it is, the human being must experience the interior division that is a result of our sin and submit to it. The human being must be led from within the living flesh in his or her own existential actuality. Only God's condescendence enables this process to be one of healing and transformation of the first creation rather than one of outright replacement of the first creation by a second that has no living connection with it.[98] This is the process that leads to the final re-creation of the cosmos (and not its de-creation). Thus the divine Word, while not conforming himself to ours sins, does indeed accustom himself to our ignorance and alienation.[99] He works in our history in slow degrees so that it may really be our freedom, darkened by sin but educated gradually in the divine light, which is set free. In his book on Mary, just now expounded, Bouyer once more reveals his Irenaean (as well as Newmanian) cast of mind. In the slow process of salvation history, which is nevertheless not without decisive, kairotic events, he sees God making us ready to receive Him and making Himself ready to receive our own condition of alienated finitude, all the while being without sin.[100] In order to be fully embodied in our history, Bouyer insists, God must be fully received by us in freedom. Mary represents the pinnacle of this slow movement of history. She is also a decisive leap forward. The Word of God comes

97 Ibid., 111–12.
98 Ibid.
99 Ibid., 112.
100 Ibid., 112–14.

into her in order to rest in the womb of humanity. The fruit of divine Wisdom radiates from her with a glory that is unparalleled by what came before. In her fiat, which is a continuous response of her whole life, including the totality of her motherhood, and is not completed until her final self-surrender of her Son at the foot of the Cross, the preparatory movement of all cosmic history is completed.[101]

This process of preparation was cosmic in scope but was centered on the history of the People of God called out in Abraham and set apart to be uniquely bearers of the divine Word. This special history within history runs from the First Adam and Eve until the Second Eve is immaculately conceived to bring the New Adam to birth, inverting the order of human origins in the garden of paradise, when the First Eve was drawn from the rib of the First Adam. Mary is, Bouyer insists in explaining Catholic doctrine, the endpoint of God's "extensive and thorough work, infinitely delicate and patient, by which grace would regain possession of humanity."[102] In Mary's perfect act of freedom, the divine and the human, the supernatural and the natural, are united—"the two planes meet, that of human freedom, the plane of history, and that of grace, that is to say, of that meta-history which is predestination."[103] Mary is the decisive, personal center for God's accomplishment of His eternal plan concretely instantiated amidst the twists and turns of the history of the People whom God called out from the beginning to be His own:

> The whole of Israel's history is but the course taken, according to God's intention and will, by sinful man, the history of fallen humanity as culminating in its resurrection, as, at one and the same time, the history of its salvation. The Immaculate Conception is simply the final outcome of that election by which God, in the sovereign freedom of His grace, set apart a branch of the fallen race to be the bearer of His promises, the executor of His plans.[104]

101 Ibid., 173.
102 Ibid., 120.
103 Ibid., 121.
104 Ibid.

The distinction between divine personhood in Christ and Mary's human personhood as affirmed by the Christological councils in the early Church has, as we have seen, a significant meaning for Bouyer. He emphasizes that it is by virtue of this distinction that Mary can be said to be even more than Christ himself the personal intersection in which "the two mysteries of human freedom and pre-existence come together, since she receives the advances of grace."[105] Mary is able to receive this preparatory grace in the fullness of human faith, while Christ, whose human soul is ever in a condition of beatific vision and perfect union with the divine Word, is above faith.[106] Mary's faith represents the definitive "yes" of creation to God's redeeming action forever willed by Him. She is an exemplary manifestation of created Wisdom, its personalization on the individual plane within history. God projects His own freedom outside of Himself in her. Mary's motherhood is the personal perfection of creation in faith and holiness. Christ's humanity, enhypostatized in the divine Word, ever remains our own only because he received it in and through the womb of his virginal mother. Her perfect exercise of maternal freedom, through which creation is fulfilled in its own order, as Mother of the Creator, makes her Mother not only of the Church but of creation *in toto*. Mary is, as an expression drawn from Byzantine liturgy tells us, "more vast than the universe." She is, we would say, following in the strict train of Bouyer's thought although set in the language that we put forth in our introduction, the individual "macrocosm" who clears the path in her perfect creaturely holiness for the consummation of the cosmos in the final filling-in of the great world of the eschatological Church.

THE BRIDE OF THE LAMB

Toward the end of the final writing published in his lifetime, *Sophia*, Bouyer says that everything in this final book should be understood as a preparation for the exegesis of the double vision of John's Apocalypse that he carries out near the end: first of Revelation 12:1 of the woman clothed with the sun, with the moon beneath her feet, and crowned

105 Ibid., 123.

106 On this point, Bouyer remains in line with the thought of Aquinas: see *ST* III.7.3 and 34.4.

with twelve stars, and, second, of the vision in Revelation 21:10 of the New Jerusalem, the Bride of the Lamb that descends from heaven to earth with the Bridegroom at the end of days.[107] This statement should not be confined to his book on Wisdom but extends to the whole of his trilogies on dogmatic theology, especially given that *Sophia* is the final volume of these and of his life's work in its totality. *Sophia* is not the only volume in the trilogies where this double vision from John is the summary consideration. It is also the case in his book on the Church as well as chapter twenty-two of *Cosmos*.[108] So far in this chapter, we have confined our focus largely to the first of these apocalyptic visions, that of the celestial woman of Revelation in chapter twelve, in that Bouyer, like Newman before him, associates this figure with the *Theotokos*. The maternal implication of this image has been our concern. We shall now focus on the spousal dimension of the Church in Revelation 21.

The image of the Church as Spouse of the Lamb complements the image of the Church as Body of Christ. The spousal image of the Church in John's revelation can be taken, in the broad perspective of the analogy of faith, to connect with chapter five of Paul's Letter to the Ephesians. In the *Church of God*, Bouyer argues that the analogy of the Church as Bride, although it emphasizes the distinction of two persons in communion, complements the analogy of the Church as Body, which emphasizes that the Church constitutes one organism with Christ.[109] He explains in this text—much more so than he does in *Cosmos*—the harmony of the two analogies as they are bound together by Paul himself in Ephesians, where Saint Paul first speaks of the Church as the Body of which Christ the savior is head (5:22) before applying the mystery of the spousal union between husband and wife to that of Christ and his Church (5:32). The two, Bride and Bridegroom, "will become one flesh" (5:31) in a communion of love, each spouse loving the other as him or herself (5:33).

This nuptial analogy, Bouyer argues, brings to the fore the Pneumatological character of the Church, rooted in its Christological foundation. It leads to the recognition that our unity in Christ is a gift of the Spirit that consummates our personal existence in reciprocal distinction from

107 *Sophia*, 161.
108 *Cosmos*, 230–33.
109 *CG*, 528–29.

the One into whose life we are adopted in deification.[110] He suggests that the sophiological school of thought based on Soloviev's work had the merit of understanding "with incomparable force that the work of the Spirit is a work of unity, but of *essentially interpersonal unity*, in which—far from doing away with their distinctness—persons succeed only in being themselves."[111] There is a qualified sense in which the Church has to be understood as a personal being which develops, as Bouyer says, "over against Christ himself, and completing his humanity, which in him, is personalized only in the Divine Person of the Son."[112] Balthasar showed that on this matter Bouyer, although adding the figure of Wisdom to the story, follows in the line of thought of Cardinal Charles Journet (1891–1975) and Jacques Maritain (1882–1973), for whom God gives to the eschatological Church a "supernatural unity that is sufficiently real for God to seal it with the seal of *subsistentia*."[113] The eschatological Church, animated by the Holy Spirit, will maintain its personal being in relation to Christ. Bouyer speaks of it as analogous to the Trinity. The one personality of God subsists in the three persons. The "personality" (*la personnalité*) of the perfect Church, by comparison, will subsist in its individual members without them blending into one another in their perfect communion.

The Oratorian places this understanding in the context of the Wisdom and nuptial motif. For him, as we have seen more than once, the eschatological Church is the personification of God's eternal Wisdom, first realized in an embodied, created individual in history in the person of Mary. And, as we have seen, he insists that it will become in the perfected Church truly a communal person, "the ultimate, definitive person of humanity, even of the entire creation."[114] Bouyer suggests that it is not enough to say that the Church of the Last Day will be the appearance of a perfect society of separate persons, for the perfection of creation in the eschatological Church will be a created analogue of the society of the

110 Ibid., 529.
111 Ibid. The italics are Bouyer's own.
112 Ibid.
113 Hans Urs von Balthasar, *Theo-Drama III: Dramatis Personae: Persons in Christ*, trans. Graham Harrison (San Francisco: Ignatius Press, 1992), 346.
114 *Sophia*, 163.

persons of the Trinity, who constitute together a single personality. As Balthasar explains, Bouyer sees in this recreated perfection of creation a shared "supernatural superexistence" (an expression drawn directly from Bouyer's *The Church of God*) in which all souls united in the Church will constitute together "an image of God's three personal unity."[115] Balthasar explains that for Bouyer, in the Trinity itself, the divine persons together "pre-exist" in the person of the Father, are projected into the person of the Son, and are recapitulated in the Father by the Holy Spirit. In the consummation of creation, when redeemed and saved humanity is at last perfectly raised individually and altogether into a living likeness of the triune God, the created hierarchy will be perfectly united with the Trinitarian exchanges. Creation will be joined with the uncreated thearchy so as to constitute together a final personality, all without doing away with the distinction between individuals.[116] The redeemed and saved will join together as a single Bride of the Lamb, in a personal unity consummated only in and through relations by way of self-outpouring *ekstasis*. Bouyer holds that this new creation, united as a single person, the final personification of created Wisdom, was pre-contained within creation in the virginal, supernatural motherhood of Mary.[117] Mary, in her virginal *potentia*, is thus a kind of reverse image of the Father. The Church has its united origin in the human personhood of Mary and in some sense will be brought together anew, in a mature personalization, if we can put it thus, in the perfect reciprocity of consciousnesses in the eschatological completion of creation.[118]

In order to explicate further this understanding, Bouyer draws on Günther's view that the movement of salvation history under the incubation of the Holy Spirit is a "counterposition" of God's Trinitarian *circumincession* of interpersonal life.[119] Humanity in its development, according to this concept, does not directly represent the triune God in the perfection of its likeness to Him but bears an indirect likeness

115 Balthasar, *Theo-Drama III*, 346. The quoted expression is taken from *Church of God*, 530.

116 Ibid., 165.

117 Ibid.

118 Ibid.

119 Cf. *Cosmos*, 194.

to God, a living likeness that is a mirror image and therefore indeed a reverse image. Bouyer explicitly suggests that we can understand this by reference to the actualized motherhood of both Mary and the Church with respect to the Holy Spirit.[120] The perfection of maternity in Mary and the Church images in reverse or in "counterposition" the perfecting of divine being by the Holy Spirit in his eternal procession as well as in his perfecting of created being by his mission in salvation history.[121] Through his particular work in the act of creation and mission in salvation history, the Holy Spirit perfects us by dwelling within us at the most interior depth of our being. As a child grows in the womb of its mother, so the Holy Spirit is enveloped by our exteriority.[122] In the womb of the mother, the child grows in living communion with her but in dependence on her. The Holy Spirit, on the other hand, is not dependent on the one he indwells but is, instead, the principle of new creation in the human being. The Spirit does not nourish itself or grow from the being of the one who surrounds it, quite the reverse: the spiritual soul in whom the Spirit dwells is nourished with God's supernatural, personal life.[123] Moreover, the child comes to completion in the human mother only when she lets go of him or her, surrendering the child to its own eventual and rightful development. By way of counterposition, it is only in the full unity of the Holy Spirit perfectly united to us in our action and thinking that we attain to spiritual adulthood.[124]

If paternity is one with the divine essence and properly belongs only to God, maternity is the essential capacity of being in the order of creation. "The high point of motherhood," Bouyer explains in *The Church of God*, "is thus the summit of a creature's possibilities as a creature: to live by a life received."[125] Both divine paternity and creaturely maternity should be distinguished from the filiation to which we are called in Christ. Bouyer recognizes that filiation is a Trinitarian reality, fundamentally divine, but he holds that it is reflected in humanity on the natural plane, where

120 *Sophia*, 166–67.

121 We explored this manner of construing the divine procession of the Holy Spirit in chapter 8 of our study. See also *Cosmos*, 194.

122 *Sophia*, 166–67.

123 Ibid., 167.

124 Ibid.

125 *CG*, 589.

it can receive a supernatural meaning in relation to the Son of God.[126] Christ who is eternally the firstborn of creation becomes in Mary the firstborn of innumerable brothers and finally in his Resurrection the firstborn of the dead.[127] The filiation to which we are called in Christ is essentially one of fraternity or of multitude. Christ extends the singularity of his own filiation to all of those who are born anew in him by the gift of the Holy Spirit to ascend with him to the Father's celestial throne. This divine work in us cannot be understood without consideration of the mystery of divine spiration, the eternal breath of the Holy Spirit. Spiration, like paternity, is a reality of God's eternal life alone.[128] But God extends the divine filiation to us in our real adoption in the Son, through which we are assimilated to the eternal idea that He has for us individually and altogether in his divine Wisdom, only by giving us the very personal gift of communion that is the third person of the Trinity. The Spirit of God is, Bouyer says, "the grace to return with the Son, in the Son, to the Father, who becomes our own in all truth."[129]

All of the elect are brought together in the Son by the gift of the Spirit of the Father. In the final, eschatological in-gathering of creation, the divine Wisdom will be fully personalized as projected outside itself, with all of the elect coming to realize their divine idea in the eternal Wisdom generated within the Eternal Son.[130] In recognizing that Bouyer so stresses the presence of the divine plan for all things in his Wisdom, it is once again difficult to accuse him, as Garrigues does, of equating without distinction the divine ideas with created realities. In Christ, at the end of history, all of the redeemed and saved will have a common relation with the Eternal Son. "It is," Bouyer explains, "this very relation which defines the whole Church as the heart of cosmic Wisdom."[131] In the eschatological Church, we shall enter, in the unique person of the Spouse of the Lamb, into a perfect condition of relationship with the Trinity, such that our deepest human capacity as essentially relational

126 *Sophia*, 168.
127 Ibid., 169.
128 Ibid.
129 Ibid.
130 Cf. *Cosmos*, 231–32.
131 *Sophia*, 180.

beings is actualized.[132] We are called to become in our own existence what we were eternally in the thought and love of God. While requiring our cooperation on the model of Mary, this perfecting of relationality is ultimately wholly God's work in us. Wisdom descends from heaven to earth as the perfect Spouse of the Son. All the Trinity is recapitulated in its paternal source by the procession of the Holy Spirit. Personified Wisdom, counterposed to the Holy Spirit, is a recapitulation of our relational subsistence in personal unity.[133]

For now, the Church is only betrothed to Christ, as Revelation implies. It is *in via* or *in fieri*, still imperfectly one with the Incarnate Lord.[134] Bouyer insists that in the eschatological Church, Wisdom will become incarnate as Bride. In one respect, "the Church in her transcendent personality is presented to us as pre-existing the whole creation as its very principle."[135] In its historical emergence, it is born from the side of Christ, from the blood and water of the Cross. On the Last Day, it will descend from heaven from the side of the eternal God, establishing in full the communion of reciprocal, created consciousnesses united in the Son in the final Kingdom of charity.[136] The Son will then hand over the reign of his kingdom, whose content is the Church itself, to the Father. The eschatological Church in which God brings to fulfillment his own Body in the Son "will be," Bouyer says, "the last word uttered in history of this primary and unique thought of God about all things."[137] In the consummation of creation in the perfect Church, in the personalization of Wisdom, history will cease, and its inner reality and meaning made transparent by the radiant fire of divine glory.

132 Ibid.

133 Ibid.

134 *CG*, 528. Marie-David Weill points out that the Book of Revelation employs two different words, γυνὴ (fiancée) and νύμφην (bride) that Bouyer took up in his later writings. The Church in the present dispensation is betrothed to Christ. In its eschatological perfection, it will be wed to him. She points the reader to Rev. 19:7, 21:2, and 22:17, where we find the language of betrothal, and to Rev. 21:9, where we find the language of both betrothal and bride. See Weill, *L'Humanisme Eschatologique de Louis Bouyer*, 568.

135 *CG*, 530.

136 Ibid., 591–93.

137 Ibid., 592.

COSMIC WISDOM AND GLORY

In coming to the end of Bouyer's "Retrospections" in *Cosmos*, one has to remember the goal that Bouyer sets for theology and why he prefers "positive theology" over more speculative approaches that are grounded in *a priori* philosophical considerations. He thinks that theology should maintain a kind of rigorous, phenomenological openness to divine revelation in its ever-surpassing meaning, which our concepts can never fully grasp. This is an especially important consideration to keep in mind as we try to assess, as we shall do in our next and final chapter, the potential impact and relevance of his cosmology. Cosmology, whether philosophical or theological, has to give an account of the *logos* of physical being, of matter, of the body. Bouyer's positive theology, with its resolutely economic and soteriological focus, enables us to think about the physical cosmos in the density of its historical actuality in terms of ultimate meaning. He shows that the meaning of matter and the body is fully disclosed only in God's self-revelation in Christ, who comes to us in Mary and exists in the plenitude of his Body only in the Parousia.[138] The qualitative understanding of matter that Bouyer recovers, as discussed in our previous chapter, fits into this phenomenological, soteriological, and economic paradigm. From the standpoint of Christian faith, which a theologian must embrace in order truly to be a theologian, one can grasp the dynamic metaphysical structure of the cosmos as meaningful in light of the final, consummating truth that God definitively promises to give us in existential fullness in the Parousia of the Eternal Son, of which we have anticipatory signs even now, in Scripture and liturgy.

Although the light of Christian faith shows us that it is only with respect to the view of our absolute end that the meaning of our being can be discerned, one can certainly explore this issue on a broader footing. The philosopher can inquire as to whether we can know our absolute end but must surely admit frustration at the task, which would indicate a real achievement of philosophical insight. Plato understood that only true myth could disclose our ultimate origin as well as end and realized that he did not himself have access to the true myth. We require divine

138 Bouyer's cosmology thus develops an insight from *Gaudium et Spes* 22, from the Second Vatican Council: "The truth is that only in the mystery of the incarnate Word does the mystery of man take on light."

disclosure in order to have some real glimpse of our ultimate purpose, and the theologian stakes his or her claims on the veracity of the Eschatological Adam, the bearer of the true story of the cosmos and its end that he shall bring about himself in his final coming. Saint Paul's definitive summary in his famous cosmic hymn from the Letter to the Colossians is decisive for the theologian:

> He [Christ] is the image of the invisible God, the first-born of all creation; for in him all things were created, in heaven and on earth, visible and invisible, whether thrones or dominions or principalities or authorities—all things were created through him and for him. He is before all things, and in him all things hold together. He is the head of the body, the church; he is the beginning, the first-born from the dead, that in everything he might be preeminent. For in him all the fullness of God was pleased to dwell, and through him to reconcile to himself all things, whether on earth or in heaven, making peace by the blood of his cross.[139]

Bouyer, in his great fresco of salvation history in its cosmic proportions, cleaves to the basic pattern of this imagistic, narrative, mythic foundation of Christian understanding as communicated through the biblical Word received in sacred liturgy. He incorporates into this picture the bridal and maternal imagery from John's Revelation as well as from Paul. Bouyer tells a coherent story of the whole creation, drawing on vast swaths of the entire history of cosmology, attuned to modern questioning, and this is one of the great merits of his cosmology. He realizes that cogent narrative provides the ultimate basis for issuing intelligible propositions regarding human and cosmic meaning. Cosmology ultimately needs access to the true story of being if it is to be a discipline ordered by *logos*. It seems to us that Marie-David Weill is correct to say that the "perfectly unified, contemplative vision of the divine design over man, predestined from all eternity to recapitulate in his body of flesh all the cosmos, in order to lead it into God... [is] one of the most profound

139 Col. 1:15–20.

contributions of Fr. Bouyer to contemporary theological thought."[140] The ontological considerations that we explored in the previous chapter regarding matter and spirit must be seen in the light of this unified, ultimately theodramatic (not to be absolutely opposed to narrative) vision of the whole. Bouyer's qualitative concept of the physical universe fits with his sense of our need for historical intelligibility in cosmology. Modern cosmology seeks to understand matter, but its positivistic science is ultimately of no avail to grasp matter on the level of purpose or *telos*, because it cannot speak of the meaning of cosmic origins on the basis of the end for which we were foreordained. Ancient mythic intelligence operated from the intuition that the physical cosmos cannot be rendered intelligible without a sense of the true story of being, but the content of the myths was ultimately nihilistic, seeing the physical universe as fundamentally agonistic and without genuine hope for a non-destructive liberation. Balthasar is correct to suggest that Bouyer's demonstrations showing this are exemplary.[141]

In the Christian theodrama that Bouyer embraces, centered on the apocalypse of Wisdom of which Paul speaks in Ephesians, we learn that the cosmos is indeed ordered toward liberation in and through the glorified Body of the risen Christ. Both the Transfiguration of Christ on Mount Tabor (cf. Mt. 17:1–8) and the Resurrection of Christ from the dead provide the effective models for our own resurrection in glory and disclose the meaning and capacity of our bodies. A katalogical framework of thought alone unfolds the meaning of Christ's own mission to glorify the Father. Human persons cannot be associated with the consummation of Christ's Mystical Body without the resurrection of their own bodies, and, as Bouyer shows in *Sophia* as in *Cosmos*, the resurrection of the human body is inconceivable without the resurrection of the whole creation.[142] Our author gladly takes up the Pauline theme in Romans (8:19) of creation groaning for liberation in and through the resurrection of the elect. Any lesser, more constrained view of the cosmos, any falsely

140 Weill, *L'Humanisme Eschatologique de Louis Bouyer*, 565.

141 Cf. Balthasar, "Epilogue," *Woman in the Church*, 116–17. In our introduction, we noted that Balthasar issues a similar commendation of Bouyer's work in this regard in his introduction to *MT*.

142 *Sophia*, 182.

Gnostic deviation from the Pauline view, leaves us trapped in a cosmic surd. Without a sense of the fulfillment of cosmic Wisdom in the glory of the Incarnate Christ, the meaning of the body's origin is unintelligible, at best a great cosmic accident or falling away from plenitude. We must turn to the Eschatological Adam and his Spouse to set the framework for any metaphysics of matter that truly deals with its *logos* as ordered to an intelligible end. Eschatology provides the ultimate hermeneutic framework for a Christian interpretation of modern science. Drawing many threads of biblical revelation together, Bouyer teaches that the eschatological Bride of Christ, cosmic Wisdom personified, will radiate in the fullness of splendor the glory of the triune God, establishing the cosmos at last as fully cosmos.

As we saw in chapter one, Bouyer ends *Cosmos* in its original, French edition with a "Postlude" that speaks of his visit to Hagia Sophia.[143] The effect of this encounter on him can hardly be downplayed. He saw in this Byzantine basilica a ritual space that evokes the *Shekinah* of light and life, the glorious divine presence that led the People of God through the desert as a pillar of cloud and fire, coming to rest in the flesh of the Incarnate Lord of Israel, Christ himself.[144] The aesthetic bearing of the symbolism of this cathedral, more than any other church in Bouyer's considerable experience, truly evoked for him the celestial Jerusalem. Truly, he thought, in this space of supernatural ritual, of Christian liturgy, heaven pierced and invaded the earth. Under the fiery dome of Hagia Sophia, God not only encountered the earth but animated it, fecundated it, making it more profoundly what it is in drawing it to Himself in the espousal of Wisdom. The basilica was a sign or pledge of the glory that will fill the whole creation at the day of its total renewal in the Spirit of glory. The city of pure gold of the Johannine apocalypse, transparent as crystal, will radiate in utter transparence with the very glory of the Creator as it sounds out in praise to His triune *agape*. The Hagia Sophia was in Bouyer's estimation Christendom's foremost architectural sign that the cosmos itself will one day be a perfect Eucharist of praise without end to the Creator and Redeemer.

143 *Cosmos* (Fr.), 377–79. We have already touched on this postlude in chapter 1 of our study.

144 Cf. *Gnôsis*, 45–51; *The Invisible Father*, 140–42.

In this final nuptial union God will be "all in all."[145] This is a Pauline expression that Bouyer frequently invokes. The subtitle of *Sophia*, Bouyer's final book, is "the world in God" (*le Monde en Dieu*). We have already discussed this, but we should clarify more than we have done that for Bouyer this indwelling of creation in the Creator is ultimately in some sense a promise for the future, an act of consummation in nuptial relationship that will give final consecration to the Creator-creature distinction by the perfect unity of the latter's consummate personalization in communion with God. We have always been in God, but God is not yet "all in all": "God besets the world on all sides, so to speak, to be finally all in all—just as everything, from all time, exists for him and existed only in him."[146] In the eschatological Church, the Bride of the Lamb, God will perfect the glory of creation not only by restoring it to its virginal beauty but by elevating it, completing our new creation in Christ (cf. 2 Cor. 5:17). This beauty will be the shining of divine glory over all of our flesh. The icon of God will be perfectly imprinted on us in the perfection of our adoption into Christ at the Parousia, in the descent of the celestial Jerusalem into our world, when the glory of the Spirit will radiate from every pore of our being.[147]

The eschatological Bride will be humanity living in truth and charity, bearing the marks of its historical existence glorified in the Spirit. Bouyer teaches that the eschatological spouse will be permeated with the glory of the Immolated Lamb and share in his own glorified wounds.[148] He insists that the glorification of the saints called to constitute together the Spouse of the Lamb in marital union can be realized and effected only by their full share in mortal experience. Only the paschal door of the Immolated Lamb gives the elect access to the vivifying divine light that perfects their material bodies.

There is no denying the beauty and poetic splendor of Bouyer's way of describing the condition of eschatological glory at the end of *Cosmos* and elsewhere in his writings. His eschatological proclamation in doxological

145 Col. 3:11.

146 *Cosmos*, 231.

147 Ibid., 232–33.

148 *Sophia*, 175. Bouyer draws this teaching especially from Saint Bernard of Clairvaux.

key provides a fitting denouement to the story that he realizes must be told in its totality in our day in the framework of cosmologies that challenge the Christian story on a fundamental level but are nevertheless ineffective in giving us a cosmos in which to live with meaning and purpose. He presents in an exemplary way a theology of Wisdom that enables us to think of *theosis* or human and cosmic glorification in the *lumen gloriae* of divine grace as spousal communion in the interchange of persons, uncreated and created. The deifying light of divine glory, communicated through the energies of God, will be personalized Sophia or created Wisdom actualized at the end of history. Eastern and Western strands of tradition come together in the Oratorian's view.[149] Exemplary and final causality, missing from scientific accounts of the world, are simultaneously affirmed in the context of intelligible narrative, in a dramatic key, and rigorous doxology.[150]

Will every human being share in this condition of nuptial glory, wherein human capacity is actualized by the deifying light of God in the virginal fecundity of the eschatological Church? Bouyer raises this particular question at the end of *Cosmos*, and, significantly, he does not follow Balthasar's proposal that we should hope for the salvation of all or for *apokatastasis* (ἀποκατάστασις).[151] Certainly, there is no indication that he believed in a *massa damnata* or that the great majority of human beings will be damned. His frequent lambasting in his wider theological writings of strands of Christian thought and practice that

149 Cf. Caldecott and Walker, "The Light of Glory: From *Theosis* to Sophiology," 252–264.

150 Cf. Weill, *L'Humanisme Eschatologique de Louis Bouyer*, 584–86. We cannot agree with Weill's position that Bouyer privileges an Eastern view of exemplary causality over the Thomist concern for final causality and thus is led to an aesthetic view that tends to evacuate from human existence the incarnational plenitude of its existence in time. The sophiological position that Bouyer embraces unites the divine exemplars so closely to the *telos* of human existence as willed by God in its existential actuality that it cannot truly be said that he does not give proper place to final causality. Eschatology orders protology in the Bouyerian view. The exemplary ideas for creation in God are so intimately connected to our historical existence that he understands them in living and dynamic terms centered on our respective, concrete mission in relation to Christ and his Spirit.

151 *Sophia*, 176.

would turn the Creator and Redeemer into a vengeful tyrant indicates his aversion to such rigorism.[152] Yet he does indeed raise the question near the end of *Cosmos* of what will happen to the resurrected bodies of those who suffer a "second death" (Rev. 21:8). He responds that they will have gotten what they wanted: the damned will be "trapped in time . . . precisely in time leading nowhere, since time no longer exists for all those who have accepted God's love and have entered eternity."[153] The elect will exist in the *aevum*, the share in eternal life that creation is given.[154] Those who are left outside of this share will have only themselves to blame. They will have chosen non-being over being, but such a choice can never lead to its desired end, the annihilation of being. No creature can truly accomplish this anti-goal, however much he or she may move toward it. These unfortunate souls, joined to the fallen angels, will have embraced a meaningless condition of spatio-temporal movement without purpose.[155] The second death is time without *telos*. Surely we have some inkling of this in our modern condition, when cosmology has stripped the cosmos of *telos*, in fact doing away with the cosmos and leaving us with a mere shell of a universe, consigning many of us to live in a hellish condition of temporality without apparent meaning or by purposes that we illusorily construct on the basis of the fragmentary desires that constitute our fallen subjectivity. Bouyer does not say so explicitly at the end of *Cosmos*, but he recognized explicitly in other writings that the end of history will be achieved only by the divine establishment of "new heavens and a new earth where justice dwells" (2 Pt. 3:13): "Christianity . . . asserts that human history must end in a catastrophe, that it will be interrupted by the supremely miraculous event, the return of Christ, and the universal judgment and resurrection."[156] In the meantime, the Antichrist will have come onto the scene: "It seems that this person [the Antichrist], who incarnates

152 Cf. *ES*, 346–48. These pages pertain to a criticism of Saint Anselm. Yet, as several of Bouyer's commentators have noted, they in fact have to do more with a subsequent tradition that hardens the thought of the Archbishop of Canterbury than with Anselm's own thought.

153 *Cosmos*, 232.

154 Ibid.

155 Ibid., 232.

156 *Christian Initiation*, 104.

in himself the power of Satan, is going to appear at the end of time to lead an ultimate battle against Christ and the Church."[157]

Bouyer does not see the end without connection to the battle between the Antichrist and the Son of Man, the eschatological judge. The resurrection is not achieved as the simple fulfillment of the immanent processes of nature but through a dramatic spiritual event, a "catastrophe" even. However, cosmic battle and dark or hellish evocations are not the final emphasis that he maintains in *Cosmos*. The ultimate consideration of the final main chapter of the book regards the glorified condition of the elect individually and gathered all together in the eschatological Church. Bouyer is, in his general writings, a "resurrection realist."[158] He holds that matter will have a full share in eternal life, but he maintains this position without subscribing to millenarian literalism. He does not seem ever to have embraced the thesis popular among Catholic theologians in the latter half of the twentieth century that there is an immediate resurrection in death, and that the material substratum or outer shell of the universe will not be saved. Bouyer does not explicitly say so, but his thinking on the general resurrection seems to be in accord with that of the most optimistic theologians in the Christian tradition who held that the final paradise will include the renewal of at least some non-human cosmic life.[159] This outlook would be in accord with his postulation of the non-necessary origin of animal suffering.

157 *Dictionary of Theology*, 31. This is under the heading "Antichrist." Bouyer recognizes that the theological Tradition is divided on whether the Antichrist is a particular eschatological person or instead figure realized in multiple partial manifestations throughout history.

158 His position bears comparison with that of Pope Benedict XVI; cf. Ratzinger, *Eschatology*, 108. Ratzinger opposes those who hold that matter cannot be perfected. See also Patrick Fletcher, *Resurrection Realism: Ratzinger the Augustinian* (Eugene, OR: Cascade Books, 2014), 181–246.

159 Cf. Paul O'Callaghan, *Christ Our Hope: An Introduction to Eschatology* (Washington, DC: The Catholic University of America Press, 2011), 119–20. O'Callaghan points out that Saint Jerome thought that animal and vegetative life would continue in heaven and that C.S. Lewis thought there would be especially domestic animals (dogs and cats) in heaven. See also Blowers, *Maximus the Confessor*, 253, n. 135. Blowers says: "The eschatological transformation of even non-human creatures, particularly as the outcome of the *Logos*'s immanence or incarnation in the diverse *logoi* of beings, is suggested in many texts. . . ." He then provides a list of relevant texts.

Bouyer explicitly holds that there will be, on the model of the resurrected Christ, continuity between our present individual being in its deepest unity and our transfigured being in the general resurrection. He speaks of a "spiritualization of the body," but this will be, he explains, for the sake of the perfection of the association of the body with the soul that is its substantial form united to God.[160] He does not seem to mean by this that matter will become totally indistinguishable from spirit as a limit that makes individuation possible. Yet for those who question whether matter in itself will be saved, it seems that one must keep in mind Bouyer's explanation that really existing matter is never solely "in itself": it is always in-formed. It is *potentia* for in-formation. It is thus not to be radically opposed to form or to mind. In order for matter to be glorified, the forms to which it is joined must be so as well. Bouyer in fact seeks a middle ground between false materialization of the resurrection body and false spiritualization of it.[161] He takes two statements from Saint Paul to set the boundaries for any discussion of this topic that would not veer into "hazardous speculation" in either of these two extreme directions: (1) "Flesh and blood . . . cannot inherit the Kingdom of God" (1 Cor. 15:50); (2) "The Spirit . . . will give life to your mortal bodies" (Rom. 8:11).[162] He notes that the exemplary miraculous appearances of the risen Jesus show both continuity and discontinuity of his risen body with the eviscerated body that was laid in the now-empty tomb.[163] In *Cosmos*, Bouyer describes the final condition of the resurrected cosmos in terms that are evocative of Teilhard's expression "living host": "The transfigured universe, all around the resurrected bodies, is swathed in a kind of rainbow in which the indivisible glory of God shines forth, and in which the sparkles are as numerous as the elect. There is now but a single harp and its strings hum in the breath of the

160 *Dictionary of Theology*, 388–90, s.v. "Resurrection"; the quotation is on p. 388. Bouyer's Thomist hylomorphism is consistent. He can hardly be credibly accused of Origenist spiritualization of the general resurrection.

161 Ibid., 389. Bouyer does not address the question that arises for Thomist hylomorphism of how to maintain the identity of the body before death with the risen body, given that at death a substantial change occurs, and the body takes on new forms. It does not seem to have been a pressing issue to his mind.

162 Ibid.

163 Ibid.

Spirit."[164] Note that Bouyer speaks of "bodies" in the plural here, which indicates that he thinks the perfection of supernatural society and the cosmos will be simultaneous with the individual perfection of the elect. There will be, it seems, both individuation of resurrection bodies and reciprocal interchange of resurrected bodies in the unified glory of God shining through the new creation. For Bouyer, the general resurrection is the endpoint of Jesus Christ's recapitulation of creation in the human nature he assumed. The whole of creation is fulfilled through man, but only in the God made man.[165] The cosmos is in some sense a process fulfilled in human nature summed up in Christ. Physical creation exhibits increasing complexification, and consciousness exists in a new way with matter in human personal existence. However, it is only on the basis of an ontological mutation that this new union of mind and matter is achieved, as we have seen. The final "leap" to the perfection of this union is the ultimate transfiguring event of grace in the Parousia.

The focus in *Cosmos* at the end, as throughout the two trilogies that it sums up, is the nuptial completion of creation in the eschatological Church, which the Oratorian clearly awaited with great hope, urging us to pray in the spirit of the *Maranatha*—"Come, Lord Jesus"—that concludes John's Apocalypse (22:20). Fittingly, the last words in the final main chapter of *Cosmos* are a prayer, a vesperal hymn, which we feel inspired to reproduce here:

> Joyous light of the eternal Father's hallowed glory:
> heavenly, holy, and blessed Jesus Christ,
> reaching the sunset hour,
> as we gaze at the evening light

164 *Cosmos*, 233.

165 In this, there may be a needed "Eastern" supplementation to Western, Augustinian theology in Bouyer's cosmology. The Augustinian tradition may indeed separate a "matter in itself" from the vocation of humanity. If this is true, it seems to us that the Eastern tradition that Bouyer recovers is to be preferred on this point, because it shows the unified meaning of creation in a clearer light. Cf. Joseph Fletcher, *Resurrection Realism*, 216–17. Fletcher wishes to show that Ratzinger ultimately has rejected the view that creation is fulfilled in man in favor of a view that the salvation of the world is an *accidental* part of the final joy of the elect.

we praise you, O God, Father, Son, and Holy Spirit.
You are indeed worthy of the holy voices' everlasting song,
O Son of God, giver of life,
wherefore does the whole world
ceaselessly extol and magnify Your name.[166]

166 Ibid. Grintchenko observes that Bouyer reverses stanzas two and three of this hymn, found in the breviary (Grintchenko points to the French *Liturgie des Heures*, 1:672), in order to give the final word to creation chanting the glory of God.

CHAPTER 12

The Return of the King

THE PRESENT STUDY WAS, FOLLOWING THE order of Louis Bouyer's *Cosmos* itself, divided into two main parts. It is important, as a final step in our exposition and analysis, to provide our own retrospective overview showing further the link between the two parts of *Cosmos* as well as of our study and make some final suggestions in regard to the uniqueness and importance of our author's cosmovision, particularly spelling out further its metaphysical dimension. The very existence of this dimension is, as we shall see, in dispute. In the first part of the study, we expounded and contextualized the larger, genealogical portion of *Cosmos*, where Bouyer gives a history of cosmology from the standpoint of Christian faith. The focus in this part was on the inescapable importance of mythopoetic thinking and creative imagination as foundational for all human intellectual pursuits as well as on the transfiguration of mythic images and symbols in and through biblical revelation. We began by exploring our author's connections to Cardinal Newman, to the Inklings, and to Sergei Bulgakov. All of these great modern theologians and writers thought that the response of faith to the cosmological question as posed in the modern age in the tumultuous wake of Galilean science requires attentiveness to the power of imagination in shaping our understanding of reality. Imagination draws together in a sacramental way the material and the spiritual planes of being and gives us a higher sense of the unity of creation. Rationalist theological approaches were castigated by our author for their propensity to fragment the unity of God in His inner life and His economy. The resources for renewal in theology that can bring about a new, truly Christian humanist vision of the cosmos and a recovery of the cosmos as such were found to be present in the more hidden, poetic strands or trends of modern thought that some of the most ecclesially-influential Catholic theologians largely ignored or attempted to repudiate in the nineteenth and twentieth centuries. The most important of these, the sophiology exemplified in the brilliant writings of the Russians Soloviev, Florensky, and Bulgakov, was critically embraced by Bouyer. This is surely one of the unique and abiding features of his thought. He is a pioneer in twentieth-century Western theology simply for having taken this current of Eastern Christian thought and that of its forebears in the West as seriously as he did and developing it. In the second part of our study, we detailed the more metaphysical, speculative, "Retrospections" section of *Cosmos*. In order to draw out much that is entailed in the brief and dense pages in this part of the book, we set these reflections in the context of the Oratorian's larger Trinitarian, Christological, Marian, and ecclesiological writings. We saw that these retrospections are communicated in a narrative form that mimics the theological style of certain Church Fathers and of Scripture itself. This approach may in fact serve to hide the speculative impact of our author's reflections, although we suggested that his work is indeed full of impact on this level.

In order to understand the organic connection between the two main parts of *Cosmos*, it is important to return again to the theme that explicitly guides the first part of the book: the transfiguration of the royal myth by biblical revelation summed up in the Paschal Mystery of Christ. In this concluding chapter, we shall first endeavor to set the whole of *Cosmos* in the light of this theme, as we were not able to do in directly expounding the chapters of the text. We shall explore in a first section, in summary fashion, with regard both to *Cosmos* and to Bouyer's wider writings, the theological and cosmological importance for him of the symbol and dramatic image of the suffering love of the thaumaturgical and wounded King of creation, which is connected in Bouyer's thought to a theology of the eternal Wisdom of the transcendent and sovereign God. In the remainder of the chapter, beginning in a second section, we shall respond to an important critique of Bouyer, regarding whether his work is sufficiently grounded in a metaphysical anthropology. The second section will be an endeavor to explicate an implicit connection of the Oratorian's theology of creation to wider currents of modern thought regarding imagination and metaphysics, which relate to the sophiological theme of created Wisdom. In a third section, paradoxically related accents in Bouyer's cosmology will be highlighted in connection first with the metaxological metaphysics discussed by the philosopher William Desmond (b. 1951) in relation to Lev Shestov (1866–1938) as well as Soloviev and then with the dual perspectives of analogy and anticipation in metaphysical exploration that contemporary theological cosmology has brought to the fore. The concluding fourth section contains a final discussion of Bouyer's anthropology of human vocation in the cosmos as a form of royal activity summed up in liturgy. We shall suggest in the end that the metaphysical dimension in Louis Bouyer's book on cosmology, as in all his work, serves a larger purpose to inspire the reader to live liturgically, in rigorous, hopeful expectation, enlivened by charity, for the victorious return of the Risen King, who is our Creator and Redeemer. Bouyer's theological cosmology unites analogy with anticipation and elucidates the human vocation as reconciler of the cosmos by the virtue of religion when humanity is perfected on the eschatological site of the Church's Eucharistic celebration. The Church assists the sovereign Creator and Redeemer in preparing the way through its own royal activity for the glorious King's Parousia and the elevated renewal of cosmic liturgy.

SYMBOLS AND IMAGES OF THE SUFFERING KING

As we look back from our own retrospective standpoint on all that has been written in the preceding pages of this book, it seems that we need to return anew to the central image that is the center around which the overall thought as well as the cosmology of Bouyer revolves: the Immolated Lamb who, as the principle of the world, is its representative King, the one who is destined to bring all of the redeemed and saved world into the royal palace of the Father. Bouyer follows the train of thought of Saint Paul, Saint Irenaeus, and Saint Maximus the Confessor in holding that the Mystery of Jesus Christ is the center of divine Wisdom, revealing in person the eternal purpose or meaning of creation. He understands conceptually the relationship of God's eternal freedom in relation to our finite freedom in a way that aligns with Aquinas's doctrine of "presentiality." He envisions God's eternal knowledge of creation as inclusive of the whole of creation in its concrete, historical actuality, past, present, and future. This knowledge includes as its central object God's decisive response in the Incarnation and Paschal Mystery of Christ to the disobedience of finite freedoms to His will. The sacrifice of Christ was thus woven into the garment of creation from the very beginning. Eschatology is the ordering factor for protology in Bouyer's sapiential or sophianic cosmology.

This theological cosmology is presented, as we have seen, in a form that is at once narrative and dramatic. Bouyer eschews compartmentalization and orders theological understanding around the symbol and image with which conceptual reason must always work in unison. The French Oratorian does not view the biblical or sacramental symbol or image to be a mere stepping-stone to the development of the precise theological concept. He gives marked preference to certain biblical images and stories, first of all to the Immolated Lamb, the expiatory King, but in a related way to Sophia, the Second Adam, the Son of Man, the Suffering Servant, the Woman Crowned with the Sun, and the Celestial Jerusalem. He develops his concepts and logical analyses with close attentiveness to the meaning of these images and symbols, which reoccur (with others) across his writings. His theology as a whole is rabbinical in form, in that it is a kind of extensive scriptural commentary, albeit shaped by the liturgical context of Christian experience. He treats the scriptural images as irreducible to concepts, and he interprets them in their proper sacramental and liturgical *Sitz im Leben*.

None of this is to say that our author absolutely denigrates the concept. Grintchenko has summarized well the balance that Bouyer maintains in this regard:

> The symbolic character [of Bouyer's thought] disconcerts our intelligence, less exercised to see resources in the symbolic universe. In basing himself without complex on the evocative force of biblical images, Louis Bouyer deploys their rationality without emptying them and without underestimating the power of the concept essential to logical thought that the image can actualize and refine.[1]

Images and concepts are mutually related and enriching necessities of intellectual advancement. The concept itself bears a symbolic function for our author, in that he holds that human knowledge does not terminate in the concept but in the realities themselves that are known in and through

1 Grintchenko, "*Cosmos*: Une Vision Liturgique du Monde," *La Théologie de Louis Bouyer: Du Mystère à la Sagesse*, 121.

concepts.[2] He clearly understands that there is a sense in which concepts go beyond images, enabling us to clarify and give precision to what it means to speak, for instance, of God as Father, Son, and Holy Spirit (the Trinity), or to grasp what it means to speak of the "eternal humanity" of the Word of God, or to explore the meaning of matter and spirit in relation to the human person. At the same time, he realizes, in line with Henri de Lubac, that biblical, liturgical, and iconic images and symbols contain "wheels within wheels" of meaning that can never be exhausted by any and all possible sets of concepts.[3] Andrew Louth's description of Saint Maximus the Confessor's theological style fits that of Bouyer: "characteristic of Maximus' approach to theology is his tendency to work out his ideas in relation to specific images, or icons."[4] This could be said for Bouyer, but with the proviso that he balances recovery of Patristic spiritual interpretation of Scripture and liturgical images with an attempt to take into account some of the concerns of modern historical scholarship to which he was introduced as a young Protestant ministry student in Paris and Strasbourg. We hope that our chapters dedicated to the genealogical portion of *Cosmos* gave some indication of this concern. Bouyer in fact holds that this modern turn to history does not have to be considered antithetical to rabbinical and Patristic approaches to Scripture, because these approaches often expressed an abiding desire to stay grounded in the literal meaning of Scripture or the biblical text when taken in "its narrowest and most immediate sense."[5]

2 *Cosmos*, 236, n. 7. Bouyer says that according to authentic Aristotelian-Thomist epistemology "we have no knowledge of concepts as such, but only of objects in or precisely through concepts."

3 The expression "wheel within wheel" is taken from Ezekiel 1:16 and the figure of the *Merkabah* beloved by Bouyer. See Henri de Lubac, "The Problem of the Development of Dogma," in *Theology in History*, trans. A.E. Nash (San Francisco: Ignatius Press, 1996), 348–80. We thank Fr. Guy Mansini for drawing this imagery in this context to our attention.

4 See Jill Raitt, ed., *Wisdom of the Byzantine Church: Evagrius of Pontus and Maximus the Confessor, Four Lectures by Andrew Louth* (Columbia, 1997), 21. See also Andrew Louth, *Maximus the Confessor* (London: Routledge, 1996), 94; C.A. Tsakiridou, *Icons in Time, Persons in Eternity: Orthodox Theology and the Aesthetics of the Image* (Burlington, VT: Ashgate, 2013), 167–92, 175. The passages in these books explore the Confessor's theological style.

5 Cf. *Rite and Man*, 202. In other words, modern scientific study of Scripture, an outcome of the development of so-called "historical consciousness" in its modern form, can help to discover the literal meaning of scriptural texts.

Bouyer's cosmology is well integrated. It always moves from and toward the attainment of unity of vision. His overarching theology is inherently sacramental and liturgical all the way through. The liturgical and ritual dimension of his thought penetrates down to the philosophical level, in which is found, as we suggested with Michaël Devaux in our first chapter, a "quest for wisdom by the imagination."[6] The Pauline Mystery, the key of divine Wisdom and the root of our very language of *sacramentum*, is at the center of his thought even on the level of fundamental theology. He orders all of the elements of Christian truth around the Mystery of the Cross and the Resurrection of the Incarnate Lord, who, as risen, bears the wounds of his triumphant suffering and death. In order to reawaken Christian reflection to the unity of the mystery of faith, Bouyer engages modern thought not on the grounds of rationalist conceptualism as much as on the terrain of imagination.

The Oratorian envisages the possibility of developing a theological hermeneutic of the cosmos that engages modern science in light of the epistemological advance in twentieth-century philosophy in which it was recognized that all human thought is necessarily symbolic.[7] The turn to myth and the transfiguration of mythic symbols with their attendant cosmological meanings is not antithetical to dialogue with modern science, inasmuch as this dialogue requires recognition of both the necessity and differentiated function of symbolic discourse across the fundamental pursuits of human culture: religious, poetic, and scientific.[8] Humanity, as the unity of spirit and matter, necessarily makes use of the symbol in order to understand its place in the world. This is first and most basically accomplished in mythic narratives or stories. Bouyer, great scholar of the legend of the Holy Grail and of Sacred Scripture that he was, gave priority to the royal myths pointing to the coming of the Kingdom of God in Christ with its conjoined anthropological and cosmological significance.[9]

6 Devaux, "'Le Problème de l'Imagination et de la Foi' chez Louis Bouyer," 159.

7 *MT*, 256, 108–9. Bouyer, as we have seen, takes Whitehead's philosophy of the symbol to be especially decisive on this point.

8 Ibid.

9 *MT*, 255–63. In these pages, Georges Daix interestingly summarizes the whole of Bouyer's theological work with reference to his historical and creative writings on Arthurian Legend, *Le Lieux Magiques du Graal* and *Prélude à*

It has become common in recent years in moral philosophy in the English-speaking world to stress the irreducible importance of narrative or of the "narrative imagination" in shaping human moral endeavor. Narrative, it is increasingly understood, is necessary in order to explain human action and to grasp the identity of human life. Alasdair MacIntyre, whose work has been especially significant in this regard, says that "the unity of a human life is the unity of a narrative quest."[10] The movement of self-conscious beings is made possible and governed by a *telos* that unites the past, present, and future of the subject in its movement in time. The unity of a human life is instantiated in story, which exists in a relation of "mutual presupposition" to a person's metaphysical identity.[11] The philosophy of human action requires attentiveness to narrative meaning in order to be rendered coherent. It is not moral philosophy alone that benefits from recognition of the significance of narrative unity for metaphysical understanding. Douglas Hedley, following the Inklings, points out that not only the human self but the whole world is situated by story or narrative.[12] C. S. Lewis in particular, he notes, showed how important story is in conveying abstract principles with cosmic significance. Our abstract concepts are ultimately situated in the context of our living embodiment and inherent connection to the drama of the whole world, which, in turn, centers on our human drama with all of its imagistic valences. Cosmic meaning is intertwined with human moral meaning in the moral imagination. Hedley explains that only by way of the imagination can "the antagonisms and tensions of immediate, lived experience of weal and woe and the 'higher law' of universal and persisting ideals of justice and goodness be reconciled and attuned."[13] Narrative and drama do not have to be taken to be radically opposed to one another. Some have argued that theology should be dramatic rather than narrative in form, because the latter tends to fall into the distortion of epic objectivism. From this narrative vantage point, the human subject takes himself to be the observer

---

l'Apocalypse.

10 Alasdair MacIntyre, *After Virtue* (London: Duckworth, 1981), 203. Quoted by Douglas Hedley, *The Iconic Imagination*, 92.

11 Hedley, *The Iconic Imagination*, 92.

12 Ibid., 94.

13 Ibid.

of purely objective events in natural history that he presumes unfold in necessary sequence. Narrative omniscience cancels out the irreducible freedom of singular human actions and particular experiences. However, narrative cannot be completely tossed out, as even one proponent of a dramatic construal of history has maintained: "I do not believe that we can avoid at least some narrative description, but such description needs to be self-aware inasmuch as it recognizes the tendency for it to be taken over by the genre of the epic."[14] The narrative of creation unfolds not only through inevitable processes but through events of grace.

It is surely a task of great importance for theologians engaged in cosmological questioning to explore the history of creation in theological terms, and this cannot be done without explicit engagement of imagination or the mythopoetic mind with its use of symbols and images as well as its sense of the coherent meaning intrinsic to the advance of history. The effective accomplishment of this task involves bringing to light the communion of *anthropos* with cosmos. Our explicit or implicit understanding of the former's horizon of perfection shapes our understanding of the causal forces that drive the latter's total history. Bouyer's discussion of "importance" in chapter two of *Cosmos* entails this point.[15] What symbols do we mark out as central for ordering our understanding of the world and its story as well as the concrete pursuits in the stories of our daily lives? What images, tied to our own sense of self-development, govern our sense of the meaning of the world—if there is one—and its historical becoming?

In *The Idea of a University* (1852), Cardinal Newman argued that natural theology, which he understood to include consideration of the relationship of nature to culture, is the architectonic or ordering science, and that its removal from the universities would create a sapiential void that other, lower branches of science would try to fill, even though they would be methodologically inadequate to the task. This is evident in our own day in the emergence of evolutionary psychology and attempts to explain human history and the human person in light of the ideal of quantitative description, as if the human and cosmic drama can be understood by means of algorithms alone. Biology and physics have become,

14 Celia Deane-Drummond, *Christ and Evolution: Wonder and Wisdom*, 51.
15 *Cosmos*, 16.

to many minds, despite the significant resistance that we indicated at the very beginning of our introduction, the architectonic sciences to which all of the other disciplines in the university must be compelled to order or contextualize their understanding. The infra-personal domain of objective being is taken to be basic and the sole cause of all that exists. Yet even in this situation the irreducible importance of narrative has emerged. Evolutionary psychology, a particularly ambitious and expansive colonizer of intellectual disciplines, has moved beyond its native disciplinary domain of biology in recognition that the physical cosmos is inherently developmental and therefore "narrative to the core."[16] Because physical nature is ever-developing and is not yet complete, who else but the scientists who study the physical evolution of its past are proper experts to tell us about the real nature of cosmic reality and all that it contains, including the human mind? Who better than the evolutionary psychologist, casting aside the mythological worldview of the old-fashioned folk psychology, to inform us about the nature and development of human personality as well as its future potential? We might well notice, as Bouyer did, that in this scientific understanding the instincts of mythopoetic consciousness and imagination are in fact surreptitiously consecrated. In order to understand physical nature itself we need a story that embraces the whole in its past, present, and future, albeit without seeing it as a *necessary* movement of evolutionary advance into which specifically human history is absorbed. Yet the evolutionary psychologists have narrated the whole for us in precisely this reductionist manner, consolidating our scientifically-naturalist, cosmic imaginary, applying to themselves the vocation of poets of cosmogonic myth in the form of the evolutionary epic: narrative does indeed become epic in their hands. On all strata of public education throughout the world in our globalized era we increasingly turn for wisdom to these evolutionary cosmogonists, these at-times scientist-poets who are unafraid to marvel in secular praise at the power of material processes to produce the illusion of consciousness and all other attributions that we used to classify erroneously as distinctly "personal."[17]

16 Haught, *The New Cosmic Story*, 7.

17 E. O. Wilson clarified the agenda in this regard: "I consider the scientific ethos superior to religion. . . . The core of scientific materialism is the

Richard Dawkins's story of the "selfish gene" is at once radical and exemplary of the confusion in this regard. Dawkins takes the gene to be the ultimate unit of progress in the cosmos.[18] The gene, he thinks, utilizes organisms for the sake of its own propagation. It uses humans as simply "robot vehicles" for its expansion. "Inclusive fitness" or the capacity to attain success in genetic copying is taken to be the filter for evolution. For human organisms, certain adaptive characteristics are preserved that enable this success. These include aesthetic, religious, and moral patterns or presumptions present within culture, which may be sloughed off as genetic advance continues forward. The underlying motor of evolution is an amoral urge on the level of the selfish gene to propagate itself. Human culture even in its most elevated concern to reach for truth, beauty, and goodness is thought to be in substance a series of fictions devised to promote behavior that is amenable to the colonizing empire of the gene.

It is clear that Dawkins and those who follow him are captured by the tendency to image the world according to their own cultural views of personal perfection or in projecting outwardly onto nature the power of the moral volition that drives human society. We have here exemplification of that reifying tendency of the mind that Bouyer has found throughout history, including in the advent of nineteenth-century evolutionary biology, which we discussed in chapter seven. Dawkins may want to explain the personal and qualitative dimensions of being on the basis of so-called "value-neutral," impersonal processes that are amenable to mathematical description, but he ends up surreptitiously projecting onto matter personal characteristics, those of the ideal liberal

---

evolutionary epic . . . the evolutionary epic is probably the best myth that we will ever have. It can be adjusted until it is as close to truth as the human mind is constructed to judge truth. And if that is the case, the mythopoeic requirements of the mind must somehow be met by scientific materialism so as to reinvest our superb energies. . . . Every epic needs a hero, the mind will do. . . . Scientific materialism is the only mythology that can manufacture great goals from the sustained purpose of pure knowledge." Wilson, *On Human Nature* (Cambridge University Press, 1978), 201, 203, 207. Quoted by Mary Midgley, *Science and Poetry* (New York: Routledge, 2006), 269.

18 Richard Dawkins, *The Selfish Gene* (Oxford: Oxford University Press, 1976).

individualist, the self-interested man or woman of ambition who seeks power and expanse of his or her personal influence. Dawkins's unwitting personalization of the gene is an inviting target for the critique of Darwinism and Neodarwinism leveled by the French Structuralists in that part of their work that Bouyer commends.[19]

There is great theological and evangelistic importance in offering a counter-story of cosmic development, as Bouyer does, which is nourished by a very different and well-explicated imaginary that relies on a very different image of human perfection and surrounding symbols than Neodarwinism provides. Post-evolutionary science has shown us that the world is not only fundamentally dynamic and developmental but agonistic, often dominated by violence. We project different images of our own ideal of perfection onto the process of the world in order to find meaning amidst its brutal processes. Self-projection onto the world is not necessarily the root of false science. What is needed is a better form of self-projection, truer to the dignity of our own being in the image of God. As Philip Sherrard demonstrated, our human image and world image are not in the end dissociable.[20] In order to interpret the *telos* of the world in its dramatic advance, we need a self-image that accords with an understanding of human perfection that befits the depth and breadth of our incarnate spirituality as given through God's superabundant generosity in the Trinitarian act of creation and redemption.

This is a key lesson in Bouyer's cosmology. As a theologian, he points to the eternal image of God, to the icon of the Invisible Father become incarnate in our world, as the key to unlock the mystery of the whole of creation, especially with regard to the mysterious conjunction of suffering and love that he recognizes is inscribed in the materiality of the cosmos in its total process.[21] This undermines ideals of human perfection that valorize pure power and self-assertion in humanity and in nature.

19 *IF*, 50.

20 Sherrard, *Human Image, World Image*, 1–10.

21 William Hurlbut, "Saint Francis, Christian Love, and the Biotechnological Future," *The New Atlantis* 13 (Winter Spring 2013): 92–99. See also Hurlbut, "From Biology to Biography," *The New Atlantis* 3 (Fall 2003): 47–66. Dr. Hurlbut has described to us in conversations his personal tutelage with Bouyer while the Oratorian taught at the University of San Francisco. He was deeply impressed with the singularity in breadth and depth of Bouyer's knowledge.

The self-seeking economic and cultural libertine, the superficially tolerant postmodern cosmopolitan, and the transhumanist cyborg of Silicon Valley fantasies do not present images of perfection that are adequate to face the living reality of suffering intrinsic to cosmic history. William Hurlbut, a philosopher of biology at Stanford University who was a student of both Bouyer and René Girard, shows that it is only in Christ and his exemplary followers that the moral meaning of the cosmos, in this unity of suffering and love, is at last revealed in plenitude:

> There within the human form with its capacity for genuine understanding and empathy, moral truth was revealed in matter; the true Image of God was borne within a body. In the face of Jesus was made evident the face of Love, and most specifically in His suffering on the Cross. Those who looked upon Him felt His pain, yet recognized His righteousness and knew the injustice of His plight; His was the ultimate, defining act of altruism.[22]

Alighting upon the eternal image of God in the Son of God incarnate, we come face-to-face with the Person in whose historical action God transformed suffering through a supreme act of love. "The evolutionary struggle," Hurlbut continues, "the seeming futility of suffering and sacrifice and death itself, was raised to the possibility of participation in a higher order of being."[23] All of creation and its history were in some sense a revelation of divine Love in and through suffering. Even suffering for the sake of the good of the species in the animal kingdom dimly foreshadows Christ's sacrifice on the Cross. Yet only in light of this latter intervention that ruptured our settled, agonistic ways, were our eyes opened to the intrinsic meaning of creation, to the very divine design for all things in God's redemptive, transforming embrace of creation in its deepest sorrow and pain, whose purpose ultimately is to liberate the whole of the world from captivity to the destructive personal beings who have caused this plight.

The image of God in the fullness of the Mystery of the Cross and Resurrection should be for Christians the source of their self-image and

22 Hurlbut, "Saint Francis, Christian Love, and the Biotechnological Future," 99.

23 Ibid.

world-image in the face of the suffering and death that are ubiquitous in the postlapsarian condition of creation.[24] The myths speak of the cosmic interlinking of suffering and love, but their understanding is tied to a sense of the world that never truly goes beyond eternal return, with death bringing about the cyclical arising of life as it has always been. The royal myths of antiquity embody this understanding of the cosmos, trapped, as they portray it, in endless destructions and renewals. Bouyer shows in exemplary fashion that in the true royal myth of the Gospel, the endless cycle of return is broken and a new possibility for liberating transcendence is opened up in religious history. Cosmic liberation is achieved through the once and for all suffering and death of the One, True King. As King and Priest, Christ the Eternal Son and Immolated Lamb is the one, true reconciler of a divided world. He embodies the eternal Father's governing Wisdom and performatively unveils the Mystery at its heart.

The myths have an enduring significance for cosmological inquiry. No other mode of cultural activity captures in the way that the narrative image that is their stock-in-trade can do the unity of cosmos with *anthropos* and the ordering of both to transcendence. Royal myths bespeak our desire for social and cosmic perfection, not to mention the unity of humanity and the world joined to an exemplar of the human species. The mythical king is the chief protagonist through which humanity in its historical formation impresses its stamp upon the world.[25] The royal myths seem dimly to intuit that the perfection of this cosmic humanization can be achieved only through the intervention of a celestial human who unites heaven with earth in thaumaturgical and expiatory action.[26] God gave us the prototypical image of kingship and human perfection by personally intervening in our history in the enfleshment of His Son. He thereby communicated the light of His glory to us. The image of the King in the Son, the Immolated Lamb who has promised to return to us in the royal glory of his Spirit, was emblazoned in our history through his effective personal deeds, culminating in his Cross, descent into Hell, Resurrection, and Ascension into Heaven. The light of this supernatural image shining

24 Cf. *Cosmos*, 27. We discussed this in chapter 2.

25 Cf. *IF*, 133.

26 *MT*, 206. Bouyer sees the Arthurian myths, with their deep roots in ancient culture, in this light.

in history can be the source of conceptual exploration of the drama of creation that refuses to reduce the power of nature to blind and voracious will, to self-expansion at the cost of alterity, or, alternatively, to the promotion of a leveling cosmopolitanism at the cost of individual freedom.

This account of the economy of creation and salvation that Bouyer develops, in harmony with mythopoetic thinking, is joined with a theological rendering of the divine life in sophiological perspective through a vision of the whole that unites the *logoi* of creation in the divine Wisdom. God's Wisdom, as we have seen, exists in the heart of the Eternal Son, who is the icon of the Invisible Father. Wisdom is a royal attribute. God, the one, true King, alone is wise. He possesses the Mystery around which His wise ordering of creation revolves in the secret of His inner life until which time, from our perspective, He chooses to accomplish His eternal design for our liberation in and through history. Through His Mystery and Wisdom revealed in the Cross, God makes His superabundant love known to us and draws us toward the inner recesses of Trinitarian life in freedom. Bouyer's positive theology issues in *Cosmos* in a narrative and theodramatic vision of the whole creation which consecrates imagination and the mythopoetic mind. It is filled with reflections of a more explicitly metaphysical character on what it means to speak of kingship, wisdom, and glory in the triune essence of God as well as in the world, and we have tried to highlight and develop these reflections in this study.

In the divine Wisdom, the ideas for creation exist in God's self-revelation in the Son. They are unified in the Son who is the image of the Father and whose splendor shines forth in his Spirit, resting on him eternally. There is, as we have seen, liveliness to this world of ideas. It seems to us that sophiology has helped Bouyer to read this understanding of the divine ideas into the best theologians in the Catholic tradition and thereby to develop their thought. It is particularly notable that he draws out of this Tradition a Wisdom-centered approach to God's relation to the world that allows us to see divine Wisdom precisely as a kind of power of divine Imagination, even if Bouyer does not use the expression "imagination" directly in this context. Divine Wisdom is, for him, the locus through which God creatively gives being and responds in the eternity of His essence as *actus purus* to the free actions of His creatures in the dramatic flow of their actual history. In Wisdom, God responds to creation in its presentiality. It is in reference to His living Wisdom or, we could say with Stratford Caldecott, "Imagination," that God could be described in the way that William Norris Clarke has done as "the Great Jazz Player."[27] It is concordant with Bouyer's thought to think the melody of the divine composition of creation in Wisdom as marked by variations respondent to our history *in concreto* with its crescendo, harmoniously interwoven with our fleshly responses, in the Paschal Mystery. In Bouyer's hands, themes of divine ideas, Wisdom, imagination, and divine predestination are uniquely connected so as to give a unified theological metaphysics of the relation of divine to created persons congruent with a narrative understanding that is at the same time a poetic expression of wonder at the glory of the divine deeds on our behalf. This theological approach enables us to think of the God-world relationship in a manner attendant to our constitution as historical beings, with respect especially to the concrete actions by which God the King has reformed us through His redemptive embrace of suffering and death in the history of salvation.

CREATED WISDOM AND IMAGINATION

In spite of our efforts in this study to highlight the speculative and metaphysical importance of Bouyer's thought, not every reader will readily admit that his mind is entirely open to this level of reflection. For example, Sr. Marie-David Weill, in her largely sympathetic *magnum opus* on Bouyer's eschatological humanism, suggests at the end of her study that the theologian's fear of rationalism and intellectualism in theology in fact led him to downplay the importance of the speculative and metaphysical dimension of the discipline. This dimension has been highlighted as the distinguishing characteristic of theology in the West since the time of Saint Augustine and came to full bloom in the thirteenth century in

27 Clarke, *The One and the Many*, 241. Clarke does not speak of divine Wisdom or Imagination. See Aquinas, *ST* I.44.3. In this article, Aquinas shows that the divine Wisdom, which devised the order of the universe, is the first exemplary cause of the determinate forms of nature. The "types" or "ideas" of all things exist together in unity in the divine Wisdom. Moreover, Aquinas says that created things may also be exemplars by reason of likeness, either in regard to species or imitation. In regard to species, this would apply to angels. For Aquinas, too, it seems, the angels may represent created Wisdom.

the writings of Aquinas. Weill argues that Bouyer's work lacks the metaphysical foundation of the person that is needed for full explication of Christian faith, and, concomitantly, that he failed to distinguish clearly in the manner of Thomas Aquinas between mystical, philosophical, and theological manifestations of wisdom.[28] His thought, she suggests, is too reliant on the theology of the Church Fathers, as he contents himself to return to their affirmations "as a rock which has no other need of justification."[29] "The anthropology and theology of Louis Bouyer," Weill says, "will be even more convincing for humanity today, if they are based not only on the Bible and the Fathers, but equally on a solid metaphysics of the person such as John Paul II calls for in *Fides et Ratio*."[30]

Weill argues that in order to face the questioning that humanity today poses to the claims of Christian faith it will not do simply to expound, as Bouyer does, the views of the Church Fathers and then to consider immediately on this basis the meaning and implications of the Christian Mystery. This criticism is well targeted given the symbolist, historical, narrative character of Bouyer's thought, and also given that he so seldom pauses, even in *Cosmos*, his most speculative and personal work, to elaborate sufficiently his philosophical positioning.

Weill's critique leaves commentators with the task to locate and understand better Bouyer's metaphysics of the person. In our view, this can be done only if the point of the questioning that moves the trajectory of his mature thought is better understood. The weakness in Bouyer's approach has to do less with a lack of metaphysics than with a lack of elaboration not only of his philosophical positions but of the line of inquiry that motivates his research. As we suggested in the first chapter, the question that orients *Cosmos* has to do with how nature and spirit are reconciled in the human person. Yet Bouyer does not sufficiently spell out the place of this deeply metaphysical question at the heart of modern thought and its congruency with the questions that motivated Patristic theology. Bulgakov provides much fuller elaboration in this regard. He develops the genealogy of this question and responds to it with reference to Schelling. Although Bouyer does not take Schelling's work directly as a point

28 Weill, *L'Humanisme Eschatologique de Louis Bouyer*, 584.
29 Ibid., 583.
30 Ibid., 584.

of departure in the way that Bulgakov does, the Russian's engagement with Schelling, especially in his philosophy of economy, can shed light on Bouyer's own underlying metaphysics or metaphysical anthropology. Indeed, the Russian theologian and his French admirer are particularly close on this matter.

Bulgakov thought that Schelling's philosophy provides resources to recover and set in a new light the ancient Patristic understanding of humanity as workshop of creation, whose mission is to imprint the stamp of its image, liberated and perfected in Christ, on the whole of objective nature. Schelling's "philosophy of identity" became in Bulgakov's hands a resource to explicate with due concern for the challenges to philosophical anthropology stemming from the scientific revolution and the emergence of transcendental philosophy the human task to render the world "macro-anthropic." Bulgakov quotes Schelling:

> Nature must be the visible Spirit, and the Spirit must be the invisible nature. Thus the problem of how nature is resolved outside of ourselves is solved here, in the absolute identity of the Spirit within us and nature outside of us.[31]

Objective nature, in this understanding, is unconscious spirit inscrutably longing to be made conscious in human culture. Bulgakov argues that all human culture bears theurgic potentiality to achieve, in part, this spiritualization of the world. He orders the whole of his philosophy of economy around this idea. Economy is for Bulgakov the total activity of the human race through which the humanization of the cosmos is achieved. The Eucharistic celebration of the Church is the wellspring of this total cultural vocation. Like the Eastern Church Fathers from whom he ultimately draws the idea, Bulgakov argues that the vocation of humanity can be realized only in consequence of the Incarnation of God in Christ, who assumed human flesh and actualizes its potencies through the Eucharistic communication of his own divine-human life to all the redeemed and sanctified.[32]

31 Sergei Bulgakov, *Philosophy of Economy: The World as Household*, trans. Catherine Evtuhov (New Haven, CT: Yale University, 2000), 85.

32 Ibid., 102–5.

The Russian theologian thus reorients Schelling's philosophy of art around the theology of the Eucharist and the icon. Like Schelling, he thought that the purpose or ultimate goal of nature was to become an object reflective to itself through human rational consciousness and cultural-economic activity, and that this is especially achieved through the work of the artist.[33] Bulgakov argued that artistic images reproduce prototypical realities and belong to these, but they also belong in the world to the human being who is their creative bearer.[34] The human being is the mirror of the ideas, "the pan-icon of the world," who brings the world to its perfection in conscious self-reflexivity.[35] Human artists have the vocation to draw out the ideal form of things as manifested in their natural images, allowing the realities of the world to become more fully iconic by expressing their prototype in a proper medium.

Because there are ideal prototypes of the world reflecting the life of the Trinity and connected to the angels, the world possesses "sophianicity."[36] Creation is an image of the unified image of the world that pre-exists in God's Wisdom. The absolute divine prototype, the image above all other images, is that of the Son himself, and Wisdom is the content of the world in this hypostatic image subsisting in filial relationship to the Father.[37] The world in Wisdom revolves on the humanity of the Son. There is conformity or correlativity between God and the human that makes the Incarnation possible. The human being is innately theomorphic and is given the divinely appointed task to render the objective cosmos theomorphic. Sin blurred the divine image in the human and in creation, and this explains why the Old Testament rejected images of God as anthropomorphic and idolatrous. The Incarnation and mission of Christ transformed this situation of darkness. The New Adam re-established the pure image of the human being as mirror of God in creation. The human being became newly theomorphic and iconic in Christ and able thereby to "iconicize" the cosmos.[38]

33 Ibid., 91–92.

34 See Aidan Nichols, *Wisdom From Above: A Primer in the Theology of Sergei Bulgakov* (Herefordshire, UK: Gracewing, 2005), 291–308.

35 Ibid., 299. This is quoted by Nichols from *The Icon and its Veneration*.

36 Cf. Bulgakov, *The Bride of the Lamb*, 13.

37 Nichols, *Wisdom From Above*, 301.

38 Ibid., 302.

What Bouyer describes with Schelling and Coleridge as the "re-creative imagination"[39] is obviously of central importance in this human mission before the whole of creation as described by Bulgakov, and it has not been sufficiently noticed in secondary literature that the metaphysics of Wisdom in the modern sophiological tradition bears connection with the larger metaphysical construal of imagination in Schelling and in thinkers directly or indirectly influenced by him, such as Coleridge, Humboldt, Wordsworth, Whitehead, and some perennialist philosophers, such as Henry Corbin.[40] For these philosophers and poets, imagination is a power of poetic or artistic production within nature itself through which the earth blossoms in human voicing. The idea of the "human workshop" is present in this line of thought. Unlike Kant, this loosely-connected group of thinkers does not turn to the importance of imagination simply as an epistemological concern.[41] They see imagination operative in the very metaphysical constitution of the world, and hold that there is an analogy between the human production of art through creative imagination and the dynamic potencies of nature. In Schelling, the formal concept of imagination as applied to nature signals a construal of nature (1) as *natura naturans*, a creative activity and not simply the product of a creative activity, (2) which exists in its products or *natura naturata* as the principle of their individuation, and (3) that constitutes the world as a relational whole.[42]

The old organic and vitalist philosophies of the world that flowed from the intuitions of the mythopoetic mind are given a new lease on life by these modern cosmologists. Bouyer is in no way a proponent of immanentist dialectic, but he does give us more than enough reason in *Cosmos* to draw connections between his thought and this wider modern tradition that also has deep links with Bulgakov. Given the appropriate

39 Cf. *IF*, 75.

40 See especially Henry Corbin, *Spiritual Body and Celestial Earth: From Mazdean Iran to Shī'ite Iran*, trans. Nancy Pearson (Princeton University Press, 1977), 109–34. Corbin puts much stock in the importance of "imaginative perception" evident in ancient religious philosophies.

41 Antoon Braeckman, "Whitehead and German Idealism: A Poetic Heritage," in *Process Studies* 14.4 (1985): 265–86. The present paragraph summarizes this important study.

42 Ibid.

transpositions, the three characteristics of cosmic imagination just now listed can be applied to Bouyer's understanding of created Wisdom. Recall that our author says in *Cosmos* that the whole world "is actually the development of Wisdom from the initial stage when the distinction between it and the Word is purely rational, to the final stage when it becomes real."[43] He speaks of created Wisdom as a cosmic presence "striving throughout the entire history of the created world" to attain the final condition of eschatological communion between divine and created life by means of God's direct actualization of the mediatorial capacity of human nature in the flesh of Christ.[44] Wisdom, by the power of the Spirit, thus leads the historical advancement of creation toward the emergence of humanity, toward the Incarnation of Christ, and toward the completed Church, the eschatological Bride of Christ and celestial city of Jerusalem. In this "striving," immanent to creation, is created Wisdom not akin to the creative activity of *natura naturans*?

The second characteristic of metaphysical or cosmological imagination also applies analogously to created Wisdom. Creaturely Sophia is not a Trinitarian person or a fourth divine hypostasis. She is indeed created. At the same time, she is not a reality separable from the individual beings of the cosmos. She does not subsist as some kind of superorganism, nor do individual beings altogether constitute a superorganism, subsisting only as her accidents or cells. Created Wisdom is hypostatized in the corporeal world only in paragons of human receptiveness to divine action: Mary and the Church. Wisdom is thus indeed feminine.[45] She can be a principle of individuation of created beings because, reflecting the "pan-organism of the divine ideas," divine Wisdom, it is through created Wisdom that each individual attains its realization in the divine plan. Individuation is not only caused by material singularization in the historical advancement of creation but realized by creaturely attainment through God's operations in the world, freely received and embraced, of the unique idea that God has for each individual in His Wisdom. As for the third characteristic of imagination, we can rightly say that created

43 *Cosmos*, 231.

44 Ibid.

45 See Caldecott, *The Radiance of Being*, 271. Again, the closeness of Caldecott's view to that of Bouyer should be emphasized.

Wisdom enables us to understand the universe as a relational whole because it reflects, through the prism of finitude and in accordance with its own ontological dignity, the relational unity of the divine Trinity through which God the Father reveals Himself by communicating His eternal essence to the Son and to the Holy Spirit as Wisdom and glory.

The question naturally arises from these considerations whether created Wisdom can be properly referred to as a "world soul." Schelling described "imagination" in this alternate way, and Bulgakov thought that created Wisdom or "creaturely Sophia" is rightly described thus.[46] Bouyer, largely maintaining silence on this issue, seems to leave open the question of whether this designation, rejected by Fathers and Doctors of the Church, can be appropriately utilized in a modern context. Sophiology faces stiff opposition on this front.[47] Opponents of the idea of a world soul stress that it undermines the substantial unity and particularity of the individual human soul by subsuming it into a wider, cosmic, substantial unity and ends up paradoxically depersonalizing creation.[48] John Milbank has given an interesting defense of the idea of the world soul as it is found in Bulgakov in terms of the idea of "metaxological metaphysics" that he draws from the Catholic philosopher William Desmond.[49] He explains that Sophia, as soul of the world and filled with beauty, is the unconscious artist of the world, creatively bringing form and pattern to it. Bulgakov, on Milbank's interpretation, realized that the modern discovery of the transformation of species and the relativity

46 Bulgakov, *The Bride of the Lamb*, 79–103.

47 See Caldecott, *The Radiance of Being*, 78. Caldecott defends something along the lines of the idea of a "world soul" and speaks in liturgical terms of the intimate dependencies of all things. He calls into question Aquinas's rejection of Plato's idea that the cosmos is one animal. Cf. *ST* 1.18.1. Aquinas, he notes, did not have to contend with the unveiling by modern science of these interdependencies.

48 Slesinski, *The Theology of Sergius Bulgakov*, 154; See also Jacques Servais, "La 'Sagesse,' nombre d'or de la théologie de Louis Bouyer," *Gregorianum* 95.4 (2014): 725–47. Without reference to the explicit designation of a "world soul," Fr. Servais raises as a kind question for dialogue with Bouyer's thought whether the Oratorian ultimately downplays a little the full reality of individual freedom.

49 John Milbank, "Sophiology and Theurgy: The New Theological Horizon," in Adrian Pabst and Christoph Schneider, eds., *Encounter Between Eastern Orthodoxy and Radical Orthodoxy: Transfiguring the World Through the Word* (Surrey, UK: Ashgate, 2009), 4–90.

of time requires us to see the unity of the world in a more organic way than even Aristotle did. The universality and unity of the world is inexplicable by reference to fixed forms subsisting within the repetition of material events. Yet universality and unity there must be, otherwise, the world would be unintelligible. The unifying features of the world must now be located on a trans-temporal plane of continuous engendering by a quasi-personal principle of universal organism, which returns us to a Platonist construal of the world soul. This soul of the world is created Wisdom, a reality that can be vouchsafed only by affirming the existence of a transcendent Creator. It does not exist as a third term between God and the world. Divine Wisdom and created Wisdom each exist on the two sides of the metaphysical chasm that separates God from creation. The former exists in the latter, constituting together with it a non-hypostatic "shared essence or power to personify."[50] This unity of divine and created Wisdom in creation cannot be a subsistent reality outside of spiritual creatures themselves. Bulgakov associates the world soul as creaturely Sophia with a demiurgic transcendental subjectivity that is a unifying power of human action to which all individuals of the species give expression.[51] He claims that this view is rooted in the Eastern Patristic understanding of the unity of human nature and the common teaching in this tradition regarding the First and Second Adam. Milbank affirms "that the inner reality of created Sophia is created angelhood and humanity."[52]

Surely some of what Milbank says about Bulgakov on the rediscovery of the world soul in light of modern cosmology could be applied to Bouyer's theology of creation, although Bouyer perhaps makes the point clearer that created Wisdom is associated with the economy of grace, with a second gift beyond the gift of nature, that moves nature through created freedoms toward the end in which it was created in Christ and his redemptive mission. With Davide Zordan, we should maintain the emphasis that for Bouyer created Wisdom, as a supernatural superexistence, is ultimately an eschatological reality.[53] If we say that

50 Ibid., 65.

51 Bulgakov, *Philosophy of Economy*, 123–56.

52 Milbank, "Sophiology and Theurgy: The New Theological Horizon," 65.

53 Cf. Bulgakov, 201. This statement by Bulgakov seems antithetical to a consistent accent in Bouyer's work: "Ontologically, there is no place for new

created Wisdom does have something akin to the status of a world soul in Bouyer's thought, we have to keep in mind that he most certainly felt the full weight of objections to the idea.[54] His firm admonitions against those who would promote the view that individual persons are ontologically tantamount to "atoms of consciousness" are telling.[55] We have seen that while Bouyer affirms such concepts as the ontological link of all creation to the human soul in the unconscious mind through the senses or the reciprocity of consciousnesses in the hierarchical interpenetration of created spirits in communion with materiality construed in a dynamic and historical way, he nevertheless castigates so-called "personalists" who evaporate the individual person into a web of relations. He places a firm emphasis on the relation of the individual to God while also affirming the social constitution of finite being. He seeks to account simultaneously for the individual, social, and cosmic constitution of human personhood, urging balance in attending to all three dimensions of our being.

One of Louis Bouyer's most incisive summary statements of his conceptualization of Wisdom is found in a passage from *The Eternal Son*:

> God, without having to change or having to submit to any passivity in His vision and eternal will, includes our freedoms and their redemption after the fall, for our salvation and not for our destruction. From all eternity, "in" the Son, by knowing Himself, God knows and wills all that will be, "by" the Spirit who is the eternal and living bond of love between Himself and the Son, and who seems to call into distinct existence to complete their being, and into a special communion of participation in divine being, all these *logoi* in the one *Logos* whom we are from all eternity. Uncreated Wisdom, likewise, unfolds herself as created

creation in the all-exhaustive fullness of creation, in which God rested from his work." It seems there is more of a sense of the elevation of the old creation in the new in Bouyer's theology.

54 See Keith Lemna, "Louis Bouyer and Alfred North Whitehead: A Dialogue in Trinitarian Cosmology," in *Gregorianum* 95.4 (2014): 749–73, 764. In this earlier article, we said that Bouyer does not see created Wisdom as a world soul. It is because he demonstrates cognizance of these objections that we made this claim, which we are now more than tempted to retract.

55 See *IF*, 41. We discussed this in chapter 2.

> throughout time in her initial distinctiveness and in her supreme reunion with the Son with whom, at the end of time, she will be mutually consummated in the marriage of the Lamb.[56]

The biggest difference that Bouyer recognizes between his conception of Wisdom in the world and that of Bulgakov in his wider writings pertains to whether the presence of the world in the divine life is perfective of the latter. As we have seen, Bouyer dissociates his understanding of divine Wisdom from the idea that the kenosis of creation shapes the divine essence. This would render creation a necessary, non-contingent action through which God is made more divine by going outside Himself in loving relation with the created other, although it is likely that Bouyer misreads Bulgakov on this front in maintaining that the Russian himself held such a view.[57] At the same time, the Oratorian emphasizes that God includes our history in His own eternal perfection. In His eternal humanity, God is the very subject of our history. This is by virtue of the power of love and grace in which He both exists in Himself and gives Himself to us as Creator and Redeemer.[58] Louis Bouyer's Christian, Pauline, organic, and vitalist understanding of creation is at once an affirmation of the classical theist understanding of God's transcendent, eternal perfection, in which there is no *agon*, no "dark ground of Spirit" in this sense, and an attempt to develop the orthodox doctrine of creation in light of modern discovery of the singular processes of nature in their productive outpouring through which the *telos* of the world is both imprinted and concretely attained over time, ultimately explicable only in the style of drama and event. He maintains a balance between so-called "classical theism" and the modern concern to relate more adequately time and history to the divine life. His thought constitutes a kind of *metaxu* connecting Aquinas with Bulgakov. He does not make any explicit connection between

56 *ES*, 413.

57 Ibid., 407. Bulgakov in fact insists that God does not need the world. He creates necessarily, but this is because God's love moves Him in necessary freedom to go outside of Himself. Bulgakov thus chastises the very question of whether or not God had to create as "childish." Still, Bouyer construes the necessary freedom of God in a different way than the Russian and so never approaches Bulgakov's level of shock pedagogy on this question.

58 Ibid., 413.

created Wisdom and the language of "world soul." This crucial text just quoted stresses that at the end of history created Wisdom will be fully re-united with divine Wisdom, of which it is the unfolding in time. Yet the conceptuality of nuptial relation and communion is maintained, that is, the view that distinction is not done away with in this union but perfected precisely in union. John's image of the Wedding Feast of the Lamb is invoked: the vision of deification as communal personalization of the Bride with the Bridegroom is biblically established. For Bouyer, the shared supernatural superexistence of the redeemed and saved, fully actualized on the Last Day, is a power that will render them one with God and with one another in elevating perfection of their created being.

The philosophical "foundation" for this is indeed in a surpassing "meta-anthropology" understood in Balthasar's sense, as we suggested in the introduction, and we thus cannot agree with Sr. Weill that Bouyer's work lacks a solid metaphysics of the person. It is truer to say that metaphysics in the Oratorian's thought revolves around the mystery of corporeal personal being as the summit of creation in which the liturgy of creation is uplifted anew in a single voice to the Creator. The French Oratorian does not, it is true, set apart a philosophical foundation in a separate treatise on personhood upon which doctrinal considerations are then added, but this is because the human and cosmic sciences are integrated in his works, perhaps especially in *Cosmos*, in an overarching vision that is at once theological and sapiential/philosophical. The meta-anthropological foundation of this vision is within the whole, permeating it, not underneath it, and includes an interpretation of modern physical science with a rather unique pedigree in which an account of the intelligibility of scientific civilization is an essential component. The Bouyerian metaphysics is included within a sophianic meta-anthropology that emphasizes historical humanity as the summit of physical creation in a world already torn asunder, and, elevated in the New Adam and the New Eve, as the personalized receiver of God's Trinitarian self-communication by the gift of the Spirit in the Church, the Body of Christ and Temple of the Spirit. The philosophical dimension of his work exists within the total dramatic-narrative and phenomenological ordering of his "positive" presentation of the Christian vision of faith. It is rooted in thorough scriptural exegesis in light of the Church Fathers, but the total presentation is ordered by the quite modern question of the relation of nature to spirit in the vast historical unfolding in freedom of created being. This question elicits considerations that are at once metaphysical and anthropological in character, and Bouyer's response to it is aptly described as "meta-anthropological." In the context of his problematic, where nature, history, and their end(s) are of paramount concern, it is appropriate to think through the meaning of all that is in play in competing eschatological (or cosmogonic) myths that alone can reveal the destiny toward which we are all being led. This brings the object of faith into the open field to be contemplated through both the *logos* of philosophy and the *logos* of theology. There is room in Bouyer's thought to consider one and the same supernatural object of attention, God's revelation in Christ, in these two ways, either philosophically or theologically, either in terms of phenomenological possibility or in terms of actuality as affirmed by the light of faith.[59]

METAXOLOGY, ANALOGY, ANTICIPATION

It may be helpful, in order to understand more deeply the metaphysical or meta-anthropological dimension of Bouyer's thought, to place it in light of the seminal essay by William Desmond on Lev Shestov (1866–1938) and Vladimir Soloviev that has been the inspiration for John Milbank and some others to explicate sophiology in terms of Desmond's metaxological analysis of the history of philosophy.[60] Given our exposition in the previous section, we may have given the impression that in our final estimation Bouyer, existing in the line of philosophers of nature who made imagination a topic of metaphysical concern, is incorrigibly Romantic, perhaps inebriated with an imaginative perception of the All-unity of being in Wisdom, but Bouyer's thought in fact seems to

59 Cf. Emmanuel Falque, *Crossing the Rubicon: The Borderlands of Philosophy and Theology*, trans. Rueben Shank (New York, NY: Fordham University Press, 2016), 15–28. In his seminal work on the relation between philosophy and theology, Falque notes three ways in which philosophy and theology can be distinguished in regard to consideration of a common object: with respect to possibility and actuality, heuristic or didactic method, and from below or from above. The first term in each pair designates the proper approach of philosophy.

60 Desmond, "God Beyond the Whole: Between Solov'ëv and Shestov," in *Is There a Sabbath for Thought: Between Religion and Philosophy*, 167–99.

share characteristics with both of these very different Russian thinkers expounded by Desmond, one Romantic or Idealist, the other an enemy of the Idealist corruption of Christian thought. There is in fact in the work of our eschatological humanist a tension between accents that one might describe as almost Neo-orthodox and anti-philosophical in character and accents that do indeed point to a "dangerous" perception of the unifying presence of Wisdom in creation. This tension may have been, in the end, not so much the indication of an incoherent or superficial thinker as productive of real metaphysical insight.

Desmond compares in his essay the univocal, holistic theology of "All-unity" in Soloviev with what he takes to be Shestov's equivocal, dualistic opposition of divine transcendence to creaturely immanence. Shestov, he explains, inveighed against modern philosophers following Spinoza, including Soloviev, who embrace a doctrine of "All-unity," where God, the world, and humanity are joined together in pantheist confusion. Shestov castigates these thinkers for absolutizing worldly intelligibilities and rational necessities, thereby placing God in the dock of univocal determination. He champions in response to them a religious doctrine of equivocity, which maintains the absolute difference between God and humanity, God and nature, as well as nature and humanity. He was never taken in by the allure of the systems of the Idealist dialectical philosophers of All-unity and fulminated against what he took to be their rationalist usurpation of the God of revelation. Desmond explains that Shestov understood himself to stand on the side of Paul, Luther, Pascal, Kierkegaard and—to Shestov's mind—Nietzsche, in protest against the God of the philosophers and in defense of the God of revealed religion.[61]

Was Soloviev, as Shestov thought, in fact an idolater of the Idealist Absolute? Desmond proposes an alternative interpretation. Soloviev's devotion to Sophia, on this counter view, demonstrates a mind that is willing to go beyond univocity without for all that falling into equivocation. He is uniquely open to what Desmond describes as the *metaxu*, that is, the medium between cosmic immanence and divine transcendence that joins them together without confusing them. In spite of certain real ambiguities, Soloviev's thought presents a genuine clearing to the

61 Ibid., 173–77.

transcendent God who exists beyond the whole of experience but from within the widest vantage point of this whole, without whose mediation God could not be known by us. Soloviev is not, on this interpretation, identifiable with those dialecticians whose manner of thinking beyond the oppositional duality of equivocity ultimately conflates the divine with the human. Soloviev's Sophia, at the center of Desmond's consideration, is not the immanent whole or the totality of the absolute in a condition that cannot be transcended. She is instead precisely a *metaxu*, "a pointer to a *middle between* utter transcendence and an otherwise godless immanence . . . between God as the ultimate other and the finite between as the milieu wherein the community of the divine and the human is coming to be."[62] This is not to say that Soloviev did not feel attraction to the idols of speculative metaphysics prevalent in his day and age. Toward the end of his life, he was purged of this attraction, perhaps by visions of the Devil, and his striking tale of the Antichrist written at this time is a sign of clear awakening to the duplicity of immanentist visions of the whole that disavow the *metaxu* and therefore the transcendent God who makes Himself known in it.[63] The Antichrist in Soloviev's mysterious final tale makes his appearance within the façade of a counterfeit Church that has wedded itself to immanentist dialectic. "The problem of the Antichrist," Desmond explains, "is the production of the counterfeits of God and the usurpation of reconciliation in the form of communities that have all the appearance of being ultimate and unsurpassable."[64]

As we have seen throughout our study, Louis Bouyer did not have to wait until the end of his life to be concerned about these sorts of deceptions. His eschatological humanism was from the beginning a cry of warning against counterfeit loci of reconciliation of the human race, those "pretended integrations" that we discussed in chapter six that can, in light of Desmond's essay, be associated with univocal dialectics. Keeping faithful to the character of Christian eschatological *apokalypsis*, he recognized that in the end the perfect reconciliation of God with cosmos and *anthropos* will come only through the final trial of the Church

62 Ibid.

63 Ibid., 183, n. 13. Bulgakov defines divine-humanity or creaturely Sophia as a *metaxu*. Cf. *The Bride of the Lamb*, 123.

64 Desmond, "God Beyond the Whole: Between Solov'ëv and Shestov," 183.

with the Antichrist, a deceiver bearing the mask of Christ. This theme is not played up in *Cosmos*. However, in his apocalyptic novel published in the same year as *Cosmos*, the Antichrist does, as one would expect, figure as the ultimate antagonist. The Antichrist is associated there with a counterfeit Church dominated by the ideal of a worldly and materialist ecumenism that seeks to bring about the false unification of religion.

This eschatological and apocalyptic sensitivity to the destructive presence in our world of counterfeit Christianity is demonstrated throughout Bouyer's writings, and what we referred to in this book as the "gloomier" chapters in *Cosmos* show that it is an important dimension of his cosmological treatise as well, even if the Antichrist is not called out by name in this latter book. On the basis of what some might derisively refer to as his apocalyptic Biblicism, Bouyer may indeed strike the reader at times as being ranged with Shestov against the God of the philosophers. That this reading of him is more than possible may be hinted at in Weill's criticism of his anti-rationalism and aversion to speculative metaphysics. Yet whatever urges Bouyer possessed to defend religion by way of equivocity, they were always balanced by a sophianic instinct for the *metaxu*. His protests against idolatry go hand-in-hand with a mystical theology of the *Shekinah-Yahweh*, of correlativity between the divine and human. His triadology emphasizes that God's transcendence is agapeic self-donation in which the divine Wisdom for creation is eternally present in the necessary freedom of divine being. The triune God of love is more intimate to us than we are to our own self, willing to unite us to His own life through filial adoption in Wisdom, although the Cross is the inescapable nucleus of this mediating Wisdom.

Another way of understanding this dual dimension in Bouyer's work, that is, its concomitant sense of divine transcendence irrupting into the world through unanticipated revelation and its sense of the eternal presence of creation in God, is to set his cosmology in dialogue with cosmologists who nowadays problematize the Catholic metaphysics of analogy at the basis of metaxology by comparing it unfavorably with a so-called "metaphysics of anticipation" that they think follows more directly than analogical thought from a properly biblical, eschatological standpoint. The work of John Haught, an emeritus professor of theology from Georgetown, best exemplifies this contemporary approach.

According to Haught, the traditional Catholic theology of analogy is based on a static cosmology that sees the world as only scenery or a backdrop to human life, or as a ladder that needs to be kicked aside once individual humans have attained to the height of the eternal present by salvation in God.[65] Analogy does not, he urges, enable theologians to give a proper theological response to the discovery that the cosmos is evolutionary, always coming-to-be, with its perfection in the future and not in a golden age of the past or in an eternal present beyond the world. Theologians of analogy, he insists, think of the cosmos as an accidental shell of little importance except as a container. Material being has no intrinsic ontological meaning in this view but has merely symbolic importance, pointing us to the beyond of perfect life. The point of salvation for analogical thought is to get out of the world, to escape the future. Haught, rooted in the cosmology of Teilhard de Chardin, proposes as an alternative theological metaphysics a way of anticipation that he thinks truly welcomes the future. This approach would, he insists, recognize that the deep time of billions of years of cosmic history is a gradual drama leading the way to the emergence of a future perfection that has not yet been achieved. Humanity should not be understood to be the center of a hierarchically-differentiated cosmos, as in analogy, but as the leading edge of the cosmic arrow of time. The human being is not the reconciler of a qualitatively-ordered chain of created being that extends from the celestial heights of the Cherubim and Seraphim down to the *materia quantitate signata* of quantum potentiality but the emergent agency responsible for the direction of future cosmic progress. Much like Teilhard, Alexander, Whitehead, or Bergson, Haught is engaged in an impassioned endeavor to recast the whole of human thought, and especially theology, in evolutionary terms. He criticizes theological projects rooted in analogy for failing to attend to the fact that the cosmos is a process whose future will bring more being into being than that which the present expanse of the whole currently possesses.

If we follow Desmond's categorizations, Haught seems to (mis-)interpret the theology of analogy alternatively as a theology of equivocity

65 Haught, *The New Cosmic Story*, 34–35; cf. John Haught, *Resting on the Future: Catholic Theology for an Unfinished Universe* (New York: Bloomsbury, 2015), 1–8; Ilia Delio, *Christ in Evolution* (New York, NY: Orbis Books, 2008).

for which the being of the world is of questionable subsistence or in terms of Parmenidean univocity. Yet does his own view not, in the end, exemplify a genuinely dialectical univocity for which God comes into being in and through the whole? Does the God that he maintains to be more Omega than Alpha exist beyond the whole in any meaningful sense?[66] In spite of these questions, the discussion that Haught has elicited on the basis of this dichotomy may help us to clarify further potential contributions of the thought our own author. The uniqueness of the Oratorian's speculative thought may ultimately reside in its way of implicitly joining anticipation with analogy. Bouyer is, as we have seen, a proponent of analogy, indeed, his anthropology of angelic ascent may seem to be a most extreme version of what Haught targets, but he is at the same time attentive to the creative event and concerned to account for the drama of deep time, a proponent of a sort of "way of anticipation." He quite obviously joins the theologians and philosophers of anticipation in recognizing the inseparability of human and cosmic meaning as well as the dramatic character of the latter. However, he realizes that the anticipatory standpoint is made possible because God, in the perfection of divine Wisdom, moves us to the fullness of our being in the plenitude of Christ. The future will bring the fullness of Christ in his Body. As a theologian of analogy as well as anticipation, he recognizes that the One who is coming into fullness in his human members, in created Wisdom, is paradoxically "the same yesterday, today, and forever."[67]

It may be helpful to explicate our meaning by attending briefly to an aspect of Bouyer's sacramental theology that we have not yet singled out and which clearly shows the mediating character of his thought in this regard. In *Liturgical Piety*, he argued that the Christian sacraments were never meant to be understood in terms of "some static

66 Haught, *Resting on the Future*, 121. Haught advocates for what he describes as "cosmic hope." He says on this page: "Cosmic hope . . . associates God not with an unchanging completeness untouched by the flow of perishable events but with a future fulfillment of all time, history, and creation." In spite of certain statements to the contrary in his work, his view seems similar in the end to that of Whitehead, according to which God has both an antecedent and a consequent nature.

67 Hebrews 13:8.

kind of being."[68] The Christian mysteries are, he argued, the vehicles through which the one Mystery of Christ is communicated to us in our being and action. If the sacraments are not measured by the transient existence of the present moment, and do indeed give us contact with the eternal present, they are also not ultimately intelligible simply by reference to the once-for-all accomplishment of Christ in his Paschal Mystery in the historical past. The true direction to which we must turn in order to grasp the intelligibility of the sacraments is the future. The Eschaton gives us the key to decipher the meaning of the sacraments: "only the fullness of Christ when His whole Mystical Body is complete, when the whole city of God has been offered to the Father, when, in its final resurrection, God will be all in all; only this ultimate actuality of the Mystery can be what permanently sustains the actuality of the Mystery in the sacramental order."[69] The sacramental logic of Christian faith is thus informed by the eschatological forevision elicited in the human soul by the supernatural virtue of hope. Christian attention should be directed not only to an eternal present beyond time or to a past that decisively shapes our being but to the future completion of human and cosmic existence in and through time.

The historical character of his overall theology likewise indicates that Bouyer, as a thinker of analogy, is no less insistent on the importance of anticipation. In *The Invisible Father*, he suggests that a theological "phenomenology of the Spirit" is needed that embraces the creative events through which the Holy Spirit leads creation toward its future completion. He counsels that anticipation of the event should replace *a priori* deductive logic on the model of Hegelian system-building as the proper theological way to comprehend the course of cosmic history. True openness to the future is openness to the God who is beyond the whole. He advocates for the establishment of open systems, decidedly phenomenological in character, which come only at the endpoint of a process in which positive analysis of sufficient rigor has been carried out:

68 *Liturgical Piety* (South Bend, IN: University of Notre Dame Press, 1955), 183.

69 Ibid., 184.

> In other words, it is both possible and desirable for every generation to re-do a phenomenology of the Spirit, but this cannot claim, as Hegel's did, to precipitate (in every sense of the word), the process of mankind's spiritual becoming as if it precontained a summa of absolute knowledge which, as time passed, would only need to be methodically inserted in its appropriate, pre-fabricated pigeon holes. All that such a phenomenology should do—and to do more would be to surrender to an illusion which would prove disastrously idolatrous—is to draw a dynamic synthesis, and not a static system, from meditation on the experience of the Catholic faith, an experience which cannot be other than incomplete. We mean a synthesis which will direct us from where we have actually arrived towards the place of eschatological encounter and consummation.[70]

This passage well describes the methodological foundation of the theological cosmology that we have explored in this study, which, as with all the Oratorian's work, can seem to have a frustratingly provisory character. Is this not precisely the anti-rationalism in his thought that keeps him from explicating, as Weill suggests, a solid metaphysics? This passage can be read in a contrary way. It may be that Bouyer's metaphysics simply embraces in part a different set of categories than what dogmatic theologians are used to, something along the lines of Jean-Luc Marion's "saturated phenomena," including, in addition to Marion's categories of event and paradox, a construal of matter in terms of dramatic-narrativity, of imagination as a productive power of the person that brings the cosmos to a higher level of unity, and the principle of emergence. Perhaps Bouyer teaches us that a "solid metaphysics" will have to learn to incorporate these sorts of categories. These are personal categories, open-systems categories, categories that train the mind to be receptive to the surpassing gifts of divine love. They elevate metaphysics onto a meta-anthropological level and encourage humility in systematic thought. They are each present within the entire train of Bouyer's reflections, even if he does not draw out their meaning in anything more than a cursory manner.

70 *IF*, 309.

The phenomenology of the Holy Spirit that Bouyer promotes, and the dynamic syntheses that he advocates, as properly sacramental, in fact join the cyclical vision of creation common to the myths and to philosophies of analogy in ancient religious thought with a sense of the progressive arrow of time that biblical revelation instilled in human consciousness and that Teilhardians such as Haught have radicalized in an evolutionary perspective. The true myth of God's intervention in the world in Christ can be said to break the dreamy vision of eternal return that the ancient philosophers of analogy embraced.[71] But God's re-creative intervention in history in Christ does not undo the intrinsic being of creation in its natural rhythms with which these philosophers kept in contact. God effects organic movement to the future, that is, enlargement of the cosmos from within itself:

> The divine intervention in history obviously did not suddenly put a stop to its natural rhythms. If it had done so, it would have only suppressed human history, far from transforming it and bringing it to fulfillment. What the divine intervention does, therefore, is to open up, as it were, those closed cycles of time, which up to that moment had been only revolving on themselves. These cycles—to use the same image—are now in a process of development into a larger and larger spiral, and this process will reach its term at the moment when the spiral of human history loses itself, or, rather, finds its fulfillment in God's eternity.[72]

Only a vision of the future that embraces God's already-decisive interventions on our behalf, beginning with creation itself and its cycles of life, enables us to anticipate in hope the future and the coming-into-being that continues to mark our lives in a way that makes the entirety of the cosmic process intelligible. A theology or metaphysics of anticipation,

71 Haught, *The New Cosmic Story*, 207, nn. 19, 20. The sense of "analogy" here is Haught's. Haught criticizes in these notes both Seyyed Hossein Nasr and David Bentley Hart. He says in n. 19: "For the latter [Hart] and most other contemporary apologists for analogy, the new scientific cosmic story seems to be of little spiritual or intellectual relevance, not a drama to be read with both religious interest and epistemological patience."

72 *Liturgical Piety*, 196. See also *Gnôsis*, 60.

which lifts us out of the recurrent cycles of being as it has been, needs analogy in order truly to embrace the intrinsic *logos* of creation that moves the very cycles of the cosmos. These cycles are expanded through what Bouyer describes as the increasing interiorization and universalization of divine revelation received by the People of God in the Church. Our fulfillment in God is a surpassing expansion of our being, even in its temporality, and not a de-creation of it. It is in this sense, ultimately, that *logos* and *Logos* surpass myth in wisdom while consecrating the mythopoetic instinct. The eschatological, dramatic-narrative, biblical-historical framework of Bouyer's thought begins to develop a meta-anthropology of anticipation from within the Catholic analogical worldview. Far from doing away with the latter, it might help to clarify some implications of it that have not always been noticed.

APOCALYPTIC WISDOM

If our analysis is correct, Bouyer can be said to unite analogy and anticipation in light of a total cosmic vision that takes into account, in a non-concordist manner, what modern science has unveiled of the vast course of the cosmos with its increasing processes of complexification and its decisive mutations. Because he embraces analogy, he does not have to see, as some Teilhardians have, the "inside" of the universe only in terms of its potentiality for emergent consciousness. The inside of cosmic process includes God's ubiquitous presence to it as well as that of the first created spirits who are intimately involved in the work of its actualization. This cosmic history is always, although not necessarily, as we have shown, a history of sin and salvation. Because Bouyer understands the Immolated Lamb to be the principle of creation and the Bride of the Lamb to be its *telos*, he is able to unite human, cosmic, and salvation history, while acknowledging their proper distinctions. All of cosmic history is included in God's eternal Wisdom and is in some sense sacred.[73] Yet the Oratorian also recognizes that we cannot ignore the decisive saltations or leaps that transform this history from within. For Bouyer, the most decisive "leap forward" came in the Incarnation

73 See Grintchenko, *Une Approche Théologique du Monde*, 274–75. See also Zordan, *Connaissance et Mystère*, 660. Zordan says that for Bouyer all of history is "the space-time of salvation promised then realized."

and Paschal Mystery of Christ.[74] Biblical history is the recounting of a special history of divine intervention in the world. The human capacity to envision the future was irrevocably elevated as a result of God's supernatural interventions in the cosmos. Louis Bouyer's sapiential theology leaves us in no doubt, unlike that of many Teilhardians, that true openness toward the future requires openness to reception of God's decisive interventions in our past in salvation history, culminating in the life and mission of Jesus of Nazareth, whose Incarnation is the decisive divine rupturing of our self-enclosed univocities as well as the supreme locus of reconciliation of equivocal oppositions.

The nuptial ontology that Bouyer proposes clarifies how a properly sacramental logic of analogy readies us to receive the future on the basis of what God has already done for us beginning with the very act of creation. If God oversees the whole of time in an instant from the eternal present in His divine paternity, we require, for our part, transformation into a condition of Marian active receptivity in order to receive the future coming-into-being of creation in the Eschaton. Connecting with our previous chapter, we can say that if we are to tell an intelligible story of the cosmos, we shall have to relate our future to our entire past and present. Bouyer holds that the future actualizes an embryonic potentiality that pre-exists in the womb of creation, mirroring in reverse the realized, eternal *actus purus* of the Creator. This potentiality of creation is elevated onto a new plane in human history in the womb of the virginal female and is elevated and actualized on an even higher level in the supernatural history of the People of God summed up in the Mother of God and her Son. The eschatological Church, the new creation, is embryonically present in the womb of Mary, who shows forth the intrinsic meaning and potentiality of creation as it was from the beginning of space-time in its deepest history.[75] Yet the primordial potency does not explain the future actuality. The reverse is true: the principle of emergence holds sway in interpreting this process of organic development.[76]

74 *Cosmos*, 223–24.

75 Cf. *MSW*, 154–55. We quoted from these pages in chapter 11. It is from these pages that we draw this whole paragraph.

76 We refer to our chapter 7, where we referenced Bouyer on the principle of emergence in regard to evolutionary biology. The principle of emergence fits

The future brings newness, but this is not without a relation of continuity to the past, beginning with the first creation in its maternal depths. The different phases of biblical history that sum up while transcending the total history of humanity and the cosmos are not abolished in the advance of this sacred history. Instead, all of the phases of biblical history, and thus all of cosmic time and space, are inserted and concentrated in the person of Christ.[77] The distinction between Name and Wisdom in God that we discussed in chapter eight is of conceptual significance in this regard. Time, according to this distinction, is included in the Wisdom of God. The divine Name, on the other hand, is the Father's self-expression in the Son and transcends the joining of Creator with creature.[78] The suffering of God in the drama of creation is not projected into the divine Name. In affirming with the tradition of analogy the eternal present of God's Trinitarian communion in the divine Name, we discern a metaphysical rationale for the possibility for real advance in creation, for real movement toward the future in a logic that transcends stifling conditions of possibility that the human mind sets for the world when glorying in its own representational interiority. God's transcendent actuality "grounds" the very possibility of temporal advance in His creation. Because God ultimately transcends the processes of creation, He can freely bring the future to creation, while humanity, because it is fundamentally Marian and "grounded" in the Wisdom of God, is substantially ordered to realization in and through time. The ideas for creation become, in Wisdom, progressively imprinted within the sensible order in the systole-diastole or contraction and expansion of the Church in its pre-history and its concrete actualization through the Body and Blood of Christ.[79]

The advance of created Wisdom in time cleanses humanity made in the divine image, so that the redeemed and saved become all together the spotless mirror of divine Wisdom in the final achievement of the

with his "phenomenology of the Spirit" described above.

77 See Grintchenko, *Une Approche Théologique du Monde*, 279–80.

78 Ibid., 281–82. Grintchenko holds that this distinction between Name and Wisdom in Bouyer's thought is a unique contribution in his theological cosmology.

79 Cf. *CG*, 257–58.

personalization of the Church. Yet for all our talk of "active receptivity" (an expression Bouyer does not himself use) with respect to created Wisdom, are we as individuals indeed something more than passive receivers in this process? How is it, more precisely, that we await the future? How do we actively participate in the eschatological coming-to-be of creation? Is there a sense in which we do indeed stand at the edge of the cosmic arrow of time as the effective bearers of progress that is completed in the Eschaton? Bouyer's emphasis on the paradigmatic vocation of the Christian monk might seem to limit the ways in which human activity can fulfill our vocation as reconciler of the created order. Is his way of rendering the human relation to the world not in the end an acosmic otherworldliness? Does human work in this world have any bearing on the world to come?[80] In fact, for Bouyer, the path of wisdom through which creation is brought to perfection is connected not only to passive contemplation but to a royal activity that cannot be without importance for the shape of our ultimate future.[81] It involves the very art of coming to know the world and adapting to it. If Bouyer did not develop a full-scale philosophy of economy in the way that Bulgakov did, he nevertheless understood that the totality of human culture contributes in Christ and the Church to the actualization of creation. Does not the Oratorian have in mind something along the lines of Bulgakov's idea of the "iconicization" of the world when he speaks in chapter seventeen of *Cosmos* (or in the as-yet unpublished *Religio Poetae*) of the reanimation of language in poetry?[82] Bouyer emphasizes in various works, including in the *Postlude* of *Cosmos*, the human capacity to add to or complete the beauty of natural sites: in cities and in works of art, particularly in Christian icons (East and West).[83] In *Cosmos*, as we have seen, he

80 O'Callaghan, 126–27. O'Callaghan shows that this was a fundamental question raised by the incarnationalists against the eschatological humanists in the debate we briefly summarized in chapter 6.

81 Grintchenko, *Une Approche Théologique du Monde*, 278.

82 Cf. Jean Duchesne, "Les fondements, les apports et les enjeux christologie de Louis Bouyer," *Groupe de Recherche sur l'Œuvre de Louis Bouyer* (Collège des Bernardins, April 30, 2018). Available at https://media.collegedesbernardins.fr/content/pdf/formation/etudier-louis-bouyer/2018-04-30%20-%203%20-%20Intervention%20J.%20Duchesne.pdf.

83 *Cosmos* (Fr.), 377–79. See also *LV*, 67–100.

commends Baader's view that scientist, philosopher, and poet are joined in the task to set straight and join together the scattered fragments of worldly being: the scientist by way of praxis; the philosopher by way of interpretation; the poet by way of imitation and ultimately by establishing unity on a new level.[84]

At the same time, our eschatological humanist sees the human vocation with a clear vision of the darkness ubiquitous to our concrete existence in this fallen world. On this he seems to be aligned more with Newman than Bulgakov.[85] He holds that suffering, death, injustice, and evil must be confronted head-on, and that any account of human vocation that diminishes this dimension of our existence lacks credibility. Telling in this regard is his description of the vocation of the artist in *The Invisible Father*. The artist, Bouyer says there, cannot be satisfied merely to reproduce the appearances and simulacra of immediate human perception as given to consciousness in our postlapsarian condition. She must avoid the trap and snare of superficial appearances of beauty, experiencing to its farthest depths the degradation and ugliness in this fallen world in order to glimpse God's transfiguring presence: "Hence it is that the beauty they [artists] recall, the Presence they evoke is no mere reminiscence, forcing us back beyond our existence, but a beauty transfiguring the ugliness in power and a presence abolishing absence, but only gained when that absence has been experienced to the bitter end."[86] The vocation of the artist entails taking up of the Cross, not merely in remembrance of the freshness of the original creation but in active hope for the full emergence of the new creation swathed in the glory of created Wisdom when at last fully reunited with its source. In true art, the metaxological power of imagination is freed from the temptation to create idolatrous images. It may be that any human work bearing eschatological significance, preparing the way for the new creation, will have to bear a similarly prophetic character to what Bouyer

84 *Cosmos*, 136.

85 Ibid., 205. Bouyer says of Newman with regard to the cosmos: "But the same Newman who knew how to express its beauty and luminosity so incomparably is also the one thinker who saw its dark face more fearsomely than perhaps any Christian thinker at any time."

86 *IF*, 73.

describes of the artist. Centered on the Cross seen in light of the Resurrection, Bouyer's "metaphysics of anticipation," if it can be put thus, founds the theandric humanism of the tradition that he recovers firmly on the ground of Golgotha.

In the end, Bouyer emphasized as much as or more than anyone in twentieth-century theology that human vocation is received and actualized in ritual and liturgy by the gift of divine grace with which we work in synergy to humanize the cosmos in the Body of Christ. This is essential to his metaphysics and anthropology. *Cosmos* is itself integrated in a ritual and liturgical framework, if we connect the two parts and recognize the foundational character especially of chapters two and three. The French Oratorian maintains in his work that the reanimation of language in poetry, art, or theology is first of all a liturgical act. He strongly emphasizes that development of the virtue of religion, the obligation that we all have as sharers in the royal priesthood of Christ in the Church to give proper worship to God, is the first order of business in the economy of redeemed creation. He showed in his early works on liturgy in some detail how ritual and liturgy are the primary human activities through which the world is "iconized" (although he does not use that word) or rendered transparent to religious reality, and how Christian liturgy transfigures thereby our perception of space and time. His famous defense of the orientation of the altar *ad orientem* in books such as *Liturgy and Architecture* and *Rite and Man* has its largest context of meaning in this sapiential understanding of human vocation in a fallen world. Only when seen in light of the dead and risen Christ in and through the liturgical remembrance of his words and deeds can the cosmos be rendered "theomorphic." The liturgical celebration itself must be a sign of cosmicity re-created in Christ:

> [T]he symbolic orientation of the Church (the community in prayer and the building that shelters it) expresses the incompleteness on earth of every Eucharist tending toward its fulfillment in the *Parousia*. At the same time, it is the whole cosmos that is reconstituted, centered on the Lordship of the Risen Christ, leading the whole universe, human and angelic, material and spiritual, toward the Father... the Church is above all the new

> cosmos of which Christ the Pantocrator is the Lord and Prince, the sovereign Archon.[87]

The eschatological density of this passage should be highlighted. The French Oratorian's cosmovision is ordered by a liturgically-informed desire for the coming of the celestial King, the Son of Man, on the final day, who, as we have seen throughout this study, is the Immolated Lamb slain before the foundation of the world. It is conformity to Christ in his love that we should seek. The whole cosmos is reconstituted in the ecclesial "new cosmos." The liturgy is the wellspring for any human work that is truly preparation for the new heaven and the new earth, because it is the fundamental action of the Church that conforms us to God's love poured out for us on the Cross.[88] In order for our being to be rightly conformed, we require entrance into Christ's suffering and death, so that our very existence becomes a living sacrifice to God and to others. In this way, we may love in the realization of the Body of Christ as we have been loved from all eternity.[89] This transformation in love is what the virtue of religion enacted in true worship gives us. This transformation of the human person brings about transformed cosmicity. The way of preparation set forth in this spiritual and liturgical perspective is not otherworldliness but indicative of a life lived fully in the present in hope for the future, informed by charity and filled with thanksgiving for God's already accomplished marvelous deeds. The cosmos is recovered in the Body of Christ in and through the Eucharistic celebration. When Christ, our hope of glory, is fulfilled in us, the very glory of creation will shine forth in fullness of charity. If we can speak, as some Eastern theologians do, of the humanization of creation, this is only with respect to the living flesh of the humanity of the Son communicated to us in the Eucharist. The living human being, as reconciler of a divided world, is truly the glory of God at the heart of creation, but in order to be fully alive the

87 *Rite and Man*, 184, 186.

88 Cf. *Sophia*, 188–89. Here Bouyer discusses the Church and communities of work and life through which the Church inserts itself in the cosmos. Through these communities—beginning with the family—the Church becomes the center or heart of the Wisdom of creation, the principle of its animation, without for all that becoming confused or identified with the world.

89 Cf. *Introduction to the Spiritual Life*, 32.

human being has to be conformed to Christ in his Body. The Pantocrator is the Suffering Servant. His Body has not yet been brought to fullness through the suffering of its members, in all their work. The recovery of the cosmos in the sophianic and apocalyptic optic of Louis Bouyer entails a kind of open metaphysics oriented by a phenomenologically rigorous consecration of nostalgia for the definitive future that the Father's representative King promised to bring us, and of which he gives us a foretaste in sharing his broken Body and Blood with us in the Eucharist. On the Last Day his Body will be perfectly filled in and personalized in fullness as his Bride. In him, and filled with his Spirit, we shall rejoice with all of the new creation forever gathered around the throne of the Father. We shall be joined together in praise of the one true King of heaven and earth who, in His superabundant Love, Wisdom, and glory, is the one and only source and end of both divine and created life.

ACKNOWLEDGMENTS

THANKS GO FIRST OF ALL TO MY FAMILY FOR their love and support over the years. Special thanks for this project go to John Riess of Angelico Press, for allowing me to write a large study of this very important but scandalously neglected theologian; to Fr. Neil J. Roy and Stratford Caldecott for encouraging me to study Bouyer in the first place many years ago; to Jean Duchesne, Bouyer's literary executor, who is always graciously indulgent to my frequent pestering and who has taught me much about Bouyer the man; to Peter Casarella, who agreed several years ago to direct my doctoral dissertation on this mysterious and then-unstudied theologian; and to the Saint Meinrad School of Theology for giving me a job with the encouragement to research and write. I would like to thank Fr. Luke Hassler for helping to put together the bibliography. Most of all, I thank God, who, in his mysterious Wisdom, has given me life and set me on a path that I could never have imagined for myself.

BIBLIOGRAPHY

THE TRILOGY ON THE ECONOMY

Mary: The Seat of Wisdom. An Essay on the Place of the Virgin Mary in Christian Theology. Trans. Fr. A.V. Littledale. New York, NY: Pantheon Books, 1962.

Cosmos. Le Monde et la Gloire de Dieu. Paris: Les Éditions du Cerf, 1982

Cosmos. The World and the Glory of God. Trans. Pierre de Fontnouvelle. Petersham, MA: St. Bede's Publications, 1988.

The Church of God: Body of Christ and Temple of the Spirit. Trans. Charles Underhill Quinn. San Francisco: Ignatius Press, 2011.

THE TRILOGY ON THEOLOGY

The Eternal Son. A Theology of the Word of God and Christology. Trans. Sister Simone Inkel, S.L., and John F. Laughlin. Huntington, IN: Our Sunday Visitor, Inc. 1978.

The Invisible Father. Approaches to the Mystery of the Divinity. Trans. Hugh Gilbert, OSB. Petersham, MA: St. Bede's Publications, 1999.

Le Consolateur. Esprit Saint et Vie de Grâce. Paris: Les Éditions du Cerf, 1980.

THE TRILOGY ON KNOWLEDGE OF FAITH

The Christian Mystery: From Pagan Myth to Christian Mysticism. Trans. Illtyd Trethowan. Petersham, MA: Saint Bede's Publications, 1995.

Gnôsis. La Connaisance de Dieu dans l'Écriture. Paris: Les Éditions du Cerf, 1988.

Sophia ou le Monde en Dieu. Paris: Les Éditions du Cerf, 1994.

OTHER BOOKS BY BOUYER CITED

Christian Humanism. Trans. A.V. Littledale. Westminster, MD: Newman Press, 1969.

Christian Initiation. Trans. J.R. Foster. London: Burns & Oates, 1960.

Dictionary of Theology. Trans. Charles Underhill Quinn. New York, NY: Desclée and Co., 1965.

En Quête de la Sagesse: Du Parthénon à l'Apocalypse en Passan par la Nouvelle et la Troisième Rome. Jouques: Éditions du Cloître, 1980.

Erasmus and His Times. Trans. Francis X. Murphy. Westminster, MD: The Newman Press, 1959.

Eucharist: Theology and Spirituality of the Eucharistic Prayer. Trans. Charles Underhill Quinn. South Bend, IN: University of Notre Dame Press, 1968.

Introduction to the Spiritual Life. Trans. Mary Perkins Ryan and Michael Heintz. South Bend, IN: Notre Dame Christian Classics, 2013.

La Vie de Saint Antoine. Essai sur la Spiritualité du Monachisme Primitif. Saint Wandrille: Éditions de Fontenelle, coll. "Figures Monastique," 1950.
Lectures et Voyages: Compléments aux Mémoires. Paris: Ad Solem, 2016.
Le Décomposition du Catholicisme. Paris: Aubier-Montaigne, 1968.
Le Métier de Théologien. Entretiens avec Georges Daix. Genève: Ad Solem, 2005.
Les Lieux Magiques de la Légende du Graal: De Brocéliande en Avalon. Paris: OEIL, 1986.
L'Incarnation et l'Église: Corps du Christ dans la Théologie de Saint Athanase. Coll. Unam Sanctam, no. 11. Paris: Les Éditions du Cerf, 1943.
Liturgy and Architecture. South Bend, IN: University of Notre Dame Press, 1967.
Liturgical Piety. South Bend, IN: University of Notre Dame Press, 1955.
Memoirs: From Youth and Conversion to Vatican II, the Liturgical Reform, and After. Trans. John Pepino. Kettering, OH: Angelico Press, 2015.
Newman's Vision of Faith: A Theology for Times of General Apostasy. San Francisco: Ignatius Press, 1986.
Orthodox Spirituality and Protestant and Anglican Spirituality. Trans. Barbara Wall. London: Burns & Oates, 1965.
Religieux et Clercs contre Dieu. Paris: Aubier-Montaigne, 1975.
Rite and Man. Trans. M. Joseph Costelloe, SJ. South Bend, IN: Notre Dame Press, 1963.
Sir Thomas More: Humaniste et Martyr. Chambray-lès-Tours: CLD coll. "Veilleurs de la Foi," no. 2, 1984.
The Cistercian Heritage. Trans. Elizabeth Livingstone. Westminster, MD: The Newman Press, 1958.
The Meaning of the Monastic Life. Trans. Kathleen Pond. New York: P.J. Kennedy and Sons, 1955.
The Meaning of Sacred Scripture. Trans. Mary Perkins Ryan. South Bend, IN: University of Notre Dame Press, 1958.
The Paschal Mystery: Meditations on the Last Three Days of Holy Week. Trans. Mary Benoit. Chicago: Regnery, 1950.
The Spirituality of the New Testament and the Fathers: History of Christian Spirituality, Vol. I. Trans. Mary P. Ryan. New York: Desclée and Co., 1963.
The Spirit and Forms of Protestantism. Trans. A.V. Littledale. Princeton, NJ: Scepter Publishers, 2001.
Vérité des Icônes: la Tradition Iconographique Chrétienne et sa Signification. Paris: Éditions Criterion, 1990.
Woman in the Church. Trans. Marilyn Teichert. San Francisco: Ignatius Press, 1979.
Women Mystics. Trans. Ann E. Nash. San Francisco: Ignatius Press, 1993.

PSEUDONYMOUS BOOK BY BOUYER CITED

Lambert, Louis. *Prélude à l'Apocalypse ou les Derniers Chevaliers du Graal*. Limoges: Critérion, 1982.

OTHER WORKS BY BOUYER CITED

"An Introduction to the Theme of Wisdom and Creation in the Tradition." *Le Messager Orthodoxe* 98 (1985): 149–61.

"Christianisme et Eschatologie." *La Vie Intellectuelle* 16.10 (1948): 6–38.

"Développements Récent de la Psychologie en Suisse." *La Vie Intellectuelle* 15.12 (1947): 98–117.

"La Colonne et le Fondement de la Vérité De P. Florensky." *Communio* 2 (1975): 95.

"La Découverte de l'Inconscient." *France Catholique*, 1264 (1974). https://www.france-catholique.fr/LA-DECOUVERTE-DE-L-INCONSCIENT.html.

"Le Malheur d'Avoir des Disciples (à propos de deux ouvrages sur le Père Teilhard de Chardin)." *France Catholique*, 626, 627, 628 (1958). https://www.france-catholique.fr/-Articles-du-R-P-Louis-Bouyer-parus-dans-France-Catholique-de-1957-a-1987-.html?debut_articles=75#pagination_articles.

"La Notion Christologique du Fils de l'Homme a-t-elle Disparu dans la Patristique Grecque?" Editors W.F. Albright, F. Amiot, P. Auvray. *Mélanges Biblique Rédigés en l'Honneur de André Robert*. Paris: Bloud et Gay, coll. "Travaux de l'institut Catholique de Paris," n. 4, 1957, 519–30.

"La Personnalité et l'Ouevre de Serge Boulgakoff (1871–1944)." *Nova et Vetera* 53.2 (1978): 135–44.

"La Situation de la Théologie." *Communio* 1.1 (1975): 41–48.

"L'Exemple de Pic de la Mirandola." *Communio* 1.5 (1976): 94.

"Le Problème du Mal dans la Christianisme Antique." *Dieu Vivant* 6 (1946): 15–42.

"Le Seigneur des Anneaux, Une Nouvelle Epopee?" *La Tour Saint-Jacques* 13.4 (1958).

"Le Souvenir de Mary Newman." *Études* 246 (1945): 145–59.

"Mystique Cosmiques et Mystique Interpersonnelles." Editors Hans Urs von Balthasar, Louis Bouyer, Olivier Clément. *Des Bords du Gange aux Rives du Jourdain*, 2nd ed. Paris: Saint Paul, 1983, 150–53.

"Newman and English Platonism." *Monastic Studies* 1.11 (1963): 285–305.

"Newman's Influence in France." *The Dublin Review* 435 (1945): 182–88.

"Poésie, Philosophie et Sagesse Chrétienne." Unpublished manuscript. Abbey of Saint Wandrille, November 1986, 18 pp.

Religio Poetae. La Poésie Moderne à la Redécouverte du sacré. Unpublished manuscript. 170 pp.

"Royauté Cosmique." *La Vie Spirituelle* 110 (1964): 387–97.

"The Permanent Relevance of Newman." *Newman Today. Proceedings of the Conference on John Henry Newman*. Wethersfield Institute, New York, October 14–15, 1988, ed. by Stanley Jaki, 165–74. San Francisco: Ignatius Press, 1989.

"The Two Economies of Divine Government: Satan and Christ in the New Testament and Early Christian Tradition." *Letter & Spirit* 5 (2009): 237–62.

WORKS BY OTHER AUTHORS CITED

Armogathe, Jean-Robert. "Obituary for Louis Bouyer." Trans. Adrian Walker. *Communio* 31.4 (2004): 688–89.

Ashley, Benedict M. *Theologies of the Body: Humanist and Christian*. Braintree, MA: Pope John XXIII Medical-Moral Research and Education Center, 1985.

Aspray, Silvianne. "A Complex Legacy: Louis Bouyer and the Metaphysics of the Reformation." *Modern Theology* 34.1 (January 2018): 13–22.

Balthasar, Hans Urs von. "A Resume of My Thought." Trans. Kelly Hamilton. *Ignatius Insight*. March 5, 2005. www.ignatiusinsight.com/features2005/hub_resumethought_mar05.asp.

—. *Cosmic Liturgy: The Universe According to Maximus the Confessor*. Trans. Brian Daley. San Francisco: Ignatius Press, 2003.

—. *Explorations in Theology*, vol. 1. Trans. A.V. Littleday and Alexander Dru. San Francisco: Ignatius Press, 1989.

—. *Man in History: A Theological Study*. London: Sheed and Ward, 1968.

—. *Presence and Thought: Essay on the Religious Philosophy of Gregory of Nyssa*. Trans. Mark Sebanc. San Francisco: Ignatius Press, 1995.

—. *The Office of Peter and the Structure of the Church*. Trans. Andrée Emery. San Francisco: Ignatius Press, 1986.

—. *The Paschal Mystery: The Mystery of Easter*. Trans. Aidan Nichols. San Francisco, CA: Ignatius Press, 2005.

—. *Theo-Logic III: The Spirit of Truth*. Trans. Graham Harrison. San Francisco: Ignatius Press, 2005.

Barfield, Owen. *Saving the Appearances: A Study in Idolatry*. Middletown, CT: Wesleyan University Press, 1988.

Barth, J. Robert. *The Symbolic Imagination: Coleridge and the Romantic Tradition*. Princeton, NJ: Princeton University Press, 1977.

Baur, Friederich Christian. *Die Epochen der Kirchlichen Geschichtsschreibung*. Tübingen: Drud und Verlag Ludwich Friederich Fues, 1852.

Beauregard, Olivier Costa de. *La Physique Moderne et les Pouvoirs de l'Espirit*. Paris: Greco, 1981.

—. *Le Second Principe de la Science Du Temps*. Paris: Editions Du Seuil, 1963.

—. *Time, the Physical Magnitude*. Boston, MA: D. Reidel Pub. Co., 1987.

Bellah, Robert. *Religion in Human Evolution: From the Paleolithic to the Axial Age*. Harvard University Press, 2011.

Benedict XVI. *Deus Caritas Est*. Libreria Editrice Vaticana, 2006.

—. "Faith, Reason and University Memories and Reflections." [Regensburg Address] September 12, 2006.

Bennett, Jane. *Vibrant Matter*. Durham, NC: Duke University Press, 2010.

Bevan, Edwyn. *Stoics and Sceptics*. Oxford: Clarendon Press, 1913.

Birot, Antoine. "Bouyer, Entre Thomas et Balthasar." *Laval Théologique et Philosophique* 67.3 (2011): 501–29.

Birzer, Bradley J. *Sanctifying the World: the Augustinian Life and Mind of Christopher Dawson*. Front Royal, VA: Christendom Press, 2007.

Blanchette, Oliva. *Maurice Blondel: A Philosophical Life*. Grand Rapids, MI: Eerdmans, 2010.

Blowers, Paul M. *Maximus the Confessor: Jesus Christ and the Transfiguration of the World*. Oxford University Press, 2016.

Boersma, Hans. *Heavenly Participation: the Weaving of a Sacramental Tapestry*. Grand Rapids, MI: Eerdmans Press, 2011.

—. *Nouvelle Théologie and Sacramental Ontology: A Return to Mystery*. Oxford University Press, 2009.

Bouwé, Christian-Noël. *L'Union Conjugale et le Sens du Sacré: La Sacramentalité du Mariage dans la Théologie de Louis Bouyer*. Paris: Les Éditions du Cerf, 2016.

Braeckman, Antoon. "Whitehead and German Idealism: A Poetic Heritage." *Process Studies* 14.4 (1985): 265–86.

Buber, Martin. *Ich und Du*. Leipzig: Im-Insel Verlag, 1923.

Bulgakov, Sergei. *Philosophy of Economy: The World as Household*. Trans. Catherine Evtuhov. New Haven, CT: Yale University Press, 2000.

—. *Sophia, the Wisdom of God: An Outline of Sophiology*. Trans. Patrick Thompson, et al. Hudson, NY: Lindisfarne Press, 1993.

—. *The Bride of the Lamb*. Trans. Boris Jakim. Grand Rapids, MI: Eerdmans Press, 2002.

—. *The Burning Bush: On the Orthodox Veneration of the Mother of God*. Trans. Thomas Allen Smith. Grand Rapids, MI: Eerdmans Press, 2009.

—. *The Friend of the Bridegroom: On the Orthodox Veneration of the Forerunner*. Trans. Boris Jakim. Grand Rapids, MI: Eerdmans Press, 2003.

—. *The Holy Grail and the Eucharist*. Trans. and ed. Boris Jakim. Herndon, VA: Lindisfarne Books, 1997.

—. *The Lamb of God*. Trans. Boris Jakim. Grand Rapids, MI: Eerdmans Press, 2008.

Bultmann, Rudolf. "New Testament and Mythology: The Problem of Demythologizing the New Testament Proclamation." In S.M. Ogden, ed., *New Testament and Mythology and Other Basic Writings*, 1–43. Philadelphia: Fortress, 1984.

Cabié, R. "Quand on Commençait à Parler du Mystère Pascal." *La Maison Dieu* 240 (2004): 7–19.

Caldecott, Stratford. *Beauty for Truth's Sake: On the Re-Enchantment of Education*. Grand Rapids, MI: Brazos Press, 2009.

—. *The Radiance of Being: Dimensions of Cosmic Christianity*. Kettering, OH: Angelico Press, 2013.

Caldecott, Stratford, and Adrian Walker. "The Light of Glory: From *Theosis* to Sophiology." *Communio* 42 (Summer 2015): 252–64.

Casado, Carolina Blázquez. *La Gloria de Dios en la Entraña del Mundo: Olivier Clément y Louis Bouyer. Un Estudio en Perspectiva Ecuménica de Dos Cosmovisiones Cristianas. Disertación para la Obtención del Doctorado en Teología Dogmática*. Salmanca, 2014.

Cassirer, Ernst. *The Philosophy of Symbolic Forms*, vol. 2: *Mythical Thought*. Trans. Ralph Manheim. New Haven, CT: Yale University Press, 1955.

Chardin, Pierre Teilhard de. *Christianity and Evolution*. Trans. René Hague. New York: Harcourt Brace Jovanovich, 1971.

Chenu, M.D. *La Théologie au XII Siècle*. Coll. "Études de Philosophie Médiévale," n. 45. Paris: Vrin, 1957.

Chroust, Anton-Hermann. *Aristotle: New Light on His Life and on Some of His Lost Works*. London: Routledge, 1973.

Clarke, William Norris. *The One and the Many: A Contemporary Thomist Metaphysics*. South Bend, IN: University of Notre Dame Press, 2001.

—. "Living on the Edge: The Human Person as 'Frontier Being' and Microcosm." In *The Creative Retrieval of Saint Thomas Aquinus. Essays in Thomistic Philosophy*, 132–51. New York: Fordham University Press, 2009.

Clément, Olivier. *Olivier Clément, Des Bords du Gange aux Rives du Jourdain*. 2nd ed. Paris: Saint Paul, 1983.

Coleridge, Samuel Taylor. *Biographia Literaria*. Vol. 1. Ed. James Engell and W. Jackson Bate. Princeton, NJ: Princeton University Press, 1983.

—. *Shorter Works and Fragments*. Ed. H. J. Jackson and James Robert de Jager. Princeton, NJ: Princeton University Press, 1995.

—. *The Statesman's Manual*, in *Lay Sermons*, ed. R.J. White, vol. 6 of *The Collected Works of Samuel Taylor Coleridge*. Ed. Kathleen Coburn. Princeton, NJ: Princeton University Press, 1972.

Collingwood, R. G. *The Idea of Nature*. Oxford University Press: 1970.

Congar, Yves. "Le Père, Source Absolue de la Divinité." *Istina* 27 (1980).

Connolly, James M. *Human History and the Word of God: The Christian Meaning of History in Contemporary Thought*. New York: Macmillan Company, 1965.

Corbin, Henry. *Spiritual Body and Celestial Earth from Mazdean Iran to Shī'ite Iran*. Trans. Nancy Pearson. Princeton, NJ: Princeton University Press, 1977.

Costache, Doru. "Making Sense of the World: Theology and Science in St. Gregory of Nyssa's *An Apology for the Hexaemeron*." *Phronema* 28.1 (2013): 1–28.

Dawkins, Richard. *The Selfish Gene*. Oxford University Press, 1976.

Dawson, Christopher. *Progress and Religion: An Historical Inquiry*. Washington DC: Catholic University of America Press, 2001.

Deane-Drummond, Celia. *Creation through Wisdom: Theology and the New Biology*. Edinburgh: T & T Clarke, 2000.

—. *Wonder and Wisdom: Conversations in Science, Spirituality, and Theology*. Philadelphia: Templeton Foundation Press, 2006.

Debaise, Didier. *Nature as Event*. Durham, NC: Duke University Press, 2017.

De Boer, Martin C. "Paul's Mythologizing Program in Romans 5:8." In *Apocalyptic Paul: Cosmos and Anthropos in Romans 5–8*, ed. Beverly Roberts Gaventa, 1–20. Waco, TX: Baylor University Press, 2016.

Delio, Ilia. *Christ in Evolution*. New York, NY: Orbis Books, 2008.

de Lubac, Henri. *The Drama of Atheist Humanism*. Trans. Marc Sebanc. San Francisco: Ignatius Press, 1995.

—. *Theology of History*. Trans. A.E. Nash. San Francisco: Ignatius Press, 1996.

—. "The Problem of the Development of Dogma." *Theology in History*, Trans. A. E. Nash. San Francisco: Ignatius Press, 1996.

Devaux, Michaël. "'Le Probleme de l'Imagination et de la Foi,' chez Louis Bouyer: En Lisant, en Écrivant avec les Inklings." In *La Théologie de Louis Bouyer: Du Mystère à la Sagesse*, ed. Bertrand Lesoing, Marie-Hélène Grintchenko, and Patrick Prétot,

141–62. Paris: Parole et Silence, 2014.

Duchesne, Jean. "L'Enracinement dans le Judaïsme du Mystère Chrétien." In *La Théologie de Louis Bouyer: Du Mystère à la Sagesse*, 179–90. Paris: Parole et Silence, 2014.

—. "Les Fondements, les Apports et les Enjeux Christologie de Louis Bouyer." *Groupe de recherche sur l'œuvre de Louis Bouyer*. Presented at Collège des Bernardins, April 30, 2018.

—. *Louis Bouyer*. Perpignan: Artéges Spiritualité, 2011.

Dupont, Dom Jacques. *Gnosis*. Louvain, 1949.

Dupré, Louis. "Newman and the Neoplatonic Tradition in England." In *Newman and the Word*, ed. Terrence Merrigan and Ian T. Ker, 137–54. Grand Rapids, MI: Eerdmans, 2000.

Eckhart, Meister. *Parisian Questions*. Trans. Armand A. Maurer. Toronto: Pontifical Institute for Medieval Studies, 1974.

Eliade, Mircea. *Images and Symbols: Studies in Religious Symbolism*. Trans. Philip Mairet. Princeton University Press: Princeton, NJ, 1991.

—. *Myth and Reality*. Trans. William R. Trask. San Francisco: Harper and Row, 1963.

—. *The Myth of the Eternal Return*. Trans. Willard R. Trask. Princeton, NJ: Princeton University Press, 2005.

—. *The Quest: History and Meaning of Religion*. Chicago: University of Chicago Press, 1969.

Ellenberger, H.F. *The Discovery of the Unconscious: The History and Evolution of Dynamic Psychiatry*. New York: Basic Books, 1970.

Faivre, Antoine. "Theosophy." *Encyclopedia of Christian Theology*, vol. 3. Ed. Jean-Yves Lacoste. New York, NY: Routledge, 2005.

Falque, Emmanuel. *Crossing the Rubicon: The Borderlands of Philosophy and Theology*. Trans. Reuben Shank. New York, NY: Fordham University Press, 2016.

Fernandez, Irène. *Mythe, Raison Ardente: Imagination et Réalité selon C.S. Lewis*. Geneva: Ad Solem, 2005.

Ferrar, Austin. *A Rebirth of Images: The Making of Saint John's Apocalypse*. Albany, NY: State University of New York Press, 1986.

Festugière, A. J. *De l'Essence de la Tragédie*. Paris: Aubier-Montagne, 1969.

—. *Socrates*. Paris: Éditions du Fuseau, 1932.

Feuillet, A. *Le Christ Sagesse de Dieu d'après les Épîtres Pauliniennes*. Paris, 1966.

Fletcher, Patrick J. *Resurrection Realism: Ratzinger the Augustinian*. Eugene, OR: Cascade Books, 2014.

Florensky, P. *The Pillar and Ground of Truth*. Trans. Boris Jakim. Princeton University Press, 2004.

Foltz, Bruce. "How Not to Speak of Sophia." *The Heavenly Country: An Anthology of Primary Sources, Poetry, and Critical Essays on Sophiology*. Ed. Michael Martin. Kettering, OH: Angelico Press, 2016.

Frankfort, Henri, et al. *The Intellectual Adventure of Ancient Man: An Essay on Speculative Thought in the Ancient Near East*. The University of Chicago Press, 1977.

Garfitt, Toby. "Newman at the Sorbonne, or the Vicissitudes of an Important Philosophical Heritage in Inter-war France." *History of European Ideas* 40.6 (2014): 788–803.

Garrigues, Juan-Miguel. "Le Christ Nouvel Adam, Mais Aussi 'Nouveau Lucifer.'" *La Théologie de Louis Bouyer*. Ed. Bertrand Lesoing, Marie-Hélène Grintchenko, and Patrick Prétot, 12–39. Paris: Parole et Silence, 2016.

—. "Le Dessein d'Adoption du Créator dans son Rapport au Fils d'après S. Maxime le Confesseur." In *Maximos Confessor. Actes du Symposium sur Maxime le Confesseur, Fribourg, 2–5 sept. 1980*, ed. F. Heinzer and C. Schöborn, 173–92. Coll. "Paradosis. Études de Littérature et de Théologie Anciennes," n. 27. Freiburg: Éditions du Universitaires, 1982.

Gavin, John. *A Celtic Christology: The Incarnation According to John Scotus Eriugena*. Eugene, OR: Cascade Books, 2014.

Geiselmann, Josef Rupert. *The Meaning of Tradition*. Trans. W.J. O'Hara. New York: Herder and Herder, 1966.

Gilbert, Hugh. *Unfolding the Mystery: Monastic Conferences on the Liturgical Year*. Herefordshire: Gracewing, 2007.

Gilson, Étienne. *The Philosophy of Saint Bonaventure*. Trans. Dom Illtyd Trethowan. New York, NY: Desclée and Co., 1965.

Grintchenko, Marie-Hélène. *Cosmos, Une Approche Théologique du Monde: Cosmos du Père Louis Bouyer*. Paris: Parole et Silence, 2015.

Hadot, Pierre. *Philosophy as a Way of Life: Spiritual Exercises from Socrates to Foucault*. Trans. Arnold Ira Davidson. Malden, MA: Blackwell, 1995.

Hart, David Bentley. *The Beauty of the Infinite: The Aesthetics of Christian Truth*. Grand Rapids, MI: Eerdmans Press, 2004.

Haught, John. *The New Cosmic Story: Inside Our Awakening Universe*. New Haven, CT: Yale University Press, 2017.

—. *Resting on the Future: Catholic Theology for an Unfinished Universe*. New York, NY: Bloomsbury, 2015.

Hedley, Douglas. *Living Forms of the Imagination*. New York, NY: T & T Clark International, 2008.

Heidegger, Martin. *Existence and Being*. Chicago: Regnery, 1949.

Heintz, Michael. "Mariology as Theological Anthropology: Louis Bouyer on Mary, Seat of Wisdom." *Mary on the Eve of the Second Vatican Council*, ed. John C. Cavadini and Danielle M. Peters. South Bend, IN: University of Notre Dame Press, 2017.

Heisenberg, Werner. *Physics and Philosophy: The Revolution in Modern Science*. New York: Harper and Brothers Publishers, 1958.

Holzer, Vincent. "Karl Rahner, Hans Urs Von Balthasar, and Twentieth-Century Catholic Currents on the Trinity." In *The Oxford Handbook of the Trinity*, ed. Gilles Emery, OP, and Matthew Levering, 314–27. Oxford University Press, 2014.

Horton, W.M. *The Philosophy of Abbé Bautain*. New York: New York University Press, 1926.

Howsare, Rodney A. *Balthasar: A Guide for the Perplexed*. New York, NY: T & T Clark, 2009.

Hurlbut, William. "From Biology to Biography." *The New Atlantis* 3 (2003): 47–66.

—. "Saint Francis, Christian Love, and the Biotechnological Future." *The New Atlantis* 13 (2013): 92–99.

Jaeger, Werner. *Aristotle: Fundamentals of the History of His Development*. Ed. Richard Robinson. 2nd ed. Oxford: Clarendon Press, 1948.

Jaki, Stanley. *The Origin of Science and the Science of Its Origin*. South Bend, IN: Gateway Editions, 1978.

—. *The Road to Science and the Ways to God: The Gifford Lectures of 1975–1976*. Chicago: University of Chicago Press, 1978.

John Paul II. *Homme et Femme Il Les Créa: Une Spiritualité du Corps*. Paris: Les Éditions du Cerf, 2010.

Käsemann, Ernst. "The Beginnings of Christian Theology." In *New Testament Questions Today*, 82–107. Trans. W.J. Montague. London: SCM Press, 1960.

Ker, Ian. *Newman on Vatican II*. Oxford University Press, 2014.

Kerr, Fergus. *Twentieth-Century Catholic Theologians: From Neoscholasticism to Nuptial Mysticism*. Malden, MA: Blackwell, 2007.

Kołakowski, Leszek. *Bergson*. Oxford University Press, 1985.

Kreeft, Peter. *The Philosophy of Tolkien: The Worldview behind* The Lord of the Rings. San Francisco: Ignatius Press, 2005.

Latour, Bruno. *Facing Gaia: Eight Lectures on the New Climactic Regime*. Trans. Catherine Porter. Medford, MA: Polity Press, 2017.

Leamy, Katy. *The Holy Trinity: Hans Urs Von Balthasar and His Sources*. Eugene, OR: Pickwick Publications, 2015.

Leclerc, Gérard. "Cosmique Théologie." *France Catholique*, 23 June 2015. https://www.france-catholique.fr/Theologie-cosmique.html.

Leeuw, G. van der. *Religion in Essence and Manifestation: A Study in Phenomenology*. Trans. J. E. Turner. New York, NY: Harper and Row, 1963.

Lemna, Keith. "Louis Bouyer and Alfred North Whitehead: A Dialogue in Trinitarian Cosmology." *Gregorianum* 95.4 (2014).

—. "Louis Bouyer's Development of Cardinal Newman's Sacramental Principle." *The Journal for Newman Studies* 13:1 (Spring 2016): 22–42.

—. "The Angels and Cosmic Liturgy: An Oratorian Angelology," *Nova et Vetera* 8, no. 4 (2010): 901–921.

—. "Trinitarian Panentheism: A Study of the God-World Relation in the Theology of Louis Bouyer." Doctoral Thesis. The Catholic University of America, 2007.

Lesoing, Bertrand. *Vers la Plénitude du Christ: Louis Bouyer et L'Oecuménisme*. Les Éditions du Cerf, 2017.

Lesoing, Bertrand, Marie-Hélène Grintchenko, Patrick Prétot, eds., *La Théologie de Louis Bouyer: Du Mystère à la Sagesse*. Paris: Parole et Silence, 2016.

Levering, Matthew. *An Introduction to Vatican II as an Ongoing Theological Event*. Washington DC: Catholic University of America Press, 2017.

Lévy-Bruhl, Lucien. *How Natives Think*. Trans. Lilian Ada Clare. New York: G. Allen & Unwin, 1926.

Lewis, C. S. *The Discarded Image: An Introduction to Medieval and Renaissance Literature*. Cambridge University Press, 1964.

—. *God in the Dock: Essays on Theology and Ethics*. Grand Rapids: William B. Eerdmans Publishing Company, 1970.

Lossky, Vladimir. *The Mystical Theology of the Eastern Church*. Crestwood, NY: Saint Vladimir's Seminary Press, 1998.

Louth, Andrew. *Maximus the Confessor*. London: Routledge, 1996.

—. *Modern Orthodox Thinkers: From the "Philokalia" to the Present*. Downer's Grove, IL: Intervarsity Press Academic, 2015.

—. "The Cosmic Vision of Saint Maximus the Confessor." In *In Whom We Live and Move and Have Our Being: Panentheistic Reflections on God's Presence in the World*, ed. Philip Clayton and Arthur Peackocke, 184–96. Grand Rapids, MI: Eerdmans Publishing Company, 2004.

Luce, A. A. *Sense without Matter; or, Direct Perception*. Edinburgh: Nelson, 1954.

Lutosławski, Wincenty. *The Origin and Growth of Plato's Logic, with an Account of Plato's Logic, with an Account of Plato's Style and of the Chronology of His Writings*. London: Longmans, Green, and Co., 1897.

MacIntyre, Alasdair. *After Virtue*. London: Duckworth, 1981.

—. *Three Rival Versions of Moral Enquiry: Encyclopedia, Genealogy, and Tradition*. South Bend, IN: University of Notre Dame Press, 1990.

Maldamé, J. M. "Sciences de la Nature et Théologie." *Revue Thomiste* 86 (1986): 283–308.

Marcel, Gabriel. *Coleridge et Schelling*. Paris: Aubier-Montagne, 1971.

Marion, Jean-Luc. "Cinq Remarques sur l'Originalité de l'Oeuvre Louis Bouyer." *Groupe de Recherche sur l'Œuvre de Louis Bouyer*, Nov. 9, 2016. https://www.collegedesbernardins.fr/formation/comptes-rendus-des-seances.

—. *The Crossing of the Visible*. Trans. James K.A. Smith. Stanford University Press, 2004.

—. "Introduction: L'Unité Organique d'une Oeuvre." *La Théologie de Louis Bouyer: Du Mystère à la Sagesse*. Paris: Parole et Silence, 2014.

—. *The Rigor of Things: Conversations with Dan Arbib*. Trans. Christina M. Gschwandtner. Fordham University Press, 2017.

Martin, Jennifer Newsome. *Hans Urs von Balthasar and the Critical Appropriation of Russian Religious Thought*. South Bend, IN: University of Notre Dame Press, 2015.

—. "True and Truer Gnosis: The Revelation of the Sophianic in Hans Urs von Balthasar." *Heavenly Country: An Anthology of Primary Sources, Poetry, and Critical Essays on Sophiology*, ed. Michael Martin, 339–64. Kettering, OH: Angelico Press, 2016.

Martin, Michael. *The Submerged Reality: Sophiology and the Turn to a Poetic Metaphysics*. Kettering, OH: Angelico Press, 2015.

Mascall, E. L. *Christian Theology and Natural Science*. Archon Books, 1965.

McCool, Gerald. *From Unity to Pluralism*. New York: Fordham Univ. Press, 1989.

McGrath, S. J. *The Dark Ground of Spirit: Schelling and the Unconscious*. New York: Routledge, 2012.

Michel, Aimé. "La Gnose de Princeton." *France Catholique* 207, no. 1487 (June 13, 1975).

—. *Métanoia: Phénomènes Physique du Mysticisme*. Paris: Albin Michel, 1973.

—. "Sur le Seuil de la Nouvelle Physique." *France Catholique*, vol. 294, no. 1613 (November 11, 1997). https://www.france-catholique.fr/SUR-LE-SEUIL-DE-LA-NOUVELLE-PHYSIQUE.html.

Midgley, Mary. *Science and Poetry*. New York: Routledge, 2006.

Milbank, John. "Sophiology and Theurgy: The New Theological Horizon." *Encounter Between Eastern Orthodoxy and Radical Orthodoxy: Transfiguring the World Through the Word*, ed. Adrian Pabst and Christoph Schneider, 45–90. Surrey, UK: Ashgate, 2009.

—. *The Word Made Strange: Theology, Language, Culture*. London: Blackwell, 1997.

Mowinckel, Sigmund. *He That Cometh*. Trans. George Anderson. New York: Abingdon Press, 1968.

Nault, Jean-Charles, OSB. "La Contribution à une Théologie de la Vie Monastique." In *La Théologie de Louis Bouyer: Du Mystère à la Sagesse*. Ed. Bertrand Lesoing, Bertrand, Marie-Hélène Grintchenko, and Patrick Prétot. Paris: Parole et Silence, 2016.

Nédoncelle, Maurice. *La Réciprocité des Consciences*. Paris, 1942.

Newman, John Henry. *Apologia pro Vita Sua*. London: Longman's, Green, and Co., 1908.

—. *Arians of the Fourth Century*. New York: Longmans, Green, and Co., 1908.

—. *An Essay in Aid of a Grammar of Assent*. London: Longman's, Green, and Co., 1903.

—. "The Mission of St. Benedict." *The Atlantis*, January 1858.

—. *The Philosophical Notebook of John Henry Newman*. Volume I: *General Introduction to the Study of Newman's Philosophy*. Ed. Edward Sillem. New York: Humanities Press, 1969.

—. *Sermons and Discourses*. New York: Longmans, Green, and Company, 1949.

Nichols, Aidan. *Wisdom from Above: A Primer in the Theology of Sergei Bulgakov*. Herefordshire, UK: Gracewing, 2005.

Nieuwenhove, Rik Van. *Jan Van Ruusbroec, Mystical Theologian of the Trinity*. South Bend, IN: University of Notre Dame Press, 2003.

Novotny, Vojtech. *Cur Homo? A History of the Thesis Concerning Man as a Replacement for Fallen Angels*. Prague: Karolinum Press, 2014.

O'Callaghan, Paul. *Christ Our Hope: An Introduction to Eschatology*. Washington, DC: The Catholic University of America Press, 2011.

Olsen, Cyrus. "Myth and Culture in Louis Bouyer: On Louis Bouyer's Theology of Participation." *Gregorianum* 95.4 (2014).

Onfray, Michel. *Cosmos: Une Ontologie Matérialiste*. Paris: Flammarion, 2015.

Pieper, Josef. *The Platonic Myths*. Trans. Daniel J. Farrelly. South Bend, IN: Saint Augustine's Press, 2011.

Rad, Gerhard von. *Old Testament Theology II: The Theology of Israel's Prophetic Tradition*. Trans. D. M. G. Stalker. Edingburgh: Oliver and Boyd, 1965.

Rahner, Hugo. *Greek Myths and Christian Mystery*. Trans. B. Battershaw. New York: Harper and Row, 1963.

Rahner, Karl. "Dieu dans le Nouveau Testament. La Signification du Mot 'Theos.'" *Écrits Théologiques* 1 (1959): 11–111.

Ramsey, Ian T. *Religion and Science: Conflict and Synthesis*. London: S.P.C.K., 1964.

Ratzinger, Joseph. *Collected Works*, vol. 11: *Theology of the Liturgy*. Trans. John Saward, et al. San Francisco: Ignatius Press, 2014.

—. *Eschatology*. Trans. Michael Waldstein. Washington, DC: The Catholic University of America Press, 1988.

—. *Milestones: Memoirs 1927–1977*. Trans. Erasmo Leiva-Merikakis. San Francisco: Ignatius Press, 1998.

Régnon, Théodore de. *Études de Théologie Positive sur le Dogme de la Trinité*. Paris: V. Retaux et Fils, 1892.

Rémur, Guillaume Bruté de. *La Théologie Trinitaire de Louis Bouyer*. Rome: Pontificia Università Gregoriana, 2010.

Riches, Aaron. "Christology and *Duplex Hominis Beatitudo*: Re-Sketching the Supernatural Again." *International Journal of Systematic Theology* 14 (January 2012): 44–69.

—. "*Theotokos*: Sophiology and Christological Overdetermination of the Secular." In *The Heavenly Country: An Anthology of Primary Sources, Poetry, and Critical Essays on Sophiology*, ed. Michael Martin, 269–84. Kettering, OH: Angelico Press, 2016.

Ries, Julien. *L'Homme et le Sacré*. Paris: Les Éditions du Cerf, 2009.

—. *L'«Homo Religiosus» et Son Expérience du Sacré*. Paris: Les Éditions du Cerf, 2009.

—. *Symbole, Mythe et Rite: Constants du Sacré*. Paris: Les Éditions du Cerf, 2012.

Riesenfeld, H. *The Gospel Tradition*. Philadelphia: Fortress Press, 1970.

Rogers, Katherin A. "Omniscience, Eternity, and Freedom." *International Philosophical Quarterly* 36.4 (1996): 399–412.

Ross, William David. *Aristotle*. London: Methuen & Co., 1923.

Rousseau, Olivier. "*Le Sens de la Vie Monastique*: À Propos d'un Ouvrage Récent." *Revue Générale Belge* (Oct. 15, 1953): 957–64.

Rowland, Christopher. *The Open Heaven*. London: SPCK, 1982.

Russell, D. S. *Divine Disclosure: An Introduction to Jewish Apocalyptic*. Minneapolis, MN: Fortress Press, 1992.

Rutledge, Dom Denys. *Cosmic Theology, the Ecclesiastical Hierarchy of Pseudo-Denys: An Introduction*. Staten Island, NY: Alba House, 1965.

Ruyer, Raymond. *La Gnose de Princeton*. Paris: Fayard 1974.

—. *Neo-Finalism*. Trans. Aloysha Edlebi. Minneapolis, MN: University of Minnesota Press, 2016.

Schipflinger, Thomas. *Sophia-Maria: A Holistic Vision of Creation*. Trans. James Morgante. York Beach, ME: Samuel Weiser, Inc., 1998.

Schooping, Joshua. "Touching the Mind of God: Patristic Christian Thought on the Nature of Matter." *Zygon* 50.3 (September 2015): 583–603.

Sécretan, Charles. *La Philosophie de la Liberté*. Volume I. Paris: Chez L. Hachette, 1849.

Séd, Nicolas-Jean. "Tradition Juive et Christianisme." In *La Théologie de Louis Bouyer: Du Mystère à la Sagesse*, 191–95. Paris: Parole et Silence, 2014.

Servais, Jacques. "La 'Sagesse,' Nombre d'or de la Théologie de Louis Bouyer." *The Gregorianum* 95.4 (2014): 725–47.

Shanley, Brian J. "Divine Causation and Human Freedom in Aquinas." *American Catholic Philosophical Quarterly* 72.1 (1998): 99–122.

Sherrard, Philip. *Human Image, World Image: the Death and Resurrection of Sacred Cosmology*. Ipswich, UK: Golgonooza Press, 1992.

Slesinski, Robert. *The Theology of Sergius Bulgakov*. Yonkers, NY: St Vladimir's Seminary Press, 2017.

Smith, Wolfgang. *The Quantum Enigma: Finding the Hidden Key*. San Rafael, CA: Angelico Press, 2005.

Solari, Gregory. *Le Temps Découvert: Développment et Durée chez Newman et Bergson*. Paris: Les Éditions du Cerf, 2014.

Soulignac, Aimé. "La Connaissance Angelique." *La Genèse au Sens Littéral*. Ed. A. Soulignac and A. Agaesse. Paris: Desclée de Brouwer, 1972.

Staniloae, Dumitru. *The Experience of God*. Vol. 1: *Revelation and Knowledge of the Triune God*. Trans. Ioan Lonita and Robert Barringer. Brookline, MA: Holy Cross Orthodox Press, 1998.

Stone, M.E. "Lists of Revealed Things in the Apocalyptic Literature." In *Magnalia Dei: The Mighty Acts of God, Essays on the Bible and Archaeology in Memory of G. Ernst Wright*, ed. F.M. Cross, et al., 414–52. New York: Doubleday & Co., 1976.

Susini, Eugene. *Franz von Baader et le Romantisme Mystique: La Philosophie de Franz von Baader*. 2 vols. Paris: Vrin, 1942.

Taylor, Charles. *A Secular Age*. Cambridge, MA: Harvard University Press, 2007.

—. *The Language Animal: The Full Shape of the Human Linguistic Capacity*. Cambridge, MA: Harvard University Press, 2016.

Thils, Gustave. *Transcendance ou Incarnation? Essai sur la Conception du Christianisme*. Université de Louvain, 1950.

Tolkien, J.R.R. *Tree and Leaf*. London: Harper Collins, 2001.

Tourpe, Emmanuel. *L'Audace Théosophique de Baader: Premiers Pas dans la Philosophie Religieuse de Franz von Baader (1765–1841)*. Paris: l'Harmattan, 2009.

Tsakiridou, C.A. *Icons in Time, Persons in Eternity Orthodox Theology and the Aesthetics of the Christian Image*. Burlington, VT: Ashgate, 2013.

Vandier-Nicolas, Nicole. *Art et Sagesse en Chine*. PUF, 1963.

Walgrave, Jan. *Newman the Theologian*. Trans. A.V. Littledale. New York: Sheed and Ward, 1960.

Wallenfang, Donald. *Dialectical Anatomy of the Eucharist: An Étude in Phenomenology*. Eugene, OR: Cascade Books, 2017.

Weill, Sr. Marie-David. , *L'Humanisme Eschatologique de Louis Bouyer: De Marie, Trône de la Sagesse, à l'Église, Éspouse de l'Agneau*. Paris: Les Éditions du Cerf, 2016.

White, Dominic. *The Lost Knowledge of Christ: Contemporary Spiritualities, Christian Cosmology, and the Arts*. Minneapolis, MN: Liturgical Press, 2015.

Whitehead, Alfred North. *The Concept of Nature*. Cambridge: Cambridge University Press, 1920.

—. *Modes of Thought*. New York: Macmillan, 1938.

—. *Process and Reality*. New York: Macmillan, 1929.

—. *Science and the Modern World*. Cambridge University Press, 1920.

Williams, Charles, and C.S. Lewis. *Taliessen through Logres: The Region of the Summer Stars and Arthurian Torso*. Grand Rapids, MI: Eerdmans Press, 1974.

Wippel, John. *Metaphysical Themes in Thomas Aquinas*. 2 vols. Washington, DC: The Catholic University of America Press, 2007.

Zaleski, Philip, and Carol Zaleski. *The Fellowship: The Literary Lives of the Inklings.* New York: Farrar, Strauss and Giroux, 2015.

Žižek, Slavoj. *Absolute Recoil: Toward a New Foundation of Dialectical Materialism.* New York: Verso, 2015.

Zordan, Davide. *Connaissance et Mystère: L'Itinéraire Théologique de Louis Bouyer.* Paris: Cerf, 2008.

—. "De la Sagesse en Théologie: Essai de Confrontation entre Serge Boulgakov et Louis Bouyer." *Irénikon* 79.3 (2006): 263–82.

INDEX

Made in United States
Orlando, FL
15 January 2025